ACTIVITIES

for

Active Learning and Teaching

ACTIVITIES

for Active Learning and Teaching

Selections from the *Mathematics Teacher*

Edited by

Christian R. Hirsch

and

Robert A. Laing

Western Michigan University
Kalamazoo, Michigan

NATIONAL COUNCIL OF TEACHERS OF MATHEMATICS

Third printing 1998

Library of Congress Cataloging-in-Publication Data:

Activities for active learning and teaching : selections from the Mathematics teacher / edited by Christian R. Hirsch and Robert A. Laing.
p. cm.
Includes bibliographical references.
ISBN 0-87353-363-1
1. Mathematics—Study and teaching. I. Hirsch, Christian R. II. Laing, Robert A. III. Mathematics teacher.
QA11.A25 1993
510′.071′273—dc20 93-23027
CIP

The publications of the National Council of Teachers of Mathematics present a variety of viewpoints. The views expressed or implied in this publication, unless otherwise noted, should not be interpreted as official positions of the Council.

Printed in the United States of America

Contents

Introduction

The *Curriculum and Evaluation Standards for School Mathematics* and its companion document, *Professional Standards for Teaching Mathematics*, have expanded, shaped, and refined the broad recommendations for curricular and instructional reform advocated in 1980 in the NCTM's *Agenda for Action*. Similarly, this compilation of activities from the *Mathematics Teacher* represents an expansion, reorganization, and refinement of a previous publication, *Activities for Implementing Curricular Themes from the "Agenda for Action,"* from 1986.

In selecting and editing the activities for this volume, we wanted to provide a resource of instructional materials that support the new curricular themes and new ways of teaching advocated in the two sets of *Standards*. This volume is organized in five sections around the following themes: problem solving, numeracy, algebra and graphs, geometry and visualization, and data analysis and probability. Collectively the activities cultivate students' abilities to explore, conjecture, reason mathematically, solve problems, see and use mathematical connections, and communicate mathematically. Selected activities engage students in the use of manipulatives to make sense of mathematics. Others feature the use of calculators and computers. Activities that were initially developed to capitalize on the graphics capabilities of computers have been reworked so that they can be used equally effectively with graphing calculators. Teachers familiar with graphing calculators may choose to modify activities involving simulation programs so that they too can be used in a graphing-calculator environment.

Each activity consists of a Teacher's Guide that includes objectives, the materials and technology needed, teaching suggestions, possible solutions, and three or four reproducible student activity sheets. The activities, primarily for grades 7 through 10, are designed to actively engage students in the process of doing mathematics. Although the activities can be completed by students working independently, a richer experience is obtained if students work cooperatively in groups and provisions are made for a whole-class summary of results and generalizations. Active teaching that entails listening to students' ideas, asking them for clarification and justification, helping them make connections, and providing them with encouragement and appropriate guidance will further enhance the quality of what students learn from engaging in these activities.

The pages of this book are perforated so that they can be easily removed and reproduced for instructional use. It is recommended that once an activity has been used, the masters be kept in a loose-leaf binder along with comments, reactions, and suggestions for future use.

Activities for

Problem Solving

Central to the current reform movement in mathematics education is the development of mathematical power for all students. This entails cultivating students' ability and disposition to use a variety of mathematical methods effectively to solve nonroutine problems. The first four activities in this section present a related series of highly motivational classroom activities that provide both an introduction to, and opportunities for practice with, several problem-solving skills and strategies. In order for these activities to be most effective, it is recommended that they be completed in the sequence given. Since problem solving is essentially a creative endeavor and thus cannot be built exclusively on routines, it is important that flexibility of thought be accepted and encouraged as students work through these activities. The final two activities in this section furnish opportunities to reinforce these stragegies in a variety of settings. The setting of the fifth activity is a mathematical game. The sixth activity examplifies how computing technology can be used as a tool to enhance problem solving.

The lead activity, "Teaching Problem-solving Skills," focuses on the problem-solving skills of *make a drawing* and *work backward.* Each skill is introduced through a total-class, teacher-directed activity, and then opportunities are provided for students to solve problems, either individually or in small groups, by applying the respective skill. "Developing Problem-solving Skills" uses a similar instructional model to introduce and reinforce the learning of two additional problem-solving skills: *make and read an organized list* and *search for a pattern.*

The activity "Extending Problem-solving Skills" employs an instructional sequence similar to that of the first two activities to highlight the problem-solving skills of *guess and test* and *simplify* and to introduce Polya's heuristics for problem solving, with particular attention to the "looking back" phase. Follow-up problem situations provide opportunities for students to design and carry out problem-solving strategies involving one or a combination of the skills developed through these first three activities. In the activity "Teaching the Elimination Strategy," pupils are introduced to the problem-solving skill *eliminate possibilities* as well as aspects of deductive reasoning and the use of indirect arguments. This activity also gives opportunities for the application of problem-solving strategies requiring the creative meshing of several skills: *guess and test, make and read an organized list,* and *eliminate possibilities.*

"The Peg Game" guides students to use the skills of *simplify, make an organized list,* and *search for a pattern* to solve problems entailing the enumeration of the moves and jumps and the identification of the sequences of moves in interchanging five each of two different-colored pegs on a game board. Questions requiring pupils to generalize from patterns together with two suggested extension activities furnish further opportunities to underscore the importance of the "looking back" phase of problem solving.

The final activity, "Problem Solving with Computers," presents two problem situations: maximizing the volume of an open-topped box and estimating expected values in marketing. Students first manipulate physical models that represent the situations and then use computer-generated data to arrive at solutions.

ACTIVITIES

TEACHING PROBLEM-SOLVING SKILLS

December 1984

By OSCAR F. SCHAAF, University of Oregon, Eugene, OR 97403

Teacher's Guide

Introduction: Recommendations for the reform of school mathematics stress problem solving as the focus of instruction. Among specific actions often suggested in support of this position are the following:

- The mathematics curriculum should be organized around problem solving.
- Appropriate curricular materials to teach problem solving should be developed for all grade levels.
- Mathematics teachers should create classroom environments in which problem solving can flourish.

Motivational classroom materials that provide direct instruction on problem-solving skills are the most important component. This activity consists of two independent subactivities that highlight and provide opportunities for practice with two problem-solving skills: *make a drawing* (worksheets 1 and 2) and *work backward* (worksheets 3 and 4).

Grade levels: 7–10

Materials: Centimeter graph paper, a set of worksheets for each student, and a set of transparencies for demonstration of problem-solving skills and discussion of students' solutions

Objective: To develop the problem-solving skills of making a drawing and working backward

Directions: Distribute copies of the worksheets one at a time to each student. Sheets 1 and 3 are designed for a full-class, teacher-directed activity. Sheets 2 and 4 provide opportunities for pupils to practice the problem-solving skills introduced. Encourage communication and cooperation among pupils as they engage in problem-solving processes.

Sheets 1 and 2: Use a transparency of sheet 1 to introduce the skill *make a drawing*. Following completion of the two sample problems, stress the usefulness of the drawing in their solutions. Pupils should then be directed to solve the problems on sheet 2 by using the same skill. Depending on the level of your class, you may wish to help them find two or three additional examples for problem 3 before they complete the problem individually. Cutouts from centimeter graph paper or a supply of Green Stamps would help pupils better see the possibilities. Emphasize that a system for finding examples might make the search easier.

Provide centimeter graph paper for students to use in solving problem 5. A transparency of sheet 2 can be used to exhibit a sample four-block route for problem 5a. With respect to problem 6, point out that the drawing shows a wall marked in sixths on the outside and tenths on the inside. These marks correspond to the minutes needed to paint the wall.

Sheets 3 and 4: A transparency of sheet 3 should be used to introduce the skill *work backward.* Solve problems 1a, 1b, and 2a with the class. Demonstrate the working-backward procedure with problem 2a. Have pupils work problems 2b and 2c individually or with a classmate.

Solve problems 3 and 4a with the class. In the process, note that the representations for problem 4 are not unique. For example, solutions for problem 4a could be either of the following:

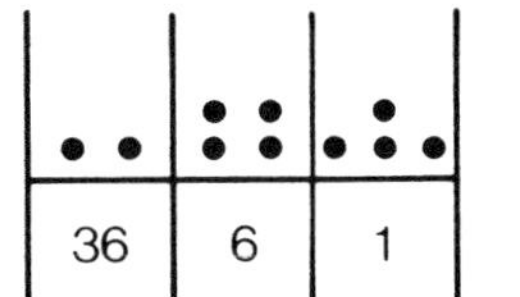

or

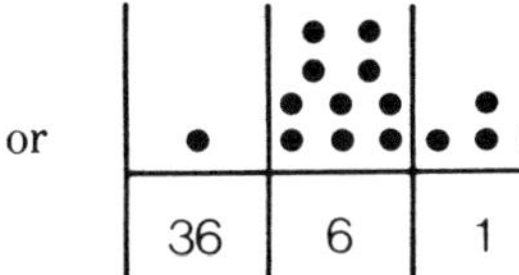

Let pupils try problems 4b and 4c on their own. Discuss their solutions and then emphasize the usefulness of working backward in solving the problems on this sheet. Next, direct students to solve the problems on sheet 4 by using this same skill.

Follow-up activities: Additional problem situations that can reinforce the problem-solving skills introduced in this activity can be found in *Problem Solving in Mathematics* (Lane County Mathematics Project 1983). This project used a similar direct approach to teaching a variety of problem-solving skills. In addition to teaching these skills directly, you can help students improve their problem-solving abilities by—

- using problem solving throughout the school year as an approach to drill and practice, laboratory activities and investigations, and the development of mathematical concepts; and
- providing pupils with appropriate nonroutine challenge problems whose solutions require the creative meshing of several skills.

A further discussion, together with examples, of this instructional approach to problem solving can be found in Brannan and Schaaf (1983).

Answers: Sheet 1: 1b. two meters, four meters, six meters; 1c. eight days. 2a. 64 feet; 2b. 64 feet, 32 feet; 2c. 368 feet

Sheet 2: 3. Nineteen different ways are possible. 4. Twenty girls. The drawing at the right should help pupils see that ten girls need to be equally spaced around each semicircle. 5b. Ten. 6. About four minutes. (If algebra is used, the solution is 3.75 minutes.)

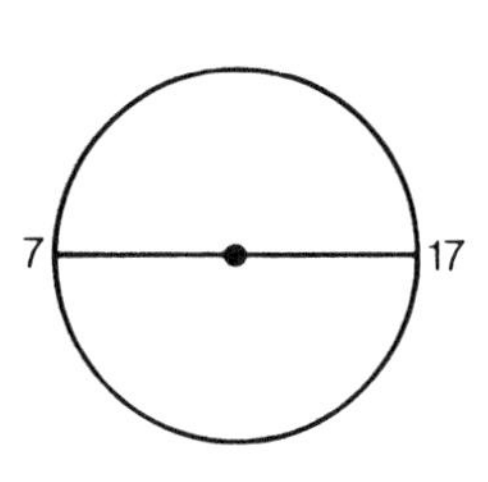

Sheet 3: 1a. fifteen; 1b. twenty-three. 2a. sixteen; 2b. twenty; 2c. twenty-seven. 3. seventy. 4. Answers may vary.

a)
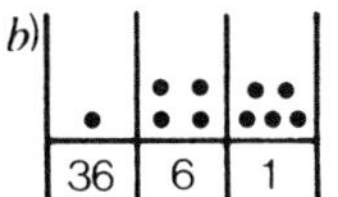

b)
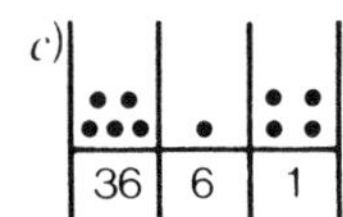

c) 36 6 1

Sheet 4: 5a. 5; 5b. 11. 6. Answers may vary.

$$\text{Equation 1: } \frac{7p - 9}{18} = 3$$

$$\text{Equation 2: } 3(17n - 135) = 54$$

7. \$49.00. 8. Pick one of the three-link pieces and cut each of the three links. Use them to join the remaining three pieces together.

REFERENCES

Brannan, Richard, and Oscar Schaaf. "An Instructional Approach to Problem Solving." In *The Agenda in Action,* 1983 Yearbook of the National Council of Teachers of Mathematics, edited by Gwen Shufelt, pp. 41–59. Reston, Va.: The Council, 1983.

Lane County Mathematics Project. *Problem Solving in Mathematics.* Palo Alto, Calif.: Dale Seymour Publications, 1983.

MAKE A DRAWING

SHEET 1

1. A frog is at the bottom of a ten-meter well. Each day it crawls up three meters. But at night it slips down two meters. How many days will it take the frog to get out of the well?

Problems of this sort can often be solved by using a drawing.

a. The drawing of the well at the right is divided into ten equal parts. Label each part in meters.

b. Use the drawing to help find how far the frog would move up the wall in—

two days ______

four days ______

six days ______

c. Now use the drawing to solve the problem.

____

____

Bottom

2. A ball rebounds 1/2 of the height from which it is dropped. Assume the ball is dropped 128 feet from a leaning tower and keeps bouncing. How far will the ball have traveled up and down when it strikes the ground for the fifth time?

The use of a carefully labeled drawing can also help in solving this problem. The drawing below shows the 128-foot drop of the ball from the tower.

a. Sketch in the first rebound and label the amount of rebound.

b. Sketch and indicate on the drawing the amounts of the second drop and the rebound.

c. Complete the drawing for the problem and give its solution. __________

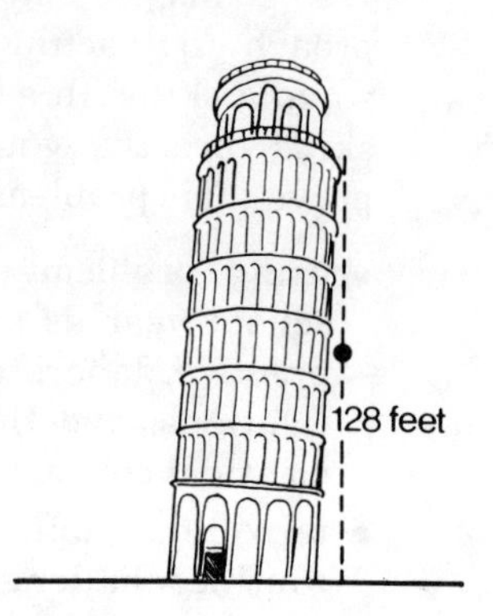

3. How many different ways can you buy four attached postal stamps? ______

   Two possible ways are shown.

   a. Make drawings to show ten additional ways.

   b. Continue this method to obtain a solution to the problem.

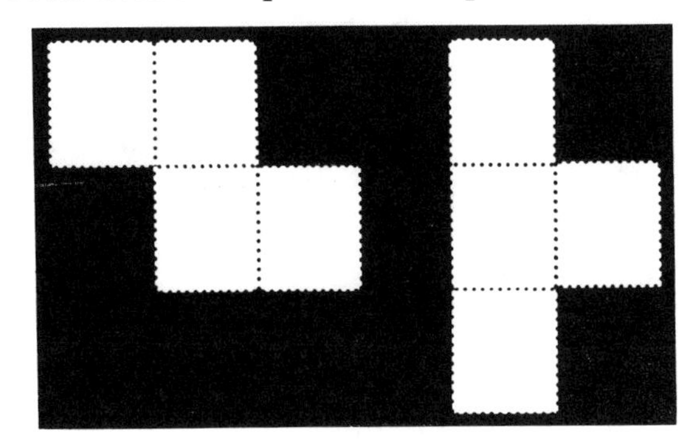

4. Some girls are standing in a circular arrangement. They are evenly spaced and numbered in order. The seventh girl is directly opposite the seventeenth girl. How many girls are in the arrangement? ______

5. a. Anne lives at point *A*. She can use six different four-block routes to walk to school. Show each possible route on a copy of this drawing.

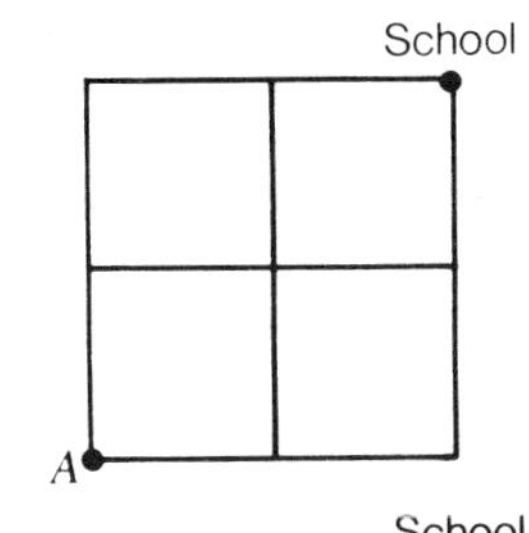

   b. Bill lives at point *B*. How many different five-block routes can he use to get to school? ______

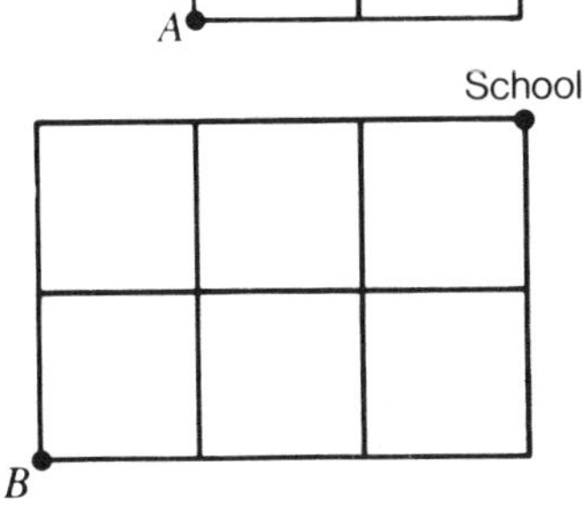

6. One painter can paint a wall in ten minutes. Another painter can paint it in six minutes. About how long will it take both painters working together to do the job? ______

Use the drawing below to help solve the problem.

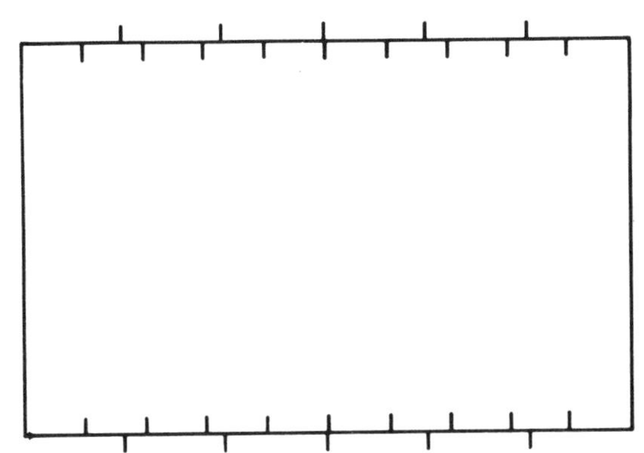

The figure at the right shows a hookup of four machines. The output from the first machine becomes the input for the next machine, and so on.

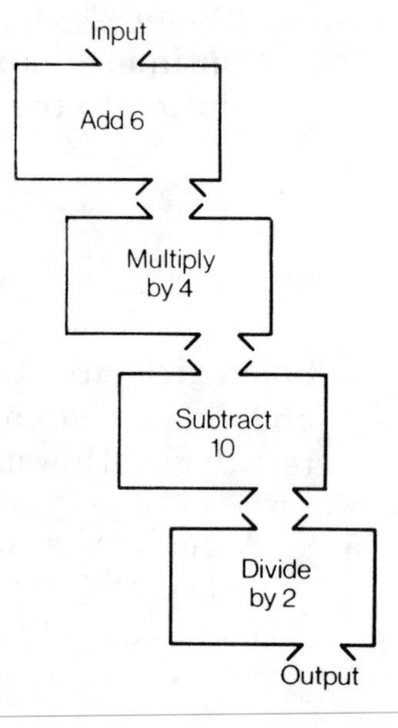

1. What is the final output number if the beginning input number is—
   a. 4? ______
   b. 8? ______

2. Work backward. Find the beginning input number if the output number is—
   a. 39 ______
   b. 47 ______
   c. 61 ______

3. Study the first counting frame shown below. It represents the number 98. What number does the second frame represent? ______

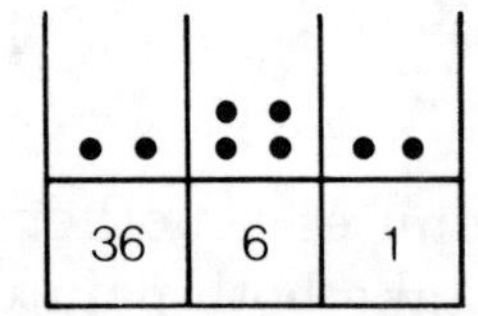

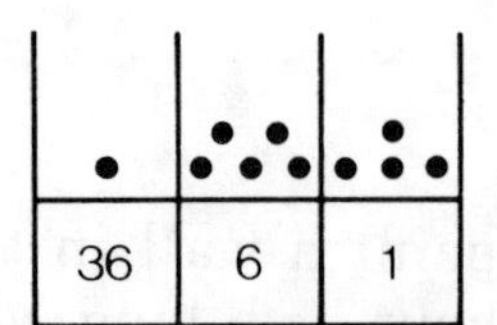

4. Work backward to complete each counting frame so that it represents the number below it.

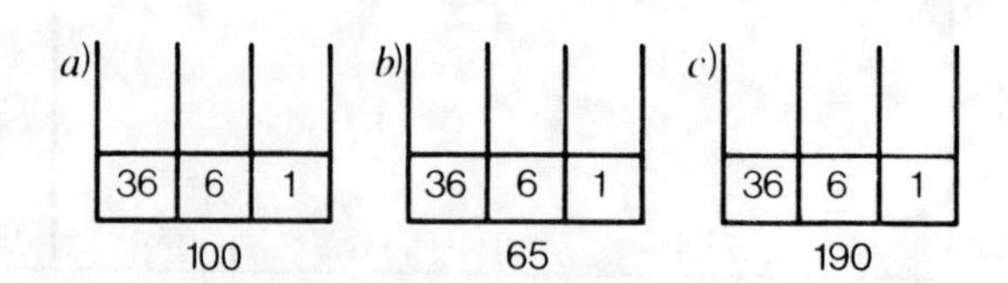

Solve each of these problems by working backward.

5. a. I'm thinking of a number. If you multiply it by 3, then subtract 5, and finally add 10, you get 20.

   What number am I thinking of? ______

   b. I'm thinking of another number. If you subtract 4 from the number, then multiply the result by 3, and then add 5, you get 26.

   What is that number? ______

6. The two equations below both have 2 as a solution:

$$\frac{7p - 2}{4} = 3 \qquad 3(17n - 30) = 12$$

Create two equations that have the same form as these but have 9 as a solution.

Equation 1 ________________ Equation 2 ________________

7. Bliksem bought a present for \$5.00, then spent 1/2 of his remaining money on jogging shoes, next bought lunch for \$2.00, and finally spent 1/2 of his remaining money on a puzzle. He had \$10.00 left.

   How much money did he start with? ______

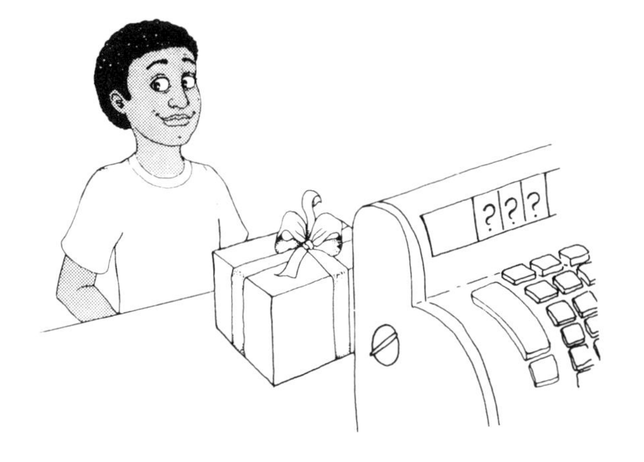

8. Below are four pieces of chain that each have three links. Explain how they can be joined into a twelve-link circular chain by cutting and rejoining only three links.

ACTIVITIES

DEVELOPING PROBLEM-SOLVING SKILLS

December 1985

By STEPHEN KRULIK and JESSE A. RUDNICK, College of Education, Temple University, Philadelphia, PA 19122

Teacher's Guide

Introduction: Recent recommendations for school mathematics suggest that the mathematics curriculum should be organized around problem solving, that classroom environments should be created in which problem solving can flourish, and that appropriate curricular materials to teach problem solving should be developed for all grade levels. The development of materials that are suitable for direct use in the mathematics classroom is, of course, a most important recommendation, since without proper instructional materials, teachers have little direction in what content to emphasize and how it might best be presented. Curricular materials previously offered in this series of activities focused on the problem-solving skills of *make a drawing* and *work backward* (Schaaf 1984) and *guess and test* and *simplification* (Laing 1985). The activities in this article are intended to provide suitable materials for introducing and reinforcing the learning of two additional problem-solving skills, namely, *make an organized list* and *search for a pattern*.

Grade levels: 7–10

Materials: Copies of the worksheets for each student and a set of transparencies for demonstration purposes

Objectives: (1) To develop the problem-solving skills of making and reading an organized list and searching for a pattern; (2) to provide practice in using the general heuristics of the problem-solving process

Directions: This activity consists of two subactivities: solving problems by making an organized list (worksheets 1 and 2) and solving problems by searching for a pattern (worksheets 3 and 4). Each subactivity will require approximately one instructional period and out-of-class work by the students. Allow ample class time for discussion of students' solutions to the problems. Although the second subactivity builds on the first, it is not necessary that the two subactivities be used in consecutive class periods.

Begin by distributing a copy of the first worksheet to each student. This sheet is designed to be used by the entire class in a teacher-centered activity. Using a transparency of sheet 1, present the first problem and explain how the list was constructed. Assume that the first tag chosen is a 3 and the second tag is a 6. This leaves 2, 5, and 1 to be distributed as shown in the first three vertical entries on the list. Now we have exhausted the 6s. The next choice should be a 3 first and a 2 second, with 5 and 1 distributed as shown in the next two vertical entries. Instruct the class to proceed in a similar manner to complete the list and then answer parts (b)–(d). After discussing the solution to this problem, review with the students how the use of an organized list enabled them to keep track of all possibilities as they worked. Note that the list *is* the answer.

Assign the students to complete the solution to problem 2 by using the list that has been started for them. Discuss their solutions. Then distribute the second worksheet and direct the students to complete problem 1. When discussing their solutions to this problem, note that Chen does not need to purchase *exactly* the 17 pounds of grass seed; it is more economical to buy 18 pounds. In this problem, the list is not the answer, but it leads directly to the answer. Instruct students to solve problems 2, 3, and 4 in a similar manner. Depending on the previous problem-solving experiences of the students, you may need to suggest that for problem 4, they begin by making a drawing.

To introduce the second subactivity, distribute sheet 3 to each student and use a transparency of this sheet to discuss the patterning rule for each exercise. Allow students as wide a selection of possible rules as you can, provided that their patterning rule fits all the examples given in the series.

Next distribute sheet 4. Allow time for the students to complete problem 1. In the discussion that follows, point out the value of the organized list in looking for the pattern and emphasize the usefulness of the discovered pattern in completing the solution. Assign problems 2 and 3 to be solved using the same skill.

Follow-up activities: The problem-solving skills of *make an organized list* and *search for a pattern* should be reinforced throughout the school year. Additional problems whose solutions are amenable to the use of these skills can be found in Dolan and Williamson (1983), Krulik and Rudnick (1980, 1984), and in the Lane County Mathematics Project (1983). Finally, note that although a problem can often be solved by algebra, an algebraic solution may not always be possible, and thus facility in making and reading an organized list is an important skill.

Solutions:

Sheet 1: 1.a.

First tag	3	3	3	3	3	3	6	6	6	2
Second tag	6	6	6	2	2	5	2	2	5	5
Third tag	2	5	1	5	1	1	5	1	1	1
Score	11	14	10	10	6	9	13	9	12	8

1.b. Every possible triple of numbers selected from {3, 6, 2, 5, 1} appears in the table. If, for example, an eleventh column was added consisting of the entries 5, 1, and 3, it would duplicate the entries in column six.

1.c. The completed list shows eight different scores.

1.d. The possible scores are 6, 8, 9, 10, 11, 12, 13, and 14.

2.

Clock 1	2:00	2:06	2:12	2:18	2:24
Clock 2	2:00	2:08	2:16	2:24	

The cuckoos come out together at 2:24.

Sheet 2: 1.

Number of 5-Pound Boxes	Cost at $6.58 a Box	Number of 3-Pound Boxes	Cost at $4.50 a Box	Total Pounds	Total Cost
4	$26.32	0	–0–	20	$26.32
3	19.74	1	$ 4.50	18	24.24
2	13.16	3	13.50	19	26.66
1	6.58	4	18.00	17	24.58
0	–0–	6	27.00	18	27.00

Chen should buy three 5-pound boxes and one 3-pound box; this combination is cheaper than buying exactly the 17 pounds of seed required.

2.

Number of Packages of 4	Number of Packages of 3	Number of Singleton Packages
3	1	0
3	0	3
2	2	1
2	1	4
2	0	7
1	3	2
1	2	5
1	1	8
1	0	11
0	5	0
0	4	3
0	3	6
0	2	9
0	1	12
0	0	15

The order can be filled in fifteen different ways.

3.

Number of Checks	Cost for Checks, @ $0.10	Total with $2.00 Service Charge	Cost for Checks, @ $0.05	Total with $3.00 Service Charge
1	$0.10	$2.10	$0.05	$3.05
2	0.20	2.20	0.10	3.10
3	0.30	2.30	0.15	3.15
4	0.40	2.40	0.20	3.20
5	0.50	2.50	0.25	3.25
6	0.60	2.60	0.30	3.30
7	0.70	2.70	0.35	3.35
8	0.80	2.80	0.40	3.40
9	0.90	2.90	0.45	3.45
10	1.00	3.00	0.50	3.50
11	1.10	3.10	0.55	3.55
12	1.20	3.20	0.60	3.60
13	1.30	3.30	0.65	3.65
14	1.40	3.40	0.70	3.70
15	1.50	3.50	0.75	3.75
16	1.60	3.60	0.80	3.80
17	1.70	3.70	0.85	3.85
18	1.80	3.80	0.90	3.90
19	1.90	3.90	0.95	3.95
20	2.00	4.00	1.00	4.00
21	2.10	4.10	1.05	4.05

At least twenty-one checks a month must be written if the new plan is to save you money.

(Solutions continued on next page)

REFERENCES

Dolan, Daniel T., and James Williamson. *Teaching Problem-solving Strategies.* Menlo Park, Calif.: Addison-Wesley Publishing Co., 1983.

Krulik, Stephen, and Jesse A. Rudnick. *Problem Solving: A Handbook for Teachers.* Newton, Mass.: Allyn & Bacon, 1980.

———. *A Sourcebook for Teaching Problem Solving.* Newton, Mass.: Allyn & Bacon, 1984.

Laing, Robert A. "Extending Problem-solving Skills." *Mathematics Teacher* 78 (January 1985):36–44.

Lane County Mathematics Project. *Problem Solving in Mathematics.* Palo Alto, Calif.: Dale Seymour Publications, 1983.

Schaaf, Oscar F. "Teaching Problem-solving Skills." *Mathematics Teacher* 77 (December 1984):694–99.

4.

Total Area	Area of Square 1	Area of Square 2	Side of Square 1	Side of Square 2	Perimeter of Square 1	Perimeter of Square 2	Sum of Perimeters
130	1	129	1	*	--	--	--
130	4	126	2	*	--	--	--
130	9	121	3	11	12	44	56
130	16	114	4	*	--	--	--
130	25	105	5	*	--	--	--
130	36	94	6	*	--	--	--
130	49	81	7	9	28	36	64
130	64	66	8	*	--	--	--
130	81	49	9	7	36	28	64
130	100	30	10	*	--	--	--
130	121	9	11	3	44	12	56

*These values will be nonintegral.

Notice that two possible answers satisfy the condition that the sum of the areas is 130 but that only one satisfies the condition that the sum of the perimeters is 64. Thus the answer is that the sides differ by 2 centimeters.

Sheet 3: 1. 5 — Patterning rule: Each term is one-half of the previous term.

2. Answers will vary. — Patterning rule: Names beginning with the letter "J"

3. 216 — Patterning rule: Each term is the cube of the number of the term.

4. Answers will vary. — Patterning rule: Names of various rock groups

5. 21 — Patterning rule: These are Fibonacci numbers. Each term is the sum of the two terms that precede it.

6. Answers will vary. — Patterning rule: Names in alphabetical order

7. ○ ○ ○ — Patterning rule: Alternating sequence of squares and circles. The number of figures of a given shape increases by one each time the shape occurs.

8. $\frac{5}{6}$ — Patterning rule: Each fraction has the denominator of the previous fraction as its numerator, and the denominator is the next integer.

9. 3750 — Patterning rule: Each term is five times the previous term.

10. — Patterning rule: The sequence of squares repeats with the upper portion of each odd-numbered term shaded and the lower portion of each even-numbered term shaded.

Sheet 4: 1.a. 4; 5

1.b.

Minutes Forward	3	7	11	15	19	23	27
Years Gained	3	8	13	18	23	28	33

By setting the timer ahead 27 minutes you will gain 33 years. The year will be 1985 + 33 = 2018.

2.

Week	Weekly Salary	Total Earnings
1	$1.00	$1.00
2	2.00	3.00
3	4.00	7.00
4	8.00	15.00
5	16.00	31.00
6	32.00	63.00
7	64.00	127.00
8	128.00	255.00
20	524 288.00	1 048 575.00

An analysis of this list suggests two patterns: the weekly salary can be expressed as 2^{n-1}, where n is the number of the week; the total earnings can be expressed as $2^n - 1$, where n is again the number of the week. From the latter pattern, it follows that the total earnings after twenty weeks of employment would be $2^{20} - 1 = \$1\,048\,575.00$.

3.

Stop #	1	2	3	4	5	6	7	8	9	10		16
Number Picked Up	5	0	5	0	5	0	5	0	5	0		
Number Dropped Off	0	2	0	2	0	2	0	2	0	2		
Number Aboard	5	3	8	6	11	9	14	12	17	15		24

The organized list enables us to see the pattern: 24 passengers will be aboard at the end of the sixteenth stop. Observe that the sequence of numbers of passengers actually consists of two subsequences: 5, 8, 11, 14, 17 and 3, 6, 9, 12, 15. ■

1. The five tags shown are placed in a box and mixed. Three tags are then drawn out at one time. If your score is the sum of the numbers on the three tags drawn, how many different scores are possible? What are the possible scores?

To keep track of the different scores, it is helpful to prepare an *organized list*. Let's begin by assuming that the first of the three tags drawn is the "3" and then listing the possibilities for the other two tags.

First Tag	3	3	3	3	3	3	6	6	6	2
Second Tag	6	6	6	2	2	5	2	2		
Third Tag	2	5	1	5	1		5			
Score	11	14	10	10	6					

a. Complete the chart.

b. Explain why this list accounts for all possible drawings of three tags where the order of selection is unimportant. ______________________

____

c. How many *different* scores are possible? ____________

d. What are the possible scores? ______________________

2. Jim's Repair-Your-Clock Shop has two cuckoo clocks that were brought in for repairs. One clock has the cuckoo coming out every six minutes, whereas the other has the cuckoo coming out every eight minutes. Both cuckoos come out at exactly 2:00. When will they both come out together again?

Let's start to make an organized list of the times that the two cuckoos come out:

Clock 1	2:00	2:06	2:12				
Clock 2	2:00	2:08					

Finish the list and solve the problem. ______________________

1. Helen Chen wants to seed her front lawn. Grass seed can be bought in 3-pound boxes that cost $4.50, or in 5-pound boxes that cost $6.58. She needs exactly 17 pounds of seed. How many boxes of each size should she purchase to get the best buy?

The use of an organized list can also help in solving this problem.

Number of 5-Pound Boxes	Cost at $6.58 a Box	Number of 3-Pound Boxes	Cost at $4.50 a Box	Total Pounds	Total Cost
4	$26.32	0	–0–	20	$26.32
3	19.74	1	$4.50		
2					
1					
0		6			

Complete the list and solve the problem. ____________________

Make an organized list to solve each of the following problems.

2. A customer ordered 15 blueberry muffins. If the muffins are packaged singly or in sets of 3 or 4, in how many different ways can the order be filled?

3. A bank has been charging a monthly service fee of $2.00 plus $0.10 a check for a personal checking account. To attract more customers it is advertising a new "reduced cost" plan with a monthly service charge of $3.00 plus only $0.05 a check. How many checks must you write each month for the new plan to save you money?

4. A piece of wire 64 centimeters in length is cut into two parts. The parts are then each bent to form a square. The total area of the two squares is 130 square centimeters. How much longer is a side of the larger square than a side of the smaller square? (Consider only whole-number solutions.)

For each of the following sets, give another element of the set. State in your own words what you think the patterning rule is.

1. 80, 40, 20, 10, ____________
   Patterning rule: ____________

2. James, Jill, Joan, John, ____________
   Patterning rule: ____________

3. 1, 8, 27, 64, 125, ____________
   Patterning rule: ____________

4. Styx, Beatles, Who, Kansas, ____________
   Patterning rule: ____________

5. 1, 1, 2, 3, 5, 8, 13, ____________
   Patterning rule: ____________

6. Alvin, Barbara, Carla, Dennis, ____________
   Patterning rule: ____________

7. □, ○, □□, ○○, □□□, ____
   Patterning rule: ____________

8. $\frac{1}{2}$, $\frac{2}{3}$, $\frac{3}{4}$, $\frac{4}{5}$, ____________
   Patterning rule: ____________

9. 6, 30, 150, 750, ____________
   Patterning rule: ____________

10. ⬒, ⬓, ⬒, ⬓, ⬒, ____
    Patterning rule: ____________

1. Scientists have invented a time machine. By setting the dial, you can move forward in time. Set it forward 3 minutes and you will be in the year 1988. Set it forward 7 minutes and you will be in the year 1993; set it forward 11 minutes and you will be in the year 1998; set it forward 15 minutes and you will be in the year 2003. If the machine continues in this manner, in what year will you be if you set the timer ahead 27 minutes? (This year is 1985.)

What do we know?	Forward 3 minutes, gain of 3 years (1988 – 1985) Forward 7 minutes, gain of 8 years (1993 – 1985) Forward 11 minutes, gain of 13 years (1998 – 1985) Forward 15 minutes, gain of 18 years (2003 – 1985)
What do we want?	Forward 27 minutes, gain of how many years?
Plan:	Make an organized list and search for a pattern.

Minutes Forward	3	7	11	15		27
Years Gained	3	8	13	18		?

a. Complete the patterning rule: Every time we move the timer ahead ________ minutes, we gain an additional ________ years.

b. Now complete the chart and solve the problem. ________________________

Solve the following problems by first making an organized list and then looking for a pattern.

2. Carlos was offered a part-time job that included on-the-job training. Because of the training, he was to be paid \$1.00 the first week, \$2.00 the second, \$4.00 the third, \$8.00 the fourth, and so on. How much money would Carlos have earned after twenty weeks of employment?

3. An empty streetcar picks up five passengers at the first stop, drops off two passengers at the second stop, picks up five passengers at the third stop, drops off two passengers at the fourth stop, and so on. If it continues in this manner, how many passengers will be on the streetcar after the sixteenth stop?

ACTIVITIES

EXTENDING PROBLEM-SOLVING SKILLS

January 1985

By ROBERT A. LAING, Western Michigan University, Kalamazoo, MI 49008

Teacher's Guide

Introduction: The mathematics curriculum must include more than the concepts and skills of mathematics to prepare students to be productive and contributing members of a rapidly changing technological society. It is essential that problem solving be an integral part of the mathematical experiences of all students.

Implementation of this recommendation requires that curriculum developers—

- give priority to the identification and analysis of specific problem-solving strategies;
- develop and disseminate examples of "good problems" and strategies and suggest the scope of problem-solving activities for each school level.

Also required is a classroom environment in which problem solving can flourish:

- Students should be encouraged to question, experiment, estimate, explore, and suggest explanations. Problem solving, which is essentially a creative activity, cannot be built exclusively on routines, recipes, and formulas.

The activities presented in this article are intended to reflect these recommendations. Two problem-solving skills, *guess and test* and *simplification*, are introduced and reinforced through a variety of motivating problem situations. Solutions of sample problems and directions for the teacher encourage the extension of problem-solving skills to include application of Polya's (1971) four phases of problem solving: understanding the problem, devising a plan, carrying out the plan, and looking back.

Grade levels: 7–11

Materials: Graph paper or square dot paper (optional), copies of the worksheets for each student, and a set of transparencies for class discussions

Objectives: (1) To develop problem-solving skills of guess and test and simplification; (2) to introduce the more general heuristics of problem solving suggested by Polya; (3) to provide opportunities for the design of problem-solving strategies using one or a combination of the following skills: make a drawing, work backward, guess and test, and state and solve a simpler problem.

Directions: Solutions of sample problems on the worksheets encourage students to devote sufficient time to understanding a problem before they begin searching for its solution. Teachers can reinforce this approach, when students claim a lack of understanding, through such questions as, "What steps have you taken to understand the problem?"

Begin by distributing sheets 1 and 2 to each student. Use a transparency of sheet 1 to present and discuss the sample problem together with the prescription for using guess and test. Following this discussion, assign students the problems on sheet 2, with problems 1 and 2 due at the next class period. Encourage pupils to keep an organized record of their guess-and-test results in tabular form. During the next class period, discuss these two problems using looking-back activities, such as those suggested in the following "Solutions" section. Assign problems 3 and 4 for solution and discussion in a similar fashion during the next class period.

You may wish to provide some experiences involving the problem-solving skill of looking for patterns prior to using sheets 3 and 4, which entail solving problems by simplification. Such activities as "Can You Predict the Repetend?" (Woodburn 1981), "Number Triangles: A Discovery Lesson" (Ouellette 1981), and "Pattern Gazing" (Aviv 1981) offer a good introduction to this skill.

Sheets 3 and 4 can be used in a manner similar to that used for the first two worksheets. However, do not expect students to complete more than one or two problems in each class period because of the increased level of difficulty of the problems and the amount of time needed to conduct the follow-up discussions adequately. Graph paper or dot paper can be distributed for use with problems 3 and 5.

Solutions and looking-back activities: An important component of every looking-back activity is the discussion of strategies that were used in solving the problem. This discussion is necessary to focus the students' attention on the processes used rather than on the characteristics of a particular problem. These processes will prove useful in later problem situations.

Answers: Sheet 2: 1. 40 m by 75 m

Area	Width (Guess)	Length	Amount of New Fence
3000	20	150	190
3000	30	100	160
3000	40	75	155
3000	50	60	160

Looking-back activities might include the following questions:

Could Mary do better by using such widths as 38.5 or 39? Use your calculators to explore this question.

Suppose Mary could use neighbors' fences for two adjacent sides of the pasture. How would this alter the problem? Let's solve the new problem together.

2. 31 and 33. Looking-back activities might include these questions:

Would this problem have an answer if the rooms were numbered with consecutive integers and the product were given as 1023? Why?

How might the concept of square root have been useful in solving the original problem?

Suppose we changed the term "product" to "sum." How must the rooms be numbered if the sum is odd? Why? For this new problem, can you find a quick method that will yield a good guess on the first trial? (Use the concept of average.)

3. 15 pigs. A table is quite useful in this problem for keeping track of guess-and-test attempts. Have different students show their organizational schemes on the chalkboard.

An interesting solution strategy described by Krulik and Rudnick (1980) is to picture the pigs standing on their hind legs and the chickens standing on one leg. If we count the number of legs in the air, we should get half of the total number of legs given in the problem. Fifty legs and thirty-five heads give fifteen more legs than heads, one contributed by each of the pigs.

4. 12 ft. and 28 ft. In looking back at this problem, the teacher might demonstrate how the inclusion of formulas in the table can be useful:

Shorter (S) (Guess)	Longer (L) (2S + 4)	Total (S + L)	Total Needed
10	20 + 4 = 24	34	40

Sheet 4: 1. Meet at the location of the innermost official. In the case of two officials, the use of some sample distances will reveal that any point between them or at either's location will yield the same minimal distance. In the case of a third official, similar reasoning will make it clear that using the location of the innermost official will result in the minimal total distance.

The general conjecture that the officials should meet anywhere between the two innermost ones in the case of an even number of officials and at the location of the innermost official in the case of an odd number of officials should be tested for the cases of four and five officials before this reasoning is applied to the case of nine officials.

2. 325 connections. Students will discover a number of patterns when simplifying this problem, making it good for sharing alternatives in class discussions. Drawing pictures of the simpler cases also reinforces this problem-solving skill.

1 classroom and office

1 connection

2 classrooms and office

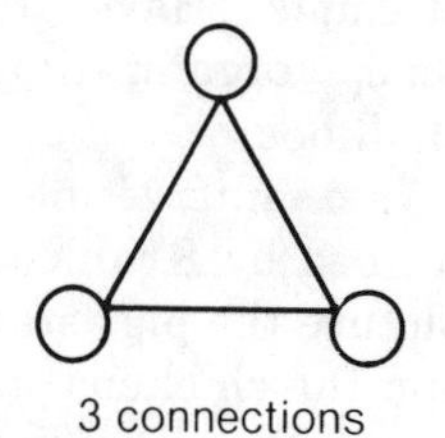

3 connections

3 classrooms and office

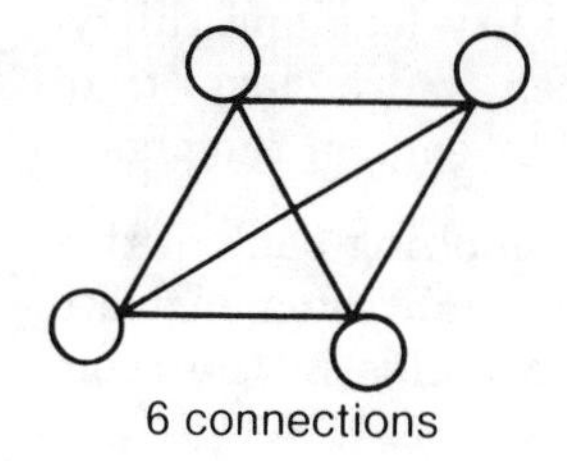

6 connections

Some pupils may recognize that the triangular numbers 1, 3, 6, 10, ... occur in the number of connections needed. The insight that results most frequently from exploring the simpler cases is that each additional classroom must be connected to each of the others, yielding the sum

$$1 + 2 + 3 + \cdots + 25 = 325$$

as the solution. You might ask students to find an easier way of finding this sum than adding all twenty-five numbers together. One discovery might be to use the average (and median, in this case) and just multiply 25 by 13. A second method would mimic a fairly standard argument found in high school textbooks for finding the sum of n consecutive counting numbers. Twice the needed sum is given by adding columns:

$$\begin{array}{r} 1 + 2 + 3 + \cdots + 24 + 25 \\ \underline{25 + 24 + 23 + \cdots + 2 + 1} \\ 26 + 26 + 26 + \cdots + 26 + 26 \end{array}$$

Therefore, the desired sum is given by $1/2 \times (25 \times 26)$.

3. 70 downward paths. In considering the simpler cases, students should discover the usefulness of the symmetry in the figures, select an appropriate notation for recording intermediate results, and note that each case is embedded in subsequent cases.

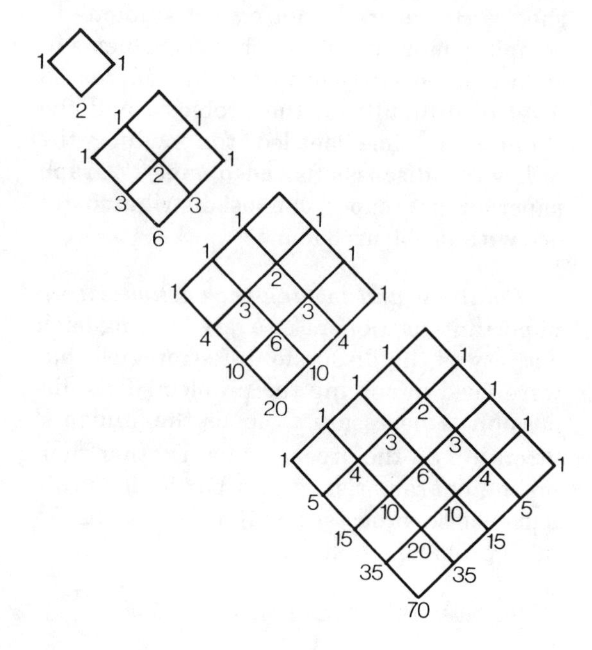

In addition to these considerations, looking-back activities should focus on number patterns that students may have discovered. These patterns can be used to extend the solution to the 5-by-5 case. The occurrence of triangular numbers, as well as of the consecutive counting numbers, should be noted. Reinforce how symmetry can be used to reduce the complexity of many problems.

4. The units digit for 7^{99} is 3. Most students soon discover that they do not have to find the entire value of each consecutive power of 7 if they are only interested in the units digit.

Power	Units Digit
7^1	7
7^2	9
7^3	3
7^4	$\underline{1}$
7^5	7
7^6	9
7^7	3
7^8	$\underline{1}$
7^9	7
7^{10}	9
7^{11}	3
7^{12}	$\underline{1}$

Although many students will discover the repeating pattern of the units digits and will see that odd powers have 7 or 3 in the units place, they may have difficulty determining which digit to use for the 99th power. The length of the repeating cycle in this example is 4 with a units digit of 1 occurring at each multiple of 4. Thus we need only to divide 99 by 4 and consider the remainder as the number of steps past the 1 that we need to go to obtain the desired digit. Follow this activity with problems that have a similar mathematical structure. The following are some examples:

Find the day of the week that is 213 days from today.

Find the time on a clock that is 100 hours from now.

5. 204 squares. In applying the simplification strategy, some students will generate the following information related to the simpler cases:

Dimension	Number of Squares
1×1	1
2×2	5
3×3	14
4×4	30
⋮	
8×8	?

A few students may discover that the differences between the consecutive cases are perfect squares and then may continue this pattern to the 8×8 case.

Another strategy that yields a more immediate solution is the discovery that the number of squares of a particular size is always a perfect square. For example, when considering the 3×3 case, the student discovers the following pattern:

Size	Number of Squares
1×1	9
2×2	4
3×3	1

$$S = 1 + 4 + 9$$
$$= 1^2 + 2^2 + 3^2$$

This pattern would suggest that the total number of squares for the checkerboard is $1^2 + 2^2 + 3^2 + \cdots + 8^2$. Circumventing the long method for obtaining this sum is difficult, since the formula for finding the sum of the squares of the first n positive integers is given by

$$S = \frac{n(n + 1)(2n + 1)}{6}.$$

In the looking-back activity, ask students how they organized their counting procedures so that they did not count the same square twice. One scheme would be to notice that each 2×2 square in the 3×3 case must have a bottom-left vertex. Because nine points could serve as a bottom-left vertex for a 2×2 square, nine such squares must exist. This problem is also one for which the answer is difficult to check; however, the validity of the process can be tested by applying it to the 5 × 5 case.

Follow-up activities: The experiences provided in this activity and that by Schaaf (1984) highlight the problem-solving skills of make a drawing, guess and test, work backward, and simplification. Solutions of the following problems will require students to develop and implement a strategy involving one or a combination of these skills. Begin to integrate problem solving into the regular curriculum by including one of these problems as part of the regular homework assignment over a period of several days.

1. A log is cut into 4 pieces in 15 seconds. At this rate, how long will it take to cut a log into 6 pieces? (Possible strategy: make a drawing; answer: 25 seconds)

2. Roosevelt wrote to 12 of his friends from music camp at a cost of $2.20 for postage. If the postage for letters is $0.23 and for postcards, $0.15, how many of each did he send? (Possible strategy: guess and test; answer: 5 letters and 7 postcards)

3. Joan decided to quit the fishing-worm business. She gave half her worms plus half a worm to Mark. Then she gave half of what was left plus half a worm to George, leaving her with a dozen worms, which she kept for herself. How many worms did Joan have when she decided to quit the business? (Possible strategy: Work backward; answer: 51 worms)

4. Fifteen people attended a party. If each person shook hands with every other person, how many handshakes were made? (Possible strategy: simplification, make a drawing, and look for patterns; answer: 105 handshakes)

5. I am a proper fraction. The sum of my numerator and denominator is 60, and their difference is 10. What is my simplest name? (Possible strategy: guess and test; answer: 5/7)

6. If 5 times a number is increased by 11 and the answer is divided by 3, the result is 32. Find the number. (Possible strategy: work backward; answer: 17)

7. To drive from Ackley to Derry, you will first pass through Booville and then through Carlton. It is five times as far from Ackley to Booville as it is from Booville to Carlton and three times as far from Booville to Carlton as it is from Carlton to Derry. If the distance from Ackley to Derry is 228 miles, how far is it from Ackley to Booville? (Possible strategy: make a drawing and guess and test; answer: 180 miles)

8. A bag of 15 silver dollars is known to contain 1 counterfeit coin, which is lighter than the other 14. What is the maximum number of comparisons that may be necessary to identify the counterfeit coin using only a two-pan balance? (Possible strategy: simplification, make a table, and look for patterns; answer: 3 weighings)

REFERENCES

Aviv, Cherie Adler. "Pattern Gazing." In *Activities from the Mathematics Teacher,* edited by Evan M. Maletsky and Christian R. Hirsch. Reston, Va.: National Council of Teachers of Mathematics, 1981.

Krulik, Stephen, and Jesse A. Rudnick. *Problem Solving—a Handbook for Teachers.* Boston: Allyn & Bacon, 1980.

Ouellette, Hugh. *"Number Triangles—a Discovery Lesson."* In *Activities from the Mathematics Teacher,* edited by Evan M. Maletsky and Christian R. Hirsch. Reston, Va.: National Council of Teachers of Mathematics, 1981.

Polya, George. *How to Solve It.* Princeton, N.J.: Princeton University Press, 1971.

Schaaf, Oscar F. "Activities: Teaching Problem-solving Skills." *Mathematics Teacher* 77 (December 1984):694–99.

Woodburn, Douglas. "Can You Predict the Repetend?" In *Activities from the Mathematics Teacher,* edited by Evan M. Maletsky and Christian R. Hirsch. Reston, Va.: National Council of Teachers of Mathematics, 1981.

Sample problem: Not Himself Today!

Today, Alan Lurt (called A. Lurt by his friends) spent 20 more minutes asleep in class than he spent awake. If the class period is 52 minutes long, how long was Alan not alert?

Problems of this sort can sometimes be solved by guessing and testing.

Solution:

What do we know? Total period is 52 minutes, and Alan spent 20 minutes more asleep than awake.

What do we want? The number of minutes asleep

Guess the answer. He slept 10 minutes.

Test your guess. 52 (total) – 10 (asleep) = 42 (awake). Awake more than asleep! Guess larger.

Guesses	Tests	Solution?
30 minutes asleep	52 – 30 (asleep) = 22 (awake) gives 8 minutes more asleep.	No
34 minutes asleep	52 – 34 (asleep) = 18 (awake) gives 16 minutes more asleep.	No
36 minutes asleep	52 – 36 (asleep) = 16 (awake) gives 20 minutes more asleep.	Yes

So A. Lurt was not alert for 36 minutes.

To solve problems using guess and test, try the following steps:

1. Read through the problem more than once.
2. Draw a picture of, or outline, the problem situation in some way until you feel you understand the problem.
3. Write down answers to these questions: (*a*) What is given? (*b*) What is wanted?
4. Make a reasonable guess.
5. Test your guess by substituting it into the problem.
6. If necessary, refine your guess and repeat the procedure until a guess satisfies the problem.

Problem 1: Good Horse Fence

Mary is asked by her dad to make a plan for fencing in a pasture for their new horses. They will require an area of 3000 m^2 for grazing and exercise. To reduce costs, Mary decides to use their neighbor's fence for the longest side of the rectangular area. What dimensions should she use for the pasture so that fencing costs are as low as possible?

Solution:

What do we know? Area (3000) = $l \times w$. Amount of new fence is $2w + l$.

What do we want? Dimensions of the pasture with area 3000 m^2, so that amount of new fence is minimal

Guess: $w = 20$ Test! Since $l \times w = 3000$, $l = 3000 \div 20 = 150$. The amount of new fence is $2(20) + 150 = 190$.

Can she do better? Complete the following table:

Area	Width (Guess)	Length	Amount of New Fence
3000	20	$3000 \div 20 = 150$	$2w + l = 2(20) + 150 = 190$
3000	30	$3000 \div 30 = 100$	$2w + l =$ ______ = ___
3000	40	______	___ = ______ = ___
3000	50	______	___ = ______ = ___

It appears that the best dimensions are approximately _____ by _____ .

Use the guess-and-test strategy to solve the problems that follow.

Problem 2: Julio's mathematics and English classrooms are next door to each other. The product of the room numbers is 1023. Find the room numbers.

Problem 3: Farmer Jones has pigs and chickens. They have a total of 35 heads and 100 legs. How many pigs does he have?

Problem 4: A 40-foot cable is divided into two sections. One section is 4 feet more than twice the length of the other section. How long is each section?

Sample problem: Who's Silly Now?

The famous football star Sam Bims (once called Silly Bims by the team's management) was offered a 4-year contract that included a bonus for each game in which he had at least 100 yards rushing. Sam was to receive a $1000 bonus for the first 100-yard game, $3000 for the second, $5000 for the third, $7000 for the fourth, and so on, throughout the contract. Sam had 40 such games. What was the total of his bonuses?

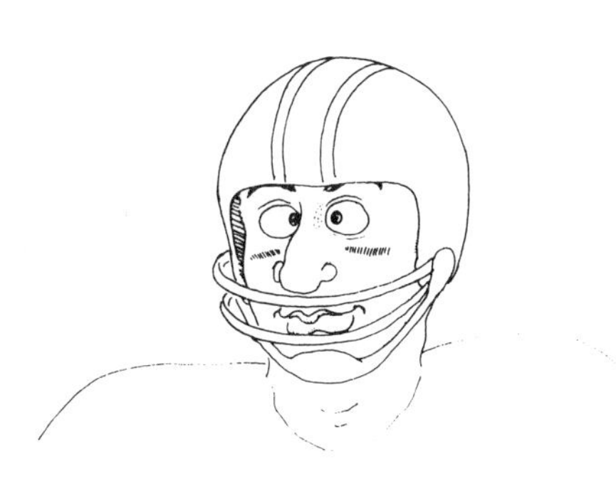

Using guess and test would not be a good choice here. Problems such as this, which are complex because of the number of conditions, are often solved more easily by solving one or more simpler problems first and looking for a pattern.

Solution:

What is known? Amount of bonus for each 100-yard game.
Sam had 40 such games.

What is wanted? Total of all the bonuses

Simplify: Let's look at some simpler problems by reducing the number of bonus games. We organize our results in a table.

Number of Bonuses	Bonuses				Total Bonus
1	1000				$1 000
2	1000	3000			$4 000
3	1000	3000	5000		$9 000
4	1000	3000	5000	7000	$16 000

Patterns: Compare the first and the last columns. Do you see a pattern? It looks like a pattern of squares. Test this conjecture: 5 bonuses would give $25 000 by the pattern. This answer checks, since $1 000 + $3 000 + $5 000 + $7 000 + $9 000 = $25 000.

Apply the pattern. 40 bonuses: $40^2 \times 1000 = 1600 \times 1000 =$ $1 600 000. Would you call him Silly Bims?

To solve problems by simplifying, try these steps:

1. Read through the problem more than once.
2. Draw a picture of, or outline, the problem situation in some way until you feel you understand the problem.
3. Write down answers to the questions: (*a*) What is given? (*b*) What is wanted?
4. Write and solve one or more simpler, but related, problems.
5. Compare the simpler situations and their solutions with the original problem.

Problem 1: Heartbreak Hill

Nine officials are positioned along a straight section of a marathon route. At what location should they meet to confer so that the total distance they travel is as small as possible?

Solution:

What is known? Nine officials who will meet at one location

What is wanted? Location of meeting to minimize total distance traveled

Simplify: Answer the question for only two officials.

O_1 O_2

Next solve the problem for three officials.

O_1 O_2 O_3

Conjecture the solution and test your reasoning with one or more smaller cases.

Solve the following problems by first stating and solving a simpler problem.

Problem 2: Fataronda Middle School is planning to install an intercom system between classrooms and the main office that permits direct conversations between any pair of classrooms, as well as between a classroom and the main office. How many room-to-room and room-to-office connections will be needed if the school has 25 classrooms?

Problem 3: How many different downward paths connect P to Q in the adjacent diagram? (One such path is shown in the diagram. Remember that each section of the path must be downward.)

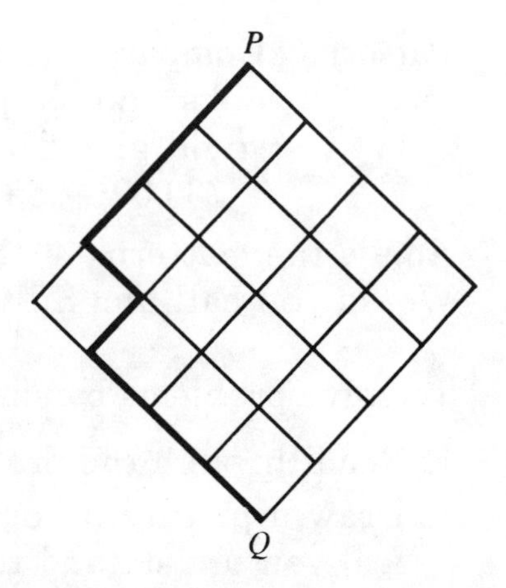

Problem 4: The units digit of 2^4 when expressed in standard form is 6. Find the units digit of 7^{99} when it is expressed in standard form.

Problem 5: How many squares are on a standard 8-by-8 checkerboard?

ACTIVITIES

TEACHING THE ELIMINATION STRATEGY

January 1986

By DANIEL T. DOLAN, Office of Public Instruction, Helena, MT 59620
JAMES WILLIAMSON, Billings Public Schools, Billings, MT 59101

Teacher's Guide

Introduction: Recent curriculum recommendations have argued that problem solving should be the focus of school mathematics. To achieve this goal, we, as classroom teachers, must create an atmosphere where problem solving flourishes, where a variety of problem-solving skills are taught in a sequential fashion that takes into account students' developmental levels, and where the skills are then applied, often in combinations, in diverse problem situations.

To establish a problem-solving atmosphere, we make the following recommendations:

- The focus of the curriculum should be on more formal and more general problem-solving approaches and strategies themselves.
- Instruction should stress the ability to select from a range of strategies and to create new strategies by combining known techniques.
- Instruction should aid in the student's transition to more abstract reasoning.

To implement these recommendations, appropriate instructional materials must be developed to teach the strategies. Previous "Activities" have provided ideas for teaching the problem-solving skills of *make a drawing* and *work backward* (Schaaf 1984), *guess and test* and *simplification* (Laing 1985), and *make an organized list* and *search for a pattern* (Krulik and Rudnick 1985).

The activities presented here are designed to provide students with an informal introduction to deductive logic through the use of the problem-solving skill of *eliminate possibilities.* Next to *guess and check, elimination* is probably the most commonly used of all problem-solving skills. We use it daily in making a variety of decisions. A consumer's decision to purchase a particular product is often made by first gathering information and then eliminating competitive products on the basis of price, service, advice of friends, or consumer information.

Most students have their first experience with formal logic in geometry at the sophomore level. The traditional junior high curriculum has virtually no introduction to logic; this lack ignores sound educational theory. Without a foundation that offers experiences with the informal application of logic in familiar situations, it is not surprising that so many students have difficulty with the application of logic to proof in a formal geometry course.

Students in grades 7–9 are just entering what Piaget calls the formal operational stage. Because levels of maturity of students in this age group vary greatly, a complete development of logic would not be appropriate. However, it should be taught at an informal and introductory level. Teaching methods of elimination at this level can tremendously help students to make the transition to the more abstract reasoning in later courses.

The process of elimination often requires the generation of a list of possible solutions from which to eliminate possibilities systematically, the use of an indirect argument, or the use of a table to organize the information. This activity therefore will be most successful if students have previously completed the aforementioned problem-solving activities focusing on these prerequisite skills.

Grade levels: 7–11

Materials: Copies of the worksheets for each student and a set of transparencies for class discussions

Objectives: (1) To develop the problem-solving skill *eliminate possibilities;* (2) to introduce students to aspects of deductive reasoning and the use of indirect arguments; (3) to provide opportunities for the application of problem-solving strategies entailing the use of elimination and those of *guess and check* and *make an organized list or table*

Procedure: The worksheets are designed to be completed in order over a period of several days, and copies should be distributed one at a time to each student. Use a transparency of each worksheet to guide the class to a solution of the first problem and then instruct the students to solve the remaining problems on the worksheet using similar approaches. Since this may be the first time many of your students are confronted with logic problems of the type presented here, encourage students to cooperate and work together. The best results will be obtained if a follow-up discussion is held with the class after each worksheet is completed. Since each worksheet builds on ideas developed in the previous worksheet, it is essential to highlight key elements during this looking-back phase.

Worksheet 1: The most important part of the first problem is the follow-up discussion. Students will probably have little difficulty solving this problem, but each clue should be analyzed to see how a suspect is eliminated as a result of the evidence. It might be helpful to discuss whether or not clue (*a*) is needed. The fact that the killer registered at the hotel does not mean that he stayed there. Frequently in the mathematical experiences of students, all the data needed to solve a problem are given in exactly the order that leads to a solution; rarely are extra data presented that are not helpful. (Solution: Yoko Red committed the crime.)

A very important aspect of problem 2 is the concept of "first clue." Throughout pupils' elementary school experiences with arithmetic, they are instructed to read a problem and then work from the beginning. In many problems whose solution requires the elimination of possibilities, it is important that the problem be carefully analyzed to see which clue might be the first or easiest to use. This is especially true when a set of possible solutions must be generated before the elimination process can be used. In this problem, clues (*b*), (*d*), and (*e*) might be the easiest to use, but if a list had to be generated, the use of clue (*c*) as a first clue would result in a shorter list than clue (*d*). (Solution: 1375)

In problem 3, it is again important to discuss which clue is the easiest to use first and why. No single answer is correct, but it is important for students to share why they feel a clue is significant. It might even be helpful to discuss the order in which clues are used so that the entire thought process used in the solution is shared. This discussion of the strategy is the important looking-back component of problem solving stressed by Polya (1971). (Solution: 1537)

Worksheet 2: The objective is to develop the idea of using a table to organize information prior to eliminating possibilities.

The first problem provides an extension of the concept of the "best" first clue introduced in worksheet 1. The main idea is first to make a list of possible solutions. The key in doing so, as indicated in the hint, is to begin with the clue(s) that will generate the shortest list. Perhaps the best choice is to list the odd multiples of 7:

Odd multiples of 7:	~~7~~	~~21~~	~~35~~	~~49~~	~~63~~	77	91
Remainders when—							
divided by 3:	1	0	2	1	0	2	
divided by 5:	2	1	0	4	3	2	

(The solution is 77.)

With respect to problem 2, the winners are circled in the following table:

Day	Opponents
1	Spikers **Setters**
2	Spikers **Servers**
3	Setters **Servers**
4	**Spikers** Servers
5	**Setters** Servers
6	Spikers **Setters**

One approach to determining the winner (loser) of each game follows. Since the Spikers never defeated the Setters, the Setters won on days 1 and 6. Each team played each of the other teams at home and away, so we can assume that the Servers played at home on days 2 and 3. Since they never lost a home game, they won on days 2 and 3. Finally, the Servers had to lose two games. They won on days 2 and 3; therefore, they must have lost on days 4 and 5. (The following is the correct solution: Spikers won one and lost three; Setters won three and lost one; Servers won two and lost two.)

Problem 3 introduces the idea of constructing a table to organize the information during the elimination process. It is very important that discussion be focused on how the table in the hint is suggested by the information contained in the problem.

Students invariably begin completing the suggested table by using clue (*b*) to enter the American as the driver of the Audi. Either clue (*c*) or clue (*d*) can be used next. The following is one of several approaches to the solution. It is important that students be allowed to explain these alternate approaches if they discover them. (*Note:* The numbers that appear in parentheses in the table correspond to the following numbered steps that can be used to obtain the solution.)

1. Since the Audi was driven by the American, the Porsche had to be driven by the Italian or the Spaniard, but from clue (*d*), we know it wasn't the Italian.
2. The Italian is the only driver left and, thus, was the driver of the MG.
3. From clue (*c*), we know that the lady from Spain drove the red car.
4. The Porsche was red, so the MG must either be black or silver (from clue (*b*)). But from clue (*a*), we know that the Italian did not drive the black car, thus the MG must be silver. The following table shows the correct solution:

Car	Audi	MG	Porsche
Nationality Color	American black	(2) Italian (4) silver	(1) Spanish (3) red

Problem 4 is similar to problem 3. The main difference is that no hint is given for the table. The following is one possible table:

Boxer	
Class	
Medal	

However, the type of table presented in the following solution is often easier to use. Be sure to discuss how the two tables are related. Again, the numbers appearing in parentheses within the table correspond to the numbered steps used to obtain the solution.

1. We know that Todd was the heaviest winner.
2. Since Grant defeated Mitchell, they must be in the same weight class. But from clue (*a*), we know that Brown is in the 118-pound class and that Todd is in the heavyweight class, so Grant and Mitchell must be in the 145-pound class and Grant won the gold medal.
3. Clue (*b*) says that Slater was the third winner.
4. Brown weighed in at 117 pounds and thus won the silver medal in the 118-pound class.
5. O'Brien is the only fighter left and, hence, won the silver medal in the heavyweight class.

The correct solution is this:

	118-Pound	145-Pound	Heavyweight
Gold Medal	(3) Slater	(2) Grant	(1) Todd
Silver Medal	(4) Brown	(2) Mitchell	(5) O'Brien

Worksheet 3: Solving a problem by eliminating possibilities often requires using an indirect argument, that is, assuming that something is true and showing that this assumption leads to a contradiction. The objective of the third worksheet is to introduce the concept of an indirect argument. The students are guided through portions of such an argument in problems 1 and 2 and then apply the techniques learned to solve problem 3.

Informal experiences with indirect reasoning such as those presented here provide an excellent introduction to the method of indirect proof that is so important in mathematics. We have found that students who have had this type of introduction often have far less difficulty with indirect proofs when they are formally introduced in geometry.

Students should be expected to write complete sentences as their answers to part (*e*) of the solution process in problem 1 and parts (*a*) and (*c*) of the solution of problem 2. The ability to express ideas clearly in writing is important and should be emphasized in all subject areas. Unfortunately, in mathematics classes, students are seldom required to express their thoughts in writing. Problems like these present an excellent opportunity for mathematics teachers to work with students in this critical area.

Problem 1 can be solved by examining each of the three possible solutions. Discussion of the solution should focus on the conclusion drawn in step (*e*) of each of the three cases.

Case 1: (*a*) Suppose that Adam spent \$0.60. (*b*) Since Adam spent four times as much as Carl, Carl must have spent \$0.15 (\$0.60/4). (*c*) Since Carl spent a third as much as Bruce, Bruce must have spent \$0.45 (\$0.15 × 3). (*d*) But Bruce also spent \$0.20 less than Adam, so Bruce spent \$0.40 (\$0.60 – \$0.20). (*e*) The results in parts (*c*) and (*d*) contradict each other. Thus, the assumption that Adam was the one who spent \$0.60 is incorrect.

Case 2: (*a*) Suppose Carl spent \$0.60. (*b*) Since Adam spent four times as much as Carl, Adam must have spent \$2.40 (\$0.60 × 4). (*c*) Since Carl spent a third as much as Bruce, Bruce must have spent \$1.80 (\$0.60 × 3). (*d*) But Bruce also spent \$0.20 less than Adam, so Bruce spent \$2.20 (\$2.40 – \$0.20). (*e*) The results in parts (*c*) and (*d*) contradict each other, so the assumption that Carl spent \$0.60 must be false.

Case 3: (*a*) Assume that Bruce spent \$0.60. (*b*) Since Carl spent one-third as much as Bruce, Carl must have spent \$0.20 (\$0.60/3). (*c*) Since Bruce spent \$0.20 less than Adam, Adam must have spent \$0.80 (\$0.60 + \$0.20). (*d*) But Adam also spent four times as much as Carl, so Adam spent \$0.80 (\$0.20 × 4). (*e*) No contradictions exist; therefore Bruce spent \$0.60. Had we considered the case of Bruce second, it would still have been necessary to check the third possibility (involving Carl) because the problem might have had multiple solutions. (Solution: Bruce spent $0.60.)

The following is a solution to the second problem. The most important parts of the solution are the indirect arguments in parts (*a*) and (*c*). (*a*) No. If any of the gumdrops in the bin were orange, then the bin would have contained both orange and black gumdrops, so it would have been labeled correctly. But this case would have contradicted the fact that all the bins were labeled incorrectly. (*b*) Black gumdrops. (*c*) No. If bin 2 contained orange and black gumdrops, then bin 1 would contain only orange gumdrops and would be labeled correctly. (*d*) Orange gumdrops; orange and black gumdrops. The correct solution is the following: bin 1—orange and black gumdrops; bin 2—orange gumdrops; bin 3—black gumdrops.

Problem 3 can best be solved by using a table to organize the information. Indirect arguments are used in steps 2, 3, and 6 of the solution, but the critical argument is that in step 2. The solution proceeds as follows:

	Hat	Coat
Jim	(3) Dan's	(2) Kevin's
Dan	(2) Kevin's	Peter's
Kevin	(1) Peter's	(6) Jim's
Peter	(4) Jim's	(5) Dan's

1. This fact was given in the problem.
2. Since Jim took the coat belonging to the man whose hat was taken by Dan, they must belong to the same person. The possibilities are Kevin's and Peter's, since no man has his own coat or hat. Suppose they belong to Peter. Then Dan would have Peter's hat and so would Kevin. But this situation is impossible, so they must belong to Kevin.
3. The only choices for the hat Jim has are Jim's and Dan's, but Jim can't have his own hat.
4. Jim is the only person whose hat has not been accounted for.
5. Dan's coat was taken by the person who took Jim's hat.
6. The only choices for the coat Kevin has are Jim's and Peter's, but it can't be Peter's, since no person took the coat and the hat of the same man.

(The solution is this: Jim had taken Dan's hat and Kevin's coat. Dan had taken Kevin's hat and Peter's coat.)

Worksheet 4: This worksheet provides additional experiences using indirect reasoning in conjunction with problem-solving strategies that entail the meshing of several skills, including that of eliminating possibilities.

In problem 1, students will no doubt put the names of the states in the table first. The forget-me-not can quite easily be placed under Alaska and the yellowhammer under Alabama. (Here we include a little geography in the mathematics lesson and assume some knowledge of these states.) The key clue now is "mistletoe." Since it is not the flower of Minnesota, it must be the flower of Oklahoma or Alabama. But flycatchers nest in mistletoe, and the yellowhammer has already been identified with Alabama. Therefore, the flycatcher and the mistletoe must be placed under Oklahoma. Loons and lady's slippers go together, so they must go with Minnesota. Thus the problem is solved. The table shows the correct solution:

State	Ala.	Alaska	Minn.	Okla.
Flower	camellia	forget-me-not	lady's slipper	mistletoe
Bird	yellowhammer	ptarmigan	loon	flycatcher

Problem 2 is a straightforward application of an indirect argument. Assume that the first or second jogger is Fred, who always tells the truth. Both cases contain a contradiction. Therefore, the third jogger is Fred. Since he always tells the truth, the jogger in front is Joe and the one in the middle is Herman. (Solution: from left to right, the joggers are Joe, Herman, and Fred.)

In solving problem 3, students must build a table, list possible solutions, and use indirect arguments to determine correct solutions. An initial table might look like this:

Linda	Mary	Kim
not artist	not police officer	police officer
not police officer		

Clue (*b*) tells us about Linda and clue (*e*) about Mary. Therefore, Kim must be the police officer and Mary the artist. Kim beat Mary and the computer programmer at handball, so Linda is the programmer. The car salesperson offended the artist and sold the programmer a car, so Kim also sells cars. The programmer and the pilot are different people, and the problem is solved. (Solution: Linda is a computer programmer and a mechanic; Mary is an artist and a pilot; and Kim is a police officer and a car salesperson.)

Supplementary problems: The following problem can be used to reinforce the problem-solving strategies involving elimination. Its solution, however, has a slightly different twist. Students must first generate possible answers to the five questions using clues (*a*), (*b*), and (*d*).

Problem: A history exam has five true-false questions.

a) The exam has more true than false answers.

b) No three consecutive questions have the same answer.

c) Juanita knows the correct answer to the second question.

d) Questions 1 and 5 have opposite answers.

e) From this information, Juanita was able to determine all the correct answers. What are they?

Solution: Using clues (*a*), (*b*), and (*d*), the possible answers to the five questions are these:

1. T T F T F
2. T F T T F
3. F T F T T
4. F T T F T

The key clue now is (*c*). Juanita knows the correct answer to question 2. If the correct answer is true, the other questions have three combinations of answers. Clue (*e*), however, says that she was able to determine *all* the correct answers. Thus, the only possible answer for question 2 is false. The correct answers for the test are T, F, T, T, F.

Additional problems that offer opportunities to practice the skills developed here can be found in Charosh (1965), Dolan and Williamson (1983), and Hunter and Madachy (1963).

REFERENCES

Charosh, M., comp. *Mathematical Challenges.* Reston, Va.: National Council of Teachers of Mathematics, 1965.

Dolan, Daniel T., and James Williamson. *Teaching Problem-solving Strategies.* Menlo Park, Calif.: Addison-Wesley Publishing Co., 1983.

Hunter, J. A. H., and Joseph S. Madachy. *Mathematical Diversions.* Princeton, N.J.: D. Van Nostrand Co., 1963.

Krulik, Stephen, and Jesse A. Rudnick. "Activities: Developing Problem-solving Skills." *Mathematics Teacher* 78 (December 1985):685–92, 697–98.

Laing, Robert A. "Extending Problem-solving Skills." *Mathematics Teacher* 78 (January 1985):36–44.

Polya, George. *How to Solve It.* Princeton, N.J.: Princeton University Press, 1971.

Schaaf, Oscar F. "Teaching Problem-solving Skills." *Mathematics Teacher* 77 (December 1984):694–99.

1. Ace Detective Shamrock Bones of the City Homicide Squad is investigating a murder at the Old Grand Hotel. Five men are being held as suspects.
   - "Giant Gene" Green. He is 250 cm tall, weighs 140 kg, and loves his dear mother so much that he never spent a night away from home.
   - "Yoko Red." He is a 200-kg Sumo wrestler.
   - "Hi" Willie Brown. He is a small man—only 130 cm tall; he hates high places because of a fear of falling.
   - "Curly" Black. His nickname is a result of his totally bald head.
   - Harvey "The Hook" White. He lost both his hands in an accident.

   Use the following clues to help Detective Bones solve this crime.

   a) The killer was registered at the hotel.
   b) Before he died, the victim said the killer had served time in prison with him.
   c) Brown hair from the killer was found in the victim's hand.
   d) The killer escaped by diving from the third-floor balcony into the river running by the hotel and then swimming away.
   e) Smudges were found on the glass tabletop, indicating that the killer wore gloves.

The killer is ______.

2. Circle the number below described by the following clues:
   a) The sum of the digits is 16.
   b) The number has more than three digits.
   c) The number is a multiple of 5.
   d) It is not an even number.
   e) The number is less than 2572.

   871 745 3625 2860 2582 1780 2315
   1937 1485 1375 1671 1455
   1075 1690 2635 2590

3. Circle the number described by these clues.
   a) No two digits are alike.
   b) The sum of the two middle digits is equal to the sum of the first and last digits.
   c) The thousands digit is the smallest.
   d) None of the digits are even.
   e) The units digit is the largest.

   9315 7351 8316
   1537 5713
   9731 1449 3714
   9371 7951
   3174 1627 7624
   1818 1592
   1569 1963

1. Andrea collects coins. She wanted to arrange her collection in a display case in rows that all contained the same number of coins. She tried putting two coins in each row, but she had one coin left over. When she tried three or five coins in a row, two coins were left over. Finally, she tried seven coins in each row and it worked! What's the smallest number of coins Andrea could have in her collection? ______

   Hint: Use the clues to make an organized list of possibilities and then use elimination.

2. The three teams in the Western Volleyball League had a season in which each team played the other two both at home and away. The season's schedule is shown in the following table. Only one game was played on a given day.

Day	1	2	3	4	5	6
Opponents	Spikers Setters	Spikers Servers	Setters Servers	Spikers Servers	Setters Servers	Spikers Setters

   a) The Spikers never defeated the Setters.
   b) Even though the Servers lost two games, they never lost a game at home.

   What was the win-loss record of each of the three teams? ______________

   ____

3. A recent international auto rally resulted in a three-way tie for first place. The three cars, an Audi, an MG, and a Porsche, were driven by three women from Italy, Spain, and the U.S. Each car was a different color. Use the following clues to determine the color of each car and the nationality of the driver.

   a) The Italian was not in the black car.
   b) The Audi, driven by the American, was not silver.
   c) The red car was driven by the woman from Spain.
   d) The Porsche was not driven by the Italian.

Hint: Set up a table like this to help organize the information.

Car	Audi	MG	Porsche
Nationality			
Color			

4. Brown, Todd, Slater, Mitchell, O'Brien, and Grant reached the final round of the Olympic boxing championships. They were the finalists in the 118-pound, 145-pound, and heavyweight classes.

   a) Brown weighed in at 117 pounds, and Todd was the heaviest winner of all.
   b) Grant defeated Mitchell, and Slater won by a knockout.

   Who were the gold and silver medal winners in each weight class? ________

   ____

1. Three boys each purchased some fruit for a snack. Adam spent four times as much as Carl. Carl spent one-third as much as Bruce, and Bruce spent $0.20 less than Adam. Which boy, if any, spent $0.60?

   Since only three possibilities exist, let's examine the possible solutions individually.

   a) Suppose Adam spent $0.60.

   b) Since Adam spent four times as much as Carl, Carl must have spent ________.

   c) Since Carl spent a third as much as Bruce, Bruce must have spent ________.

   d) But, Bruce also spent $0.20 less than Adam, so Bruce spent ________.

   e) What do the results in parts (*c*) and (*d*) suggest? ____________________

   Continue this method of *indirect reasoning* by next assuming that Carl spent $0.60; and so on.

2. Eileen was visiting her uncle's candy factory. In the storeroom, she found three bins. Bin 1 was labeled "orange gumdrops," bin 2 was labeled "black gumdrops," and bin 3 was labeled "orange and black gumdrops." Her uncle told her that she could have one gumdrop from one of the bins, but he warned her that none of the bins was labeled correctly. Eileen reached into the bin labeled "orange and black gumdrops" and pulled out a black gumdrop.

   a) Could any of the gumdrops in the bin labeled "orange and black gumdrops" be orange? ________ Why? ____________________

   b) What is the correct label for bin 3? ____________________

   c) Can the bin labeled "black gumdrops" contain orange and black gumdrops? ________ Why? ____________________

   d) What is the correct label for bin 2? ________________

   For bin 1? ________________

3. Four men attended the ballet with their wives. In their hurry to leave, each man mistakenly picked up another man's coat and the hat of yet another man and left. Jim took the coat belonging to the man whose hat was taken by Dan, and Dan's coat was taken by the person who took Jim's hat. Kevin took Peter's hat. Whose hat and whose coat had Jim and Dan taken? ______

   ____

1. A group of students were playing a trivia game involving states, state birds, and state flowers. Of four states—Alaska, Oklahoma, Minnesota, and Alabama—they knew the birds were the common loon, yellowhammer, willow ptarmigan, and scissortailed flycatcher, and that the flowers were mistletoe, camellia, forget-me-not, and pink-and-white lady's slipper. No one knew which bird or flower matched which state. A call to the local library resulted in the following clues.

   a) The forget-me-not is from the northernmost state.

   b) The flycatcher loves to nest in the mistletoe.

   c) The willow ptarmigan is not the bird for the camellia state.

   d) The yellowhammer is from a southeastern state.

   e) Mistletoe and Minnesota do not go together, but loons and lady's slippers do.

   Use the clues to fill in the table below.

State				
Flower				
Bird				

2. Three joggers named Fred, Herman, and Joe are jogging toward the country club. Fred always tells the truth. Herman sometimes tells the truth, whereas Joe never does.

   Determine the names of each runner and explain how you know. (*Hint:* First, determine which one is Fred.)

3. Linda, Mary, and Kim have each been employed in two of the following jobs: artist, mechanic, computer programmer, car salesperson, pilot, and police officer. No two ever had the same job. Use the following information to determine each of their jobs.

   a) The car salesperson offended the artist by criticizing her work.

   b) The artist and the police officer dated Linda's brother.

   c) The computer programmer and the pilot both got A's in mathematics in high school.

   d) The car salesperson sold the computer programmer a car.

   e) Mary got a ticket from the police officer.

   f) Kim beat both Mary and the computer programmer at handball.

ACTIVITIES

THE PEG GAME

January 1982

By Charles Verhille and Rick Blake, University of New Brunswick, Fredericton, NB E3B 6E3

Teacher's Guide

Grade Level: 6–12.

Materials: Peg games may be purchased commercially or made from inexpensive materials.

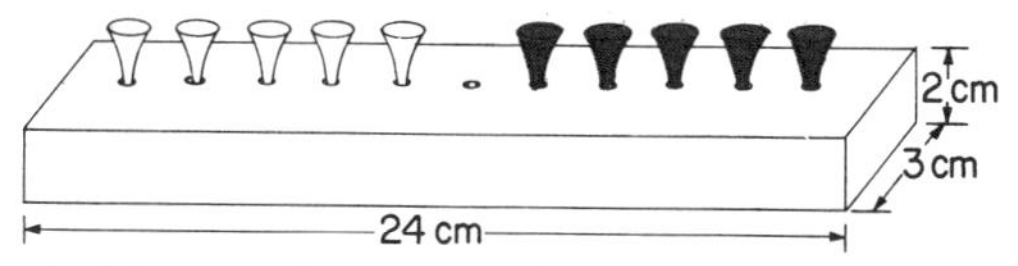

A piece of wood with the dimensions indicated can serve as the game board. Eleven equally spaced holes are drilled as illustrated. Five each of two different colored pegs are arranged, as shown, on the board. Golf tees work well for pegs.

A modified version of the game can be played using paper strips of eleven 1-inch squares for the playing board. Instead of pegs, use ten coins, five heads and five tails, to represent the colors white and black.

Objectives: To collect and organize data; to recognize, extend, and generalize patterns; and to develop a problem-solving strategy.

Procedure: The teacher should be familiar with the entire activity before attempting it with the students.

Duplicate a set of sheets for each student. The individual activities are to be done separately and in the order given. They may be presented to individual students or small groups or used in a learning center. It may be helpful to have students work in pairs so that one student can do the moving while the other counts, checks, and records.

• *Counting moves.* Distribute game boards and pegs. Students will normally complete this task at different times, allowing the teacher time to check each student's strategy before continuing. Ensure that each student demonstrates an appropriate strategy, since this is necessary for the other activities.

Some guidance may be necessary to

find patterns that lead to answers for question 5.

• *Recording the sequence of moves.* Be sure that all students understand the notation R for move a *white peg right* and L for move a *black peg left*. Note also that the sequences are set up with an R move first.

• *Counting jumps.* Here the student studies the previously developed sequences and discovers that the number of jumps needed is the square of the number of pairs of pegs at the start.

Answers:

1–3 (See answers in the table for question 7.)

4.

Number of Pairs	Number of Moves
1	3
2	8
3	15
4	24

5. *a.* 35

b. 120

The rule may come from patterns like the following:

(1) Noticing how the differences for the "number of moves" are related

3 > 5 > 2
8 > 7 > 2
15 > 9
24

(2) The next number of pairs squared minus 1: $(n + 1)^2 - 1$

(3) The number of pairs squared plus twice itself: $n^2 + 2n$

(4) The number of pairs times the number of pairs plus 2: $n(n + 2)$

7.

Number of Pairs	Sequence of Moves
1	*RLR*
2	*RLLRRLLR*
3	*RLLRRRLLLRRRLLR*
4	*RLLRRRLLLLRRRRLLLLRRRLLR*

8.

Number of Pairs	Sequence of Moves
1	1, 1, 1
2	1, 2, 2, 2, 1
3	1, 2, 3, 3, 3, 2, 1
4	1, 2, 3, 4, 4, 4, 3, 2, 1

9. 1, 2, 3, 4, 5, 5, 5, 4, 3, 2, 1

10. 7 pairs

11.

Number of Pairs	Number of Jumps
1	1
2	4
3	9
4	16

12. *a.* 25

b. 100

c. n^2

Extensions: Students might be interested in exploring the effects of changing rule C on sheet 1 of the peg game to one of these:

1. You can move a peg into an adjacent empty hole or jump over *one or two pegs* of the opposite color into an empty hole.

2. You can move a peg into an adjacent empty hole or jump over *any number of pegs* of the opposite color.

Interestingly, the number of moves needed to interchange the pegs with either of these modifications remains the same as with the original rule for a single jump.

BIBLIOGRAPHY

Duncan, Richard B. "The Golf Tee Problem." *Mathematics Teacher* 72 (January 1979):53–57.

Higginson, William. "Mathematizing 'Frogs': Heuristics, Proof, and Generalization in the Context of a Recreational Problem." *Mathematics Teacher* 74 (October 1981):505–15.

Rules for the Game

The object of this peg game is to interchange the black and white pegs using the following rules:

A. You can move only one peg at a time.
B. You can move white pegs only to the right and black pegs only to the left.
C. You can move a peg into an adjacent empty hole or jump over a single peg of the opposite color into an empty hole. You are not allowed to jump over two or more pegs.

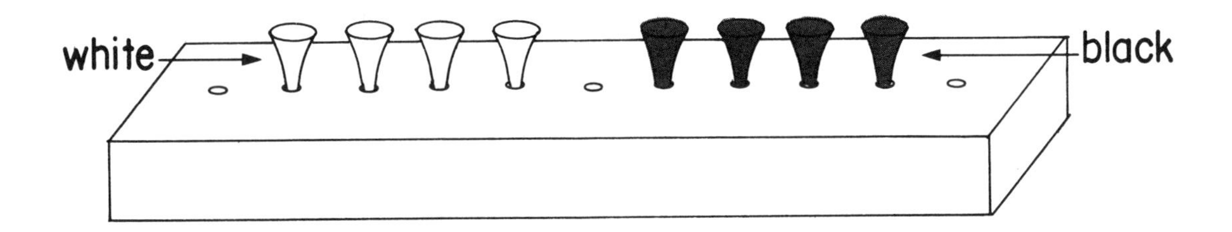

To find the number of moves needed for four pairs of pegs as shown above, first try the simpler problems that follow.

Counting Moves

1. Show how you can interchange one pair of pegs with a space between them in just three moves.

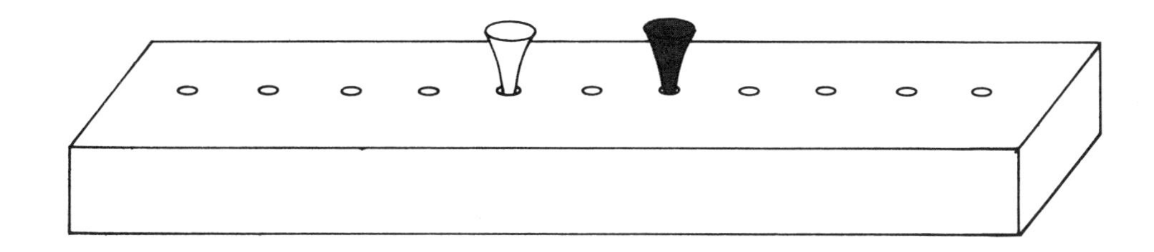

2. Now try two pairs of pegs. See if you can interchange them in eight moves.

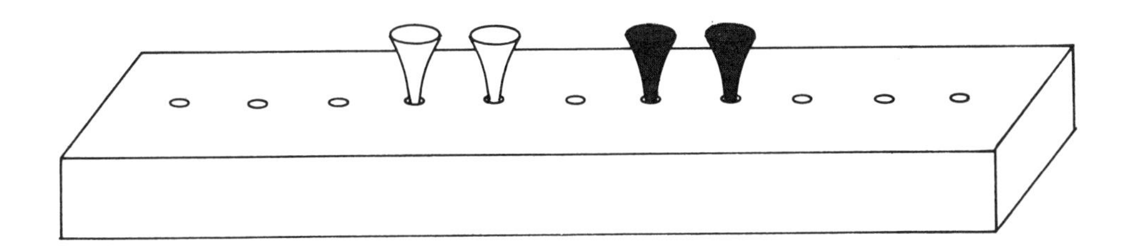

3. Next try to interchange three pairs of pegs. Look for a strategy. Count your moves carefully.

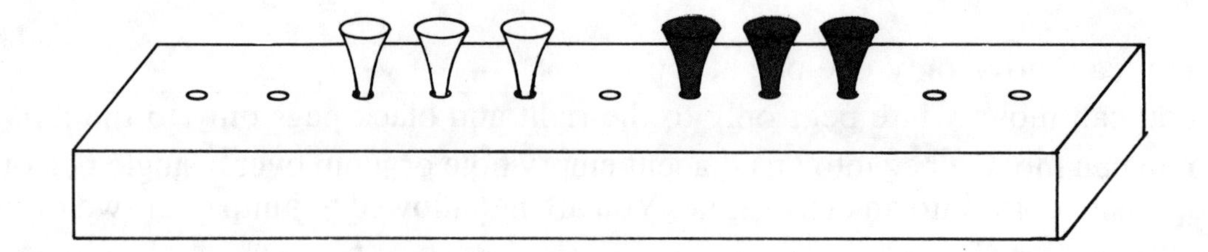

4. Complete the table below. Note that a pair consists of one peg of each color and that only one space separates the colors.

Number of Pairs	Number of Moves
1	3
2	8
3	
4	

5. Look for a pattern in your completed table that can be used to predict the number of moves for the following:

   a. five pairs of pegs?

   b. ten pairs of pegs?

   What rule can you use to find the answer for any number of pairs?

Recording the Sequence of Moves

Suppose you always begin with a move to the right. Record the consecutive moves as *R* if you move a peg to the right and *L* if you move a peg to the left. Stop when all pegs have been switched.

6. The result for two pairs of pegs is *RLLRRLLR*. Verify that this sequence of moves does interchange the pegs.

7. Complete the table.

Number of Pairs	Sequence of Moves
1	
2	*RLLRRLLR*
3	
4	

The sequence of moves can be recorded by the number of consecutive moves in a given direction, starting with a right-hand move.

Thus the sequence *RLLRRLLR* can be expressed as 1, 2, 2, 2, 1.

$$\underbrace{R}_{1}\ \underbrace{LL}_{2}\ \underbrace{RR}_{2}\ \underbrace{LL}_{2}\ \underbrace{R}_{1}$$

8. Complete the table using this numerical notation.

Number of Pairs	Sequence of Moves
1	
2	1, 2, 2, 2, 1
3	
4	
5	

9. Write the numerical sequence for five pairs of pegs.

10. How many pairs of pegs would produce this sequence of moves?

1, 2, 3, 4, 5, 6, 7, 7, 7, 6, 5, 4, 3, 2, 1

Counting Jumps

An interesting pattern emerges when you count only the jumps made in interchanging the pegs.

11. Complete the table below, recording only the number of jumps needed to switch 1, 2, 3, and 4 pairs of pegs.

Number of Pairs	Number of Jumps
1	
2	
3	
4	

12. Without using your board, how many jumps do you think will be needed for—

a. 5 pairs of pegs?__________

b. 10 pairs of pegs?__________

c. *n* pairs of pegs?__________

ACTIVITIES

PROBLEM SOLVING WITH COMPUTERS

October 1984

BY DWAYNE E. CHANNELL, Western Michigan University, Kalamazoo, MI 49008

Teacher's Guide

Introduction: Mathematics programs must equip students with the mathematical methods that support the full range of problem solving, including the use of the problem-solving capacities of computers to extend traditional problem-solving approaches and to implement new strategies of interaction and simulation.

Grade levels: 7–11

Materials: Paper (preferably centimeter graph paper), scissors, tape, and copies of the four worksheets. Access to a microcomputer is essential.

Objectives: To develop problem-solving skills involving the generation, summarization, and interpretation of computer-generated data

Directions: Distribute copies of sheets 1 and 2 to each student. The completion of these worksheets will require one or two class periods, depending on the level and background of your students.

Sheet 1: Have students read the problem carefully. To be certain that students understand the procedure used in constructing the open-topped boxes, have each student cut square corners from a piece of rectangular paper and form the resulting shape into a box. Don't insist that every student cut squares of the same size, since the nature of the problem will be much clearer if a variety of boxes are available for comparison. Allow students to complete the remainder of sheet 1.

Sheet 2: Have students enter and run the program on a microcomputer. The program was written for the Apple II using Applesoft BASIC, but it should run with little or no modification on other computers. Students should use the output generated by the program to complete exercises 5 through 7. In exercise 8, it may not be obvious to students that the optimal solution falls somewhere between 6 cm and 8 cm. It might be helpful to point out that the table

This activity consists of two problem situations, each of which illustrates how a computer can be used as a tool to assist pupils in solving mathematical problems. Following an organizational scheme used by Channell and Hirsch (1984), each situation is presented using a four-stage Polya-type problem-solving model: problem statement, analysis, solution, and looking back. The first problem, "Maximizing Volume," uses the computer to perform straightforward numerical calculations that would be tedious without the computer. The computer program does not provide the solution to the problem but rather tabular data to be analyzed by the student. The second problem, "Collecting Stickers," introduces the use of a Monte Carlo model to simulate a physical action. Again, the computer program does not give the solution to the problem. Students must collect, tabulate, and summarize data generated by the program to arrive at an approximation of the solution. Each situation illustrates important and powerful problem-solving techniques and increases students' understanding of the role computers can play in problem solving.

of volumes increases as h moves from 1 cm to 7 cm and the volumes decrease as h moves from 7 cm to 17 cm. The actual maximum volume can occur on either side of 7 cm, and so the search must be made between 6 cm and 8 cm. Students with little or no programming experience will need some assistance in completing exercise 9. Point out that a 44 cm × 36 cm rectangle and a 66 cm × 24 cm rectangle have equal areas, 1584 cm^2. However, the volume of the largest open-topped boxes are not equal. More able students might be challenged to find the dimensions of a rectangular sheet with an area of 1584 cm^2 that would produce the open-topped box with maximum volume.

Distribute copies of sheets 3 and 4 to each student. Students should be able to complete them in a single class period.

Sheet 3: Have students read the problem carefully. To be certain that the simulation process is clear to all students, simulate this problem using labeled slips of paper in a box or hat before discussing the computer program. Use the following procedure:

(*a*) Place the name of each arcade game on a separate slip of paper or cardboard, making certain that all slips are of the same size.

(*b*) Place the six slips into a container so that no one can see them.

(*c*) Draw a single slip from the container and record the name using a tally mark in a table similar to the following:

Q-Bert	Burger Time	Ms. Pac-Man
Zaxxon	Donkey-Kong	Frogger

(*d*) Replace the slip that was drawn and mix the six slips well before drawing again.

(*e*) Continue drawing, recording, and replacing slips until one for each of the six arcade games has been drawn.

(*f*) The sum of the tally marks gives one approximation to the theoretical expectation.

Discuss the program. Have students enter the program, run it on a microcomputer, and organize the output in the table provided. The program was written for the Apple II using Applesoft BASIC and will most likely need slight modification for other computers. Line 100 serves the function of the RANDOMIZE command found in many BASIC dialects. The RND(X) function used in line 110 may require an argument other than 1, or it may require no argument at all in BASIC dialects other than Applesoft.

Sheet 4: Students should use the results from sheet 3 to answer exercise 1 at the top of sheet 4. The computer program should be run four more times and an average of the results computed. Exercise 5 might best be handled as a teacher-directed activity, since it involves collecting and summarizing data generated by the entire class.

Supplementary activities: In the problem about maximizing volume, students can be asked to plot the data generated by the program at the top of sheet 2 on a coordinate system. If the seventeen points are plotted and a smooth curve sketched to connect them, a graph similar to figure 1 results. A discussion of the relationships between the tabular data and the increasing-decreasing nature of the graph may prove instructive.

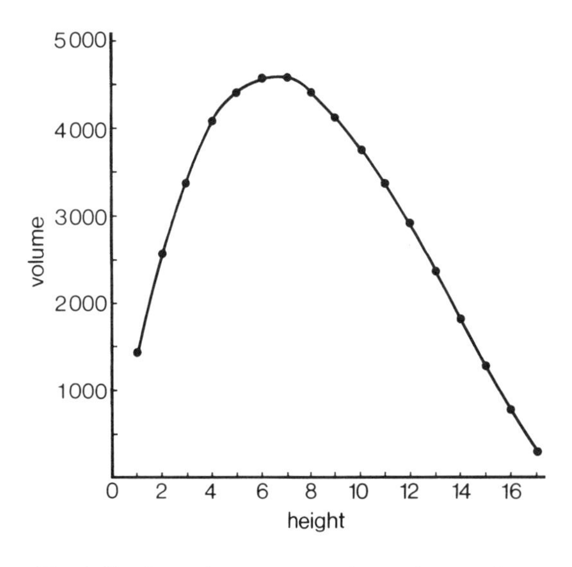

Fig. 1. Plotting points to suggest the maximum volume

For the problem about collecting stickers, students with a good programming background may enjoy writing a program that simulates this problem *and* summarizes the results for many customers. The program in table 1 summarizes results for 100 customers. In addition, it allows the user to INPUT the number of stickers in the collection. Students might enjoy running this program for the cases of 2, 3, 4, ..., 10 stickers. The results can be plotted on a coordinate system to provide a graphic representation of the increasing expected values.

TABLE 1

```
100  REM STICKER COLLECTION SIMULATION
110  HOME : POKE (203), PEEK (78)
120  INPUT "HOW MANY STICKERS TO
       COLLECT?";N
130  PRINT : PRINT "NOW BUYING MUNCHIES
       FOR 100 CUSTOMERS."
140  PRINT : PRINT "PLEASE WAIT."
150  PRINT : PRINT "EACH * REPRESENTS A
       FINISHED CUSTOMER.": PRINT
160  LET C = 0
170  FOR I = 1 TO 100
180  FOR S = 1 TO N
190  LET F(S) = 0
200  NEXT S
210  LET T = 0
220  LET S = INT (N * RND (1)) + 1
230  LET F(S) = F(S) + 1
240  LET T = T + 1
250  FOR S = 1 TO N
260  IF F(S) = 0 THEN 220
270  NEXT S
280  LET C = C + T
290  PRINT "*";
300  NEXT I
310  PRINT : PRINT : PRINT
320  PRINT "AVERAGE CUSTOMER PURCHASED
       ";C / 100;" BAGS"
330  PRINT "OF MUNCHIES TO COLLECT ";N;"
       STICKERS."
340  PRINT : PRINT : PRINT
350  INPUT "RUN THE PROGRAM AGAIN
       (Y OR N)?";A$
360  IF A$ = "Y" THEN 110
370  END
```

Answers: Sheet 1: 1. $h = 4$ cm; $l = 44 - 2(4) = 36$ cm; $w = 36 - 2(4) = 28$ cm; $V = 36 \times 28 \times 4 = 4032$ cm^3. 2. (*a*) $h = 10$ cm; $l = 24$ cm; $w = 16$ cm; (*b*) $V = 3840$ cm^3. 3. $l = 44 - 2h$; $w = 36 - 2h$. 4. Since $36 - 2h > 0$, we must have $h < 18$, and thus the maximum integral value for h is 17 cm.

Sheet 2: 5. Yes (it is hoped). 6. (*a*) 4620 cm^3; (*b*) 7 cm 7. 7 cm 8. (*a*) 6.6 cm; (*b*) The volume corresponding to a height of 6.6 cm is 4634.8 cm^3, which is larger than the 4620 cm^3 volume resulting from a height of 7 cm. (It can be shown using calculus that the maximum volume of 4634.87 cm^3 occurs with a height of 6.58 cm.) 9. Lines 110, 120, and 130 of the program should be changed as follows:

```
110  FOR H = 1 TO 11
120  LET L = 66 - 2*H
130  LET W = 24 - 2*H
```

This program gives a maximum volume of 3920 cm^3 when $h = 5$ cm. Changing line 110 to read

```
110  FOR H = 4 TO 6 STEP .1
```

gives a maximum volume of 3934.7 cm^3 when $h = 5.4$ cm.

Sheet 4: 1. Answer will vary; no; the result from one customer is probably not a good estimate of the actual solution. 2. (*a*) 6 bags; (*b*) The maximum number depends on the number of bags produced by the company. For example, if 600 000 bags were made, it is theoretically possible (but highly improbable) that a very determined, but unlucky, customer would need to buy 500 001 bags to collect all six stickers! 3, 4, 5. Answers will vary. The expected (theoretical) result for the six-sticker problem is 14.7 purchases. 6. Results will vary. The expected result for the eight-sticker problem is 21.7 purchases.

BIBLIOGRAPHY

Channell, Dwayne E., and Christian Hirsch. "Computer Methods for Problem Solving in Secondary School Mathematics." In *Computers in Mathematics Education,* 1984 Yearbook of the National Council of Teachers of Mathematics, edited by Viggo P. Hansen, pp. 171–83. Reston, Va.: The Council, 1984.

Lappan, Glenda, and M. J. Winter. "Probability Simulation in the Middle School." *Mathematics Teacher* 73 (September 1980):446–49.

National Council of Teachers of Mathematics. *An Agenda for Action: Recommendations for School Mathematics of the 1980s.* Reston, Va.: The Council, 1980.

Travers, Kenneth J., and Kenneth G. Gray. "The Monte Carlo Method: A Fresh Approach to Teaching Probabilistic Concepts." *Mathematics Teacher* 74 (May 1981):327–34.

Ms. Hawes needs several open-topped boxes for storing laboratory supplies. She has given the industrial arts class several rectangular pieces of sheet metal to form the boxes. The pieces measure 44 cm × 36 cm. The class plans to make the boxes by cutting equal-sized squares from each corner of a metal sheet, bending up the sides, and welding the edges. Ms. Hawes has asked that each box have the largest possible volume. What size of squares should be cut from the corners of the metal sheets?

Analysis

To help you visualize the problem, take a sheet of paper and cut equal-sized squares from its corners. You will get a shape like the figure shown below.

Fold the side tabs up to form a box and tape the edges together. Compare your open-topped box with those made by your classmates. Note that the size of square cut from the corners of the paper determines the height (h) of the box as well as the length (l) and width (w).

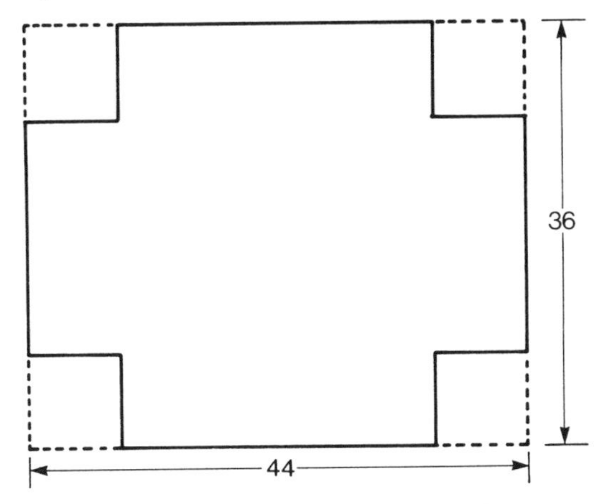

1. Assume that Ms. Hawes suggested cutting 4-cm squares from the metal sheets. What would be the dimensions of the box formed?

$h =$ ____ cm, $l = 44 - 2(4) =$ ____ cm, $w = 36 - 2(\_\_) =$ ____ cm

The volume (V) of this box is given by $V = lwh =$ ____ cm^3.

2. Assume that 10-cm squares are removed from the corners.

(*a*) Give the dimensions of the corresponding box:

$h =$ ____ cm, $l =$ ____ cm, $w =$ ____ cm

(*b*) Find the volume of the box. ______

3. If squares of size h cm are removed from the corners, then the height of the box formed would be h cm. Complete the expressions that represent the length and width of the box.

$l = 44 - 2(\_\_)$ cm and $w =$ ____ cm

4. Many different-sized squares can be cut from the metal sheets. Let h represent the size of the square cut. If h is measured to the nearest whole centimeter, the smallest possible h is 1 cm. What is the largest possible value for h? (*Hint:* The length and width of the box must both be positive values.)

____

Solution

You have found that the height h of the box can range from 1 cm to 17 cm depending on the size of the square cut from the corners of the rectangular metal sheets. The following BASIC program computes the dimensions and corresponding volumes of all seventeen boxes. Study the program, enter it into a computer, and RUN it.

```
100 PRINT "HEIGHT", "VOLUME"
110 FOR H = 1 TO 17
120 LET L = 44 - 2 * H
130 LET W = 36 - 2 * H
140 LET V = L * W * H
150 PRINT H,V
160 NEXT H
170 END
```

Looking Back

5. Examine the table of numbers produced by the program. Do the volumes corresponding to heights of 4 cm and 10 cm agree with your computations in exercises 2 and 3 on sheet 1? ____

6. (*a*) What is the largest volume listed in the table? ____
   (*b*) What is the corresponding height? ____

7. What size of square should be cut from the metal sheets to produce a box with this maximum volume? ____

8. To the nearest whole centimeter, a box with $h = 7$ cm gives the maximum volume. If the value of h is measured to the nearest tenth of a centimeter, the height of the box with maximum volume would still lie somewhere between 6.0 cm and 8.0 cm but not necessarily at 7.0 cm. Change line 110 of the program to read as follows:

```
110 FOR H = 6 TO 8 STEP .1
```

RUN this modified program and study the output.

(*a*) To the nearest tenth of a centimeter, what is the height of the box that gives the maximum volume? ____

(*b*) Is this an improvement over the whole-number solution of $h = 7$ cm?

____

9. Assume that the industrial arts class had sheets measuring 66 cm × 24 cm. Modify the computer program to help you determine the height of the open-topped box with maximum volume that could be made from these sheets.

Junk Foods, Inc., introduced a new cheese-flavored snack food called "Cheese Munchies." To encourage people to buy their new product, they included a colorful sticker depicting a popular arcade game in each large-sized bag of the snack food. The games depicted are Q-Bert, Burger Time, Ms. Pac-Man, Zaxxon, Donkey-Kong, and Frogger. Equal numbers of each of the six stickers were randomly distributed in the first production of Cheese Munchies. If you, as an average customer, wanted to collect all six stickers, about how many bags of Cheese Munchies would you have to buy? Your guess: ____

Analysis

One approach to solving this problem is to simulate, or imitate, buying a bag of Cheese Munchies and checking which one of the six stickers was included. Would you believe that a computer can be used to simulate the buying process?

Let's define a correspondence between the names of the arcade games depicted on the stickers and the numbers from 1 to 6 inclusive:

Q-Bert	Burger Time	Ms. Pac-Man	Zaxxon	Donkey-Kong	Frogger
↕	↕	↕	↕	↕	↕
1	2	3	4	5	6

The following program generates random whole numbers between and 1 and 6 inclusive:

```
100  POKE(203), PEEK(78)
110  PRINT INT(6 * RND(1)) + 1,
120  INPUT "ANOTHER NUMBER (Y OR N)?";A$
130  IF A$ = "Y" THEN 110
140  END
```

Each number generated by the program can represent the purchase of a bag of Cheese Munchies. The value of the number generated determines which sticker was in the bag. Numbers should be generated by the program until each of the numbers 1–6 has been printed.

Solution

Enter and RUN the program. Use the following table to keep a tally of the number of each different sticker collected by customer A. Stop when you have collected all six numbers. Find the total number of bags purchased.

Customer A

Sticker Number	1	2	3	4	5	6
Bag Tally						

Total bags purchased ____

Looking Back

1. How many bags of Cheese Munchies did customer A buy before obtaining all six stickers? ____ Do you think that this answer represents the exact solution to the problem? ____

2. (*a*) What is the minimum number of bags a really lucky person would have to buy to collect all six stickers? ____

(*b*) Is there a maximum number of bags that a very unlucky person would have to buy? (*Hint:* Assume the company produced 600 000 bags of Cheese Munchies in their first production run.) ____

3. The results for customer A do not necessarily represent the expectations of the "average" customer. To better approximate the number of bags purchased by the average customer, the problem should be simulated several times. RUN the program again for each of the following customers. Keep a tally of the results in the tables and then find the total number of bags purchased by each customer.

Customer B

1	2	3	4	5	6

Total bags _____

Customer C

1	2	3	4	5	6

Total bags _____

Customer D

1	2	3	4	5	6

Total bags _____

Customer E

1	2	3	4	5	6

Total bags _____

4. (*a*) What is the total number of bags of snack food purchased by all five customers (A, B, C, D, and E)? ____

(*b*) What is the average number of bags purchased by these five customers? ____

5. The last answer in exercise 4 represents an estimate of the solution to the sticker problem. A better estimate can be found by simulating the purchases of a larger number of customers and averaging the number of bags they purchased. With the help of your teacher, collect information from classmates who have worked on this problem. You will want to know how many customers they have simulated and how many bags were purchased by each of these customers. Calculate the average number of bags purchased by all these customers.
What is a reasonable solution to the problem stated at the top of sheet 3? ____

6. Suppose the company decided to include eight different stickers instead of six. (Centipede and Buck Rogers were added.) How many bags would the typical customer expect to buy to collect all eight stickers? Change line 110 of the program to read `110 PRINT INT(8 * RND(1)) + 1`. What is a reasonable estimate of the solution to this problem? ____

Activities for

Numeracy

The use of numbers to quantify situations throughout society suggests that the development of numeracy should occupy a major role in school mathematics. The importance of quantitative literacy has been magnified by the phenomenal growth of technology and the need to evaluate results of this technology in everyday-life situations. The activities in this section promote thinking and reasoning in quantitative terms with an emphasis on the use of estimation strategies and calculators.

The first activity, "Estimating with 'Nice' Numbers," develops the skill of rounding to "nice numbers" (powers of 10), when the context permits, to obtain quick estimates of products and quotients and then introduces strategies for refining these estimates. Students use these skills to judge the reasonableness of given calculations as well as to estimate answers to a variety of routine problems arising in daily life. "Estimating with 'Nice' Fractions" uses a similar approach to estimate sums of fractions and mixed numbers by rounding the fractions to 0, 1/2, or 1. Everyday-life situations requiring the estimation of fractional parts of numbers are explored using unit fractions and "compatible" whole numbers as an estimation strategy.

"Calculators and Estimation" is designed to further develop the student's ability to estimate products and quotients of whole numbers and decimals. Students first estimate a factor (divisor) so that indicated products (quotients) fall within a given range. The number of tries needed to achieve success is recorded. These experiences are then extended to situations in which a "start" number is given and then estimation and successive multiplication (division) are used to obtain a product (quotient) in a given range.

The activity "Developing Estimation Strategies" extends the "compatible-numbers strategy" (for estimating quotients involving whole numbers and products involving fractions and percents), introduces averaging and rounding to the nearest half (to estimate totals), and provides opportunities for pupils to apply these strategies in real-life contexts. "Estimate and Calculate" continues the development of these abilities. In this activity pupils form, from a given set of digits, two numbers whose product (quotient or percent the first is of the second) will best approximate a given number. Their "off-scores" are analyzed and the information used to adjust their estimates so as to improve their subsequent selections.

"Estimation, Qualitative Thinking, and Problem Solving" is designed to develop numeracy through the estimation of measurement situations occurring in daily life. The activity provides students with opportunities to design algorithms for estimating measurements appropriate for particular situations and culminates with problem-solving experiences in which data necessary to the solution of the problem must be identified and estimated in the problem-solving process.

The final activity, "Examining Rates of Inflation and Consumption," cultivates numeracy in real-world situations modeled by exponential functions. These functions are used to investigate the effects of inflation on consumer costs, the effects of consumption rates on our nation's coal reserves, and population growth rates.

ACTIVITIES

ESTIMATING WITH "NICE" NUMBERS

November 1985

By ROBERT E. REYS, University of Missouri—Columbia, Columbia, MO 65211
BARBARA J. REYS, Oakland Junior High School, Columbia, MO 65202
PAUL R. TRAFTON and JUDY ZAWOJEWSKI, National College of Education, Evanston, IL 60201

Teacher's Guide

Introduction: Computational estimation is a very important and useful skill. Its value rests on producing reasonable answers to computations, sometimes very messy computations, quickly. Since *An Agenda for Action*'s call for an increased emphasis in mathematics programs on mentally estimating results of calculations (NCTM 1980, 3), many powerful and useful estimation techniques have been identified (cf. Reys and Reys 1983; Reys et al. 1984; Schoen 1986).

This activity is derived from a lesson originally created for the National Science Foundation project "Developing Computational Estimation in the Middle Grades." It introduces one effective strategy, which we call the "nice" numbers estimation strategy. "Nice" numbers, as used here, are those numbers that allow mental computation to be performed quickly and easily. The nice numbers considered here are powers of ten (e.g., 1, 10, 100).

This activity is designed to be initiated through a teacher-directed lesson developed through overhead transparencies. The transparencies highlight the concept of nice numbers and show how they can be used to obtain estimates quickly. Examples for students to try under your direction are included on each transparency. The student worksheets offer opportunities for additional practice, along with real-world applications of estimation; these sheets should be started in class and completed as homework. As time permits, a discussion of selected exercises on the following day will promote thinking about estimation and awareness of some different ways of obtaining reasonable estimates.

Grade levels: 7–12

Materials: Calculator (preferably one modified for overhead-projection use), transparencies of sheets 1–3, and a set of worksheets (sheets 4 and 5) for each student, to be used after the transparencies have been discussed

Objective: To develop estimating skills when multiplying or dividing by numbers near a power of ten

Procedure: Prior to using this activity, review with the class multiplication and division by powers of ten. The nice-numbers strategy encourages rounding to nice numbers (e.g., 1.03 ≐ 1; 9.93 ≐ 10) and then performing the operations indicated. After these operations are performed, the estimate is adjusted. The following example illustrates the entire process:

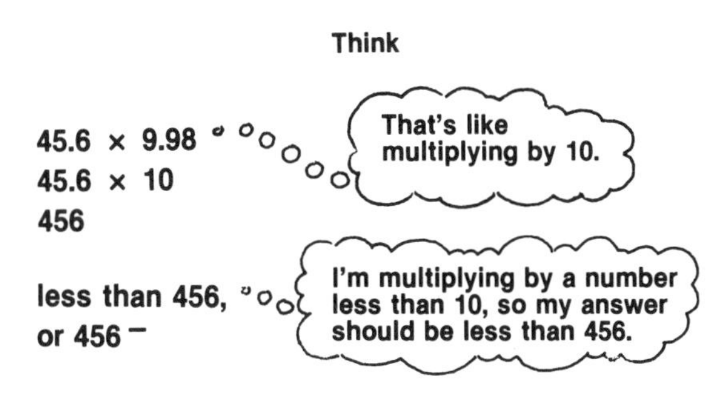

The first transparency (sheet 1) provides some examples of nice numbers. Mention that many numbers occur near nice numbers, but the acceptable range for near-nice numbers is generally determined by the context of the problem in real-life situations. For this activity, numbers near 10 will generally be between 9 and 11, whereas numbers near 100 will be between 90 and 110. As you work through this transparency with the class, encourage students to add other nice numbers to the list. Use the bottom portion of the transparency to have students select the numbers that are near nice numbers and state which nice numbers they are near.

Use a transparency of sheet 2 to begin collecting and recording estimates for such exercises as 23 × 0.97. Ask for estimates of the product from several students. These estimates should be recorded. Encourage students to think of 0.97 as "a little less than 1." For example:

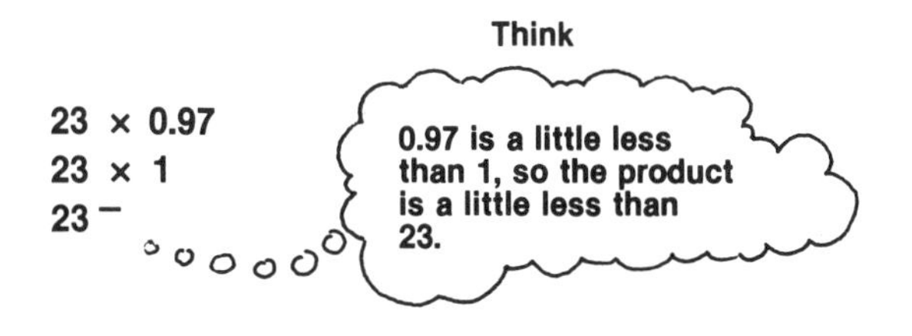

Estimate: "A little less than 23, I'll say 22."

After some estimates have been recorded and discussed, the exact answer can be found using a calculator.

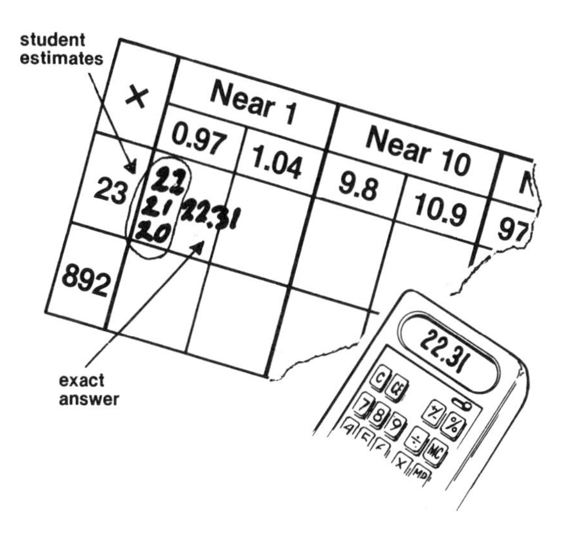

Continue in this manner, estimating 892 × 0.97. Repeat this process with several other numbers near nice numbers (e.g., 1.04, 9.8, 10.9, 97.2, 103.0). Pick and choose items from the chart as you wish. Most students will *not* need to complete the whole chart before going on.

Students should be encouraged to seek a pattern and verbalize it when observed. In general, the pattern might be described as follows:

> Given a number near n, where n is a power of ten, the product of this number and any other number is found by multiplying by n and then adding or subtracting an adjustment (compensation) determined by whether the given number was greater than or less than n.

Compensation—adding or subtracting the adjustment—is a complex process. Students need to realize that exactness is not the objective. The emphasis is on determining whether the exact answer will be more or less than the initial estimate.

Use the same process for division by near-nice numbers. Help students notice that the pattern is now reversed. When dividing by a number greater than n, the adjustment is subtracted. When dividing by a number less than n, the adjustment is added. This concept is more complex and usually requires plenty of discussion. For example:

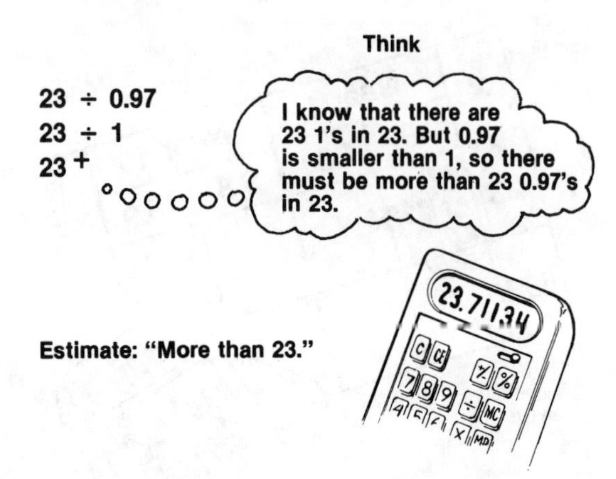

Students will need to see a variety of examples to help them clearly establish this idea.

The third transparency (sheet 3) highlights some work with nice numbers. Discuss the two examples given and then have the students try the six exercises. It is suggested that students share estimating strategies orally with the class. Specifically, they should be encouraged to verbalize how they move from their quick estimate to their adjusted estimate. Often the adjusted estimate can best be expressed by using the terms "more than" or "less than" or by using symbols "+" or "−."

Sheet 4 should be started immediately following the introductory lesson. It will give a quick check on how well the students have picked up the ideas presented. The remainder of this sheet and sheet 5 can be completed as a homework assignment.

Answers: The concept that a single situation involving estimation can have a number of different answers, all reasonable and falling within the acceptable range, will be foreign to many students. Some may demand to know "the answer" or the "best" answer. Encouraging several different estimates during class discussions can help students develop a greater tolerance for estimation.

Sheet 3

1. 485; adjusted estimate: 485^-, or about 480
2. 780; adjusted estimate: 780^+, or about 800
3. 1750; adjusted estimate: 1750^+, or about 1800
4. 5600; adjusted estimate: 5600^-, or about 5500
5. 0.862; adjusted estimate: 0.862^-, or about 0.840
6. 350; adjusted estimate: 350^-, or about 340

Sheet 4

1. 4600; 4600^-, or about 4500
2. 782; 782^+, or about 790
3. 840; 840^+, or about 850
4. 673; 673^+, or about 680
5. 5420; 5420^+, or about 5500
6. 29.8; 29.8^+, or about 30
7. 425 000; $425\,000^+$, or about 430 000
8. 86.7; 86.7^+, or about 87
9. 342; 342^+, or about 350
10. 168; 168^+, or about 170
11. 520; 520^-, or about 510
12. 8.46; 8.46^-, or about 8
13. 12.38; 12.38^+, or about 13
14. 436.7; 436.7^-, or about 436

Sheet 5

1. 559.55	6. 86.57	11. less
2. 8654.69	7. 53 990.74	12. $\$475^-$
3. 2.999	8. 6.999	13. 32.5^-mpg
4. 80.86	9. \$1.40–\$1.60	14. $\$110\,000^-$
5. 41.28	10. yes	15. more

REFERENCES

National Council of Teachers of Mathematics. *An Agenda for Action: Recommendations for School Mathematics of the 1980s.* Reston, Va.: The Council, 1980.

Reys, Barbara, and Robert Reys. *Guide to Using Estimation Skills and Strategies, GUESS, Box 1 and 2.* Palo Alto, Calif.: Dale Seymour Publications, 1983.

Reys, Robert E., Paul Trafton, Barbara Reys, and Judy Zawojewski. *Computational Estimation Materials, Grades 6, 7 and 8.* Washington, D.C.: National Science Foundation, 1984.

Schoen, Harold, ed. *Estimation and Mental Computation.* 1986 Yearbook of the National Council of Teachers of Mathematics. Reston, Va.: The Council, 1986.

Examples of Nice Numbers

$3 \times \underline{1}$ $\quad$ $40 \div \underline{10}$ $\quad$ $58 \times \underline{100}$

$85 \div \underline{1}$ $\quad$ $67 \times \underline{10}$ $\quad$ $2500 \div \underline{100}$

Why are the underlined numbers called "nice" numbers?

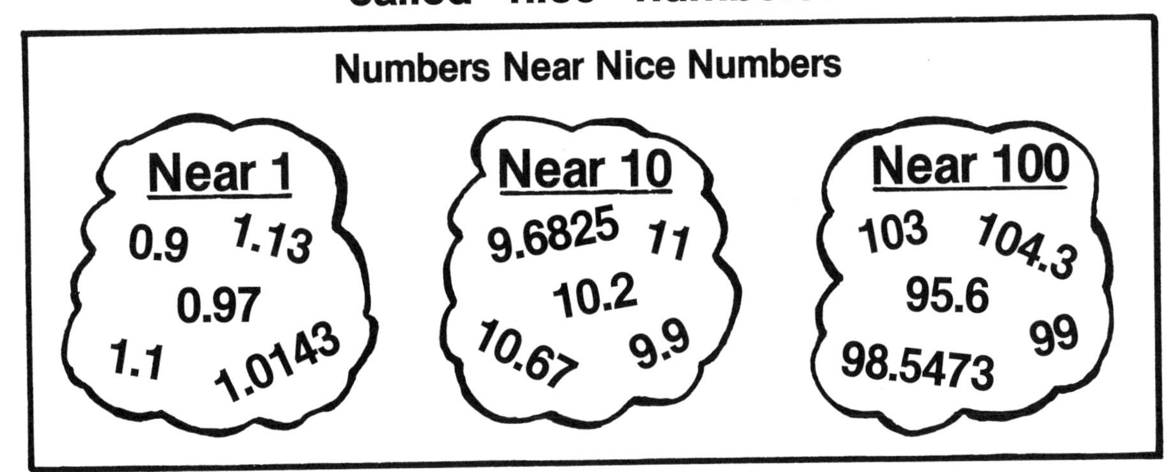

From the numbers below, select those that are near nice numbers and state which nice numbers they are near.

1.09 $\quad$ 23.4 $\quad$ 781 $\quad$ 3.6 $\quad$ 0.96

78 $\quad$ 9.86 $\quad$ 0.45 $\quad$ 102.9375

97 $\quad$ 1.127 $\quad$ 17.2 $\quad$ 10.4 $\quad$ 135

SHEET 2

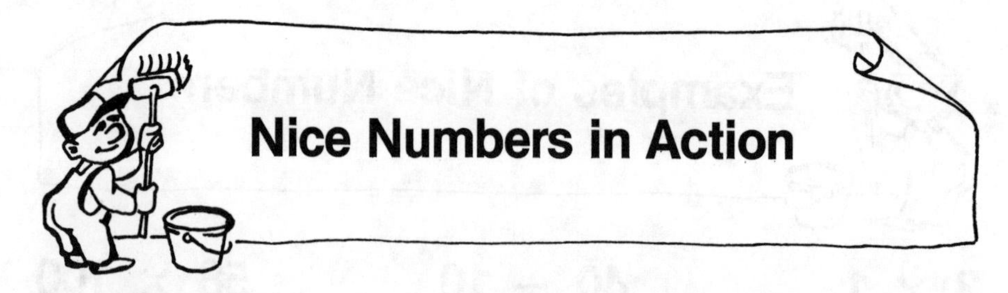

×	Near 1		Near 10		Near 100	
	0.97	1.04	9.8	10.9	97.2	103.0
23						
892						

÷	Near 1		Near 10		Near 100	
	0.97	1.04	9.8	10.9	97.2	103.0
23						
892						

To estimate 430 × 10.231

Think

$430 \times 10.231 \doteq 430 \times 10$

4300

Quick estimate

I'm multiplying by something bigger than 10, so it's more than 4300.

4300^{+}, or about 4400

Adjusted estimate

To estimate 430 ÷ 10.231

Think

$430 \div 10.231 \doteq 430 \div 10$

43

I'm dividing by a number bigger than 10, so it will be less than 43.

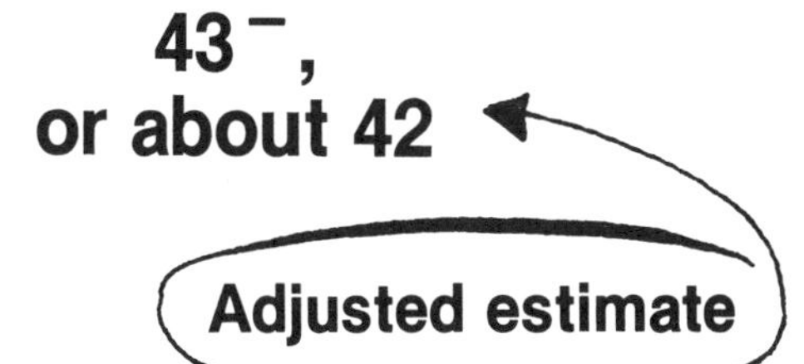

43^{-}, or about 42

Adjusted estimate

TRY THESE

1. 485 × 0.985 __________
2. 7800 ÷ 9.61 __________
3. 175 × 10.53 __________
4. 56 × 97.8 __________
5. 0.862 ÷ 1.03 __________
6. 35 000 ÷ 104 __________

Example:

We need 37 Fun Day tickets. They cost \$9.75 each. Estimate the total cost of the tickets.

Make a quick estimate for each of the following and then adjust your estimate.

	Quick Estimate	Adjusted Estimate
1. 460 × 9.75	________	________
2. 782 × 1.05	________	________
3. 84 × 10.3	________	________
4. 673 × 1.08	________	________
5. 542 × 10.29	________	________
6. 29.8 × 1.1	________	________
7. 425 × 1015	________	________
8. 86.7 ÷ 0.98	________	________
9. 3420 ÷ 9.78	________	________
10. 16 800 ÷ 99	________	________
11. 52 000 ÷ 102.7	________	________
12. 8460 ÷ 1029	________	________
13. 12 380 ÷ 991	________	________
14. 4367 ÷ 10.04	________	________

Below are some calculator-obtained results, which in some cases have been rounded. In each case, one result is correct, the other is wrong. Use estimation to identify the correct answer and then circle it.

1. 589 × 0.95	559.55	629.945
2. 89 576 ÷ 10.35	8654.69	9271.16
3. 35.75 ÷ 11.92	426.14	2.999
4. 87 ÷ 1.076	80.86	93.61
5. 387 ÷ 9.375	36.28	41.28
6. 85 397 ÷ 986.43	86.57	83.64
7. 53.6 × 1007.29	53 990.74	52 864.50
8. 6.7 ÷ 0.9573	6.257	6.999

Use nice numbers to estimate reasonable answers for these problems.

9. Your grocery store sells hamburger for $1.39 a pound. If you buy a package of hamburger that weighs 1.08 pounds, estimate what you will pay. ________
10. Suppose that 1-liter bottles of cola are on sale for 99¢ apiece and you are going to buy 24 bottles for the church picnic. Will $24.85 be enough to pay the bill? ________
11. A racer averages 9.9 miles an hour when running long distances. Will a marathon race about 26 miles long take more or less than 3 hours? ________
12. The owner of the Sound Shop has 475 tapes in inventory and makes about 95¢ on each tape sold. Estimate the profit if all the tapes are sold. ________
13. The Brocks drove 325 miles on 10.2 gallons of gasoline. Estimate their mileage. ________
14. If 9960 people bought tickets to see REO Speedwagon in concert, and the tickets sold for $11 each, estimate the total gate receipts for the concert. ________
15. The typical American eats 96.3 pounds of beef each year. If the average price of beef is $2.42 a pound, would each spend more or less than $200 a year on beef? ________

ACTIVITIES

ESTIMATING WITH "NICE" FRACTIONS

November 1986

By PAUL R. TRAFTON and JUDITH S. ZAWOJEWSKI, National College of Education, Evanston, IL 60201
ROBERT E. REYS and BARBARA J. REYS, University of Missouri—Columbia, Columbia, MO 65211

Teacher's Guide

Introduction: Interest in estimation in the school curriculum has markedly increased during the past few years. Reports on needed changes in the mathematics curriculum have called for an increased emphasis on estimation. The topic has been the focus of frequent articles and activities in journals (e.g., Ockenga and Duea [1985]; Reys [1985]; Reys et al. [1985]; Rubenstein [1985]) and of talks at mathematics conferences, and it is the major theme of the 1986 NCTM Yearbook (Schoen 1986). Estimation is being viewed in a broader and more diverse way than in the past, with particular emphasis on the many ways in which one can obtain a valid estimate.

This activity is based on material developed for a National Science Foundation project, *Developing Computational Estimation in the Middle Grades* (Reys et al. 1984). It is built around the use of "nice" fractions in estimating. The first three activity sheets focus on changing fractions to ones that are easy to work with. On the fourth sheet, students use nice fractions (unit fractions) to estimate a fractional part of a whole.

Each activity sheet should be introduced through teacher-pupil discussion of transparencies made from the sheets. After an idea has been developed, students should complete the related practice exercises.

Grade levels: 6–12

Materials: Copies of the activity sheets for each student and a set of transparencies for class discussions

Objectives: To have students learn (1) to identify fractions close to 1, 0, and $\frac{1}{2}$; (2) to estimate sums of fractions and mixed numbers by rounding fractions to 0, $\frac{1}{2}$, and 1; (3) to round fractions to easily understood fractions; (4) to estimate a fractional part of a whole with unit fractions

Procedure: Distribute the activity sheets one at a time to each student. Introduce sheet 1 by discussing how fractions can often be "messy" to work with and that recognizing fractions close to 1, 0, and $\frac{1}{2}$ can make it easier to estimate the value of a fraction and thereby make work with fractions easier. Note that a fraction is close to 1 when the numerator and denominator are about the same size, or "close" to each other. Through discussion bring out the fact that a fraction is close to 0 when the numerator is very small in comparison to the denominator. Finally, students should understand that a fraction is close to $\frac{1}{2}$ when the numerator is about half the denominator or the denominator is about twice as large as the numerator. Have students complete exercises 4, 5, and 6 on sheet 1 independently.

Following discussion of students' solutions to the exercises on sheet 1, work through examples 1–4 on sheet 2. Point out that whereas it is unlikely that one would have to compute a sum as in example 1, it is very easy to estimate the sum. Discuss the changing of each fraction to 0, $\frac{1}{2}$, or 1 and how the estimate of the sum is obtained. For the second example, the "front end" method is shown. First the sum of the whole numbers is found, and then an estimate of the sum of the fractions is added. Examples 3 and 4 introduce the idea of adjusting an initial estimate. In example 3, both fractions are rounded down. Thus the whole-number estimate, 2, is less than the actual sum and is an *underestimate.* To indicate that the actual sum is over the estimate, a superscript plus sign is written next to the 2. Similar reasoning is used to adjust the overestimate in example 4. Next assign exercises 1–8, in which students are to select the best estimate.

Sheet 3 is an extension of the work on sheet 1. Begin by having students identify fractions whose value is easily understood, such as $\frac{1}{2}$, $\frac{1}{3}$, $\frac{1}{4}$, $\frac{1}{5}$, $\frac{1}{6}$, $\frac{2}{3}$, or $\frac{3}{4}$. Now direct their attention to the problem at the top of the sheet. Point out that the fraction $\frac{6}{23}$ may not be easily understood by many people. Discuss how to find a "nice" fraction that is close to $\frac{6}{23}$. Next solve exercises 1 and 2 with the class, encouraging pupils to change both the numerator and denominator to get a fraction that can easily be simplified to a familiar one. Have students complete the remaining exercises on their own.

Sheet 4 focuses on estimating a fractional part of a whole with unit fractions. This useful, practical skill is used widely in everyday life. Two key ideas are involved:

a. Finding a fractional part with a unit fraction involves division. $\frac{1}{3}$ represents 1 of 3 equal parts. Thus $\frac{1}{3}$ of 36 is 1 of 3 equal divisions of 36, or $36 \div 3$.

b. The amount to be divided should be changed to a "compatible" number, one that is a multiple of the divisor and is easy to work with.

Discuss the example at the top of the sheet, stressing both of these ideas. Point out that $120 was selected because $120 \div 3$ is easy to compute. Next develop the first two exercises with the students. In exercise 1, $100 could also have been used and also would have produced a reasonable estimate. Instruct the class to complete exercises 3–13 individually. Students should be encouraged to adjust their estimates.

Answers

Sheet 1: 4. $\frac{3}{17}$; $\frac{1}{8}$; $\frac{2}{13}$; $\frac{6}{97}$; 5. $\frac{7}{13}$; $\frac{4}{9}$; $\frac{6}{11}$; $\frac{11}{20}$; $\frac{5}{9}$; 6. $\frac{3}{4}$; $\frac{4}{5}$; $\frac{7}{9}$; $\frac{9}{8}$

Sheet 2: 1. about 2; 2. about $\frac{1}{2}$; 3. about 1; 4. about 2; 5. over 6; 6. under 10; 7. under 8; 8. over 7

Sheet 3 (reasonable alternate answers should be accepted): 1. $\frac{1}{3}$; 2. $\frac{1}{3}$; 3. $\frac{2}{3}$; 4. $\frac{1}{2}$; 5. $\frac{1}{5}$ ($\frac{1}{4}$); 6. $\frac{1}{6}$; 7. $\frac{1}{2}$; 8. $\frac{1}{3}$; 9. Lions; 10. Tigers, Muskrats; 11. Cougars, Wildcats; 12. Bears

Sheet 4: (reasonable alternative answers should be accepted): 1. $88.00, $22.00 ($22–$25); 2. $27.00, $3.00 ($2–$3); 3. $150.00, $30.00 ($25–$30); 4. $30.00, $10.00 ($10–$11); 5. $160.00, $20.00 ($20–$22); 6. $16.00–$20.00; 7. $9.00–$10.00; 8. $11.00–$12.00; 9. $30.00–$33.00; 10. $10.00–$12.00; 11. $50.00–$60.00; 12. $10.00–$15.00; 13. $7.00–$8.00

REFERENCES

Ockenga, Earl, and Joan Duea. "Activities: Estimate and Calculate." *Mathematics Teacher* 78 (April 1985):272–76.

Reys, Robert E. "Estimation." *Arithmetic Teacher* 32 (February 1985):37–41.

Reys, Robert E., Barbara J. Reys, Paul R. Trafton, and Judy Zawojewski. "Activities: Estimating with 'Nice' Numbers." *Mathematics Teacher* 78 (November 1985):615–17, 621–25.

Reys, Robert E., Paul R. Trafton, Barbara Reys, and Judy Zawojewski. *Developing Computational Estimation Materials for the Middle Grades.* Final Report No. NSF-8113601. Washington, D.C.: National Science Foundation, 1984.

Rubenstein, Rheta N. "Activities: Developing Estimation Strategies." *Mathematics Teacher* 78 (February 1985):112–18.

Schoen, Harold, ed. *Estimation and Mental Computation.* 1986 Yearbook. Reston, Va.: National Council of Teachers of Mathematics, 1986.

FRACTIONS CAN BE "NICE" NUMBERS

SHEET 1

The numbers 0, $\frac{1}{2}$, and 1 are easy numbers to use in mental arithmetic. Many fractions can be rounded to these nice numbers to make estimates. Study the examples below to help you find a way to determine when a fraction is close to 0, $\frac{1}{2}$, or 1.

Fractions close to 1:

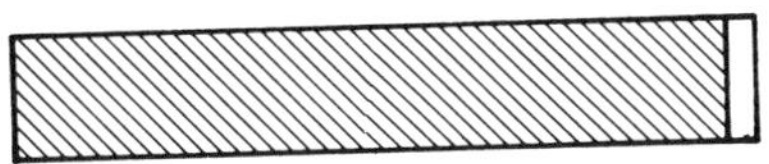

1. How can you tell when a fraction is close to 1?

____
____
____

Fractions close to 0:

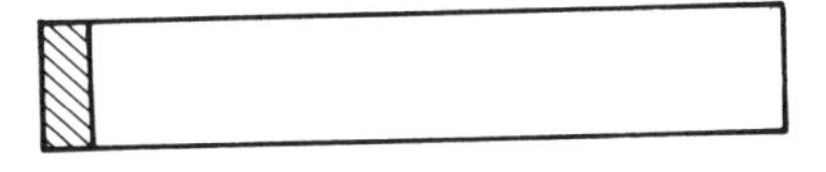

2. How can you tell when a fraction is close to 0?

____
____
____

Fractions close to $\frac{1}{2}$:

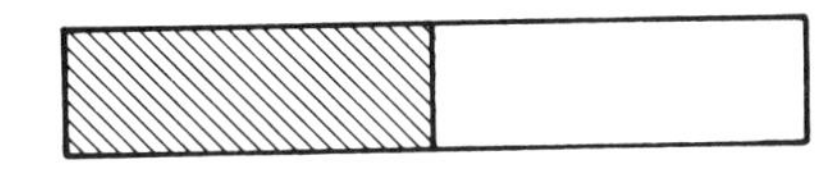

3. How can you tell when a fraction is close to $\frac{1}{2}$?

____
____
____

4. Circle the fractions that are close to 0. $\frac{3}{5}$ $\frac{3}{17}$ $\frac{1}{8}$ $\frac{2}{3}$ $\frac{2}{13}$ $\frac{6}{97}$ $\frac{4}{9}$

5. Circle the fractions that are close to $\frac{1}{2}$. $\frac{7}{13}$ $\frac{2}{7}$ $\frac{4}{9}$ $\frac{6}{11}$ $\frac{8}{9}$ $\frac{11}{20}$ $\frac{5}{9}$

6. Circle the fractions that are close to 1. $\frac{3}{8}$ $\frac{3}{4}$ $\frac{7}{12}$ $\frac{4}{5}$ $\frac{7}{9}$ $\frac{9}{8}$ $\frac{7}{20}$

It's fairly easy to add in your head with zeros, halves, and ones. By rounding fractions to these nice numbers, you can estimate their sums easily. Here are some examples.

Example 1: To estimate

$$\frac{3}{5}+\frac{9}{10}+\frac{1}{20}+\frac{16}{30},$$

Think: $\frac{1}{2}$ 1 0 $\frac{1}{2}$

Estimate: about 2

Example 2: To estimate

$$1\frac{1}{9}+2\frac{3}{4}+3\frac{1}{10},$$

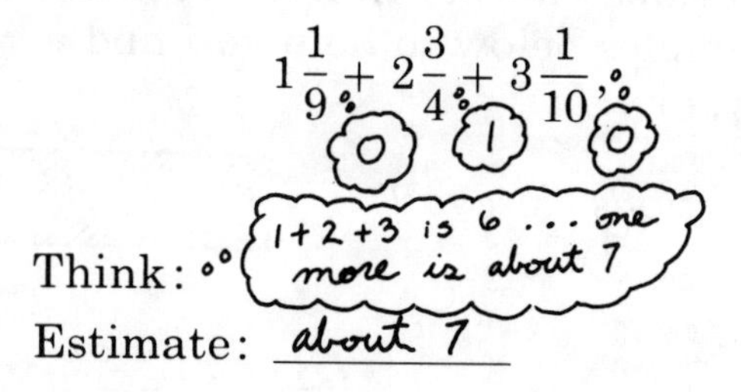

Think:

Estimate: about 7

Sometimes you can tell if your estimate is too high or too low. In such examples, you can write a "+" or a "−" to show how you would adjust your estimate.

Example 3: To estimate

$$\frac{15}{13}+\frac{9}{8},$$

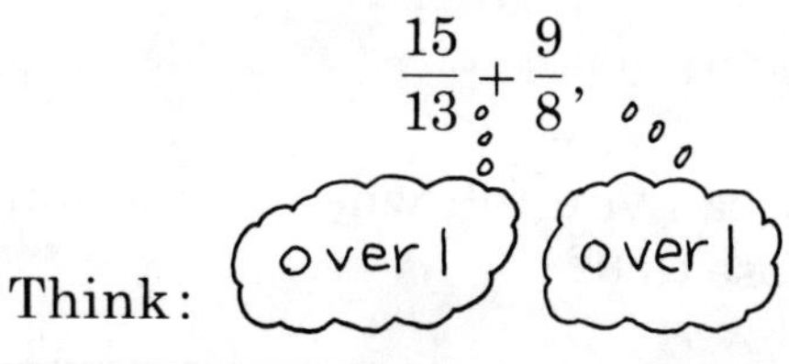

Think:

Estimate: over 2 or 2^+

Example 4: To estimate

$$3\frac{4}{5}+1\frac{7}{8}+\frac{9}{10},$$

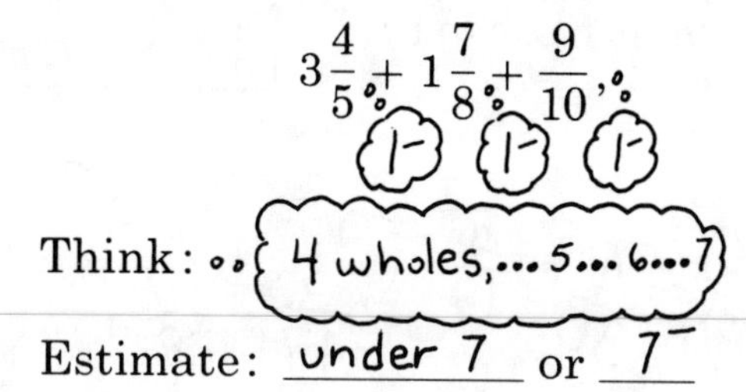

Think:

Estimate: under 7 or 7^-

Circle the best estimate of each sum.

1. $\frac{3}{4}+\frac{9}{10}$ about $\frac{1}{2}$ about 1   about 2
2. $\frac{1}{4}+\frac{3}{16}$ about $\frac{1}{2}$ about 1   about 2
3. $\frac{13}{16}+\frac{1}{8}$ about $\frac{1}{2}$ about 1   about 2
4. $\frac{8}{9}+\frac{5}{6}$ about $\frac{1}{2}$ about 1   about 2
5. $3\frac{7}{8}+2\frac{3}{10}$ over 6   under 6
6. $6\frac{1}{5}+3\frac{1}{2}$ over 10   under 10
7. $5\frac{3}{8}+2\frac{1}{4}$ over 8   under 8
8. $6\frac{5}{8}+\frac{5}{8}$ over 7   under 7

COMMUNICATING WITH NICE FRACTIONS — SHEET 3

Sometimes we round off a fraction to a nice fraction, such as $\frac{1}{2}$, $\frac{1}{3}$, $\frac{1}{4}$, $\frac{1}{5}$, $\frac{1}{6}$, $\frac{2}{3}$, or $\frac{3}{4}$, that is more easily understood.

Example:

John made 6 of 23 free throws. He made about _?_ of his shots.

$\frac{6}{23}$ is close to $\frac{6}{24}$, and $\frac{6}{24}$ equals $\frac{1}{4}$. He made about $\frac{1}{4}$ of his shots.

Think

Represent each of the following situations with a nice fraction:

1. Travis — 8 baskets, 25 shots

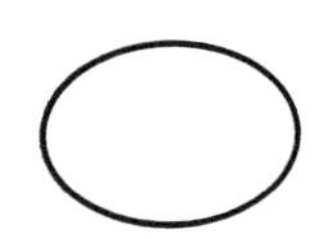

2. Mary Lou — 24 base hits, 74 times at bat

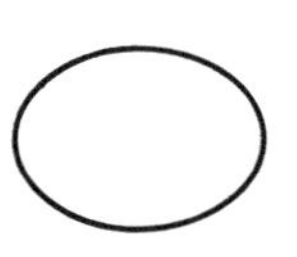

3. Laura — 11 free throws, 18 attempts

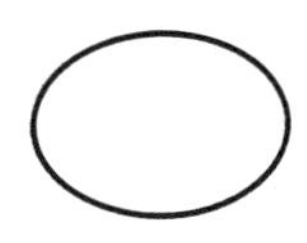

4. Jason — 21 completions, 43 passes

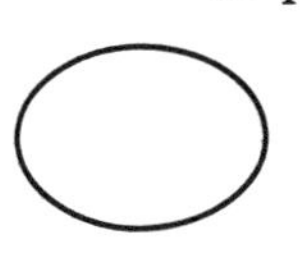

5. Alan — 15 base hits, 74 times at bat

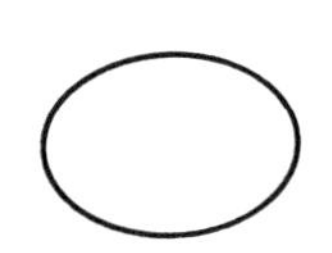

6. Tammy — 6 goals, 37 shots

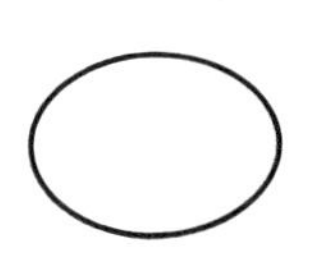

7. Jeff — 8 baskets, 17 shots

8. Terry — 11 base hits, 36 times at bat

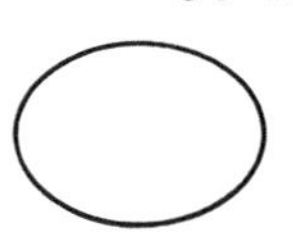

Use the table at the right to complete exercises 9–12.

9. Which team has won more than half of its games? ________
10. Which teams have won just under half of their games? ________
11. Which teams have won about one-third of their games? ________
12. Which team has won about one-fourth of its games? ________

Team	Games Played	Games Won
Tigers	21	10
Lions	18	11
Muskrats	27	13
Cougars	19	6
Bears	21	5
Wildcats	29	9

ESTIMATING WITH NICE NUMBERS SHEET 4

Example: Estimate how much you would save on the purchase of this television.

$\frac{1}{3}$ is already a nice number. To find a third of a number, you can divide by 3. However, \$118 does not divide evenly by 3. Round 118 to a number that is easy to divide by 3.

So, I would save about \$40, or a little less, $\$40^-$.

Estimate each answer by using a nice number that is easy to divide.

		Nice Number	Estimate
1. $\frac{1}{4}$ of \$89.95	$\frac{1}{4}$ of	\$88.00	______
2. $\frac{1}{9}$ of \$25.89	$\frac{1}{9}$ of	______	______
3. $\frac{1}{5}$ of \$146.15	$\frac{1}{5}$ of	______	______
4. $\frac{1}{3}$ of \$31.19	$\frac{1}{3}$ of	______	______
5. $\frac{1}{8}$ of \$163.45	$\frac{1}{8}$ of	______	______

For the sale items below, estimate how much you would save on each item by changing the price to a nice number. Write your estimate below the price tag.

10. \$46.83
11. \$237
12. \$49.95
13. \$29.19

ACTIVITIES

CALCULATORS AND ESTIMATION

February 1982

By Terry Goodman, Central Missouri State University, Warrensburg, MO 64093

Teacher's Guide

Grade level: 7–12.

Materials: One set of worksheets and a calculator for each student.

Objective: Students will estimate factors and divisors, using a calculator to check their estimations.

Directions: Students often have difficulty estimating when using very large or very small numbers. If the numbers are not whole numbers, this difficulty is more pronounced. These activities are designed to provide estimation practice for students. Since students are reluctant to make estimates if they are going to have to check them by hand, allow the students to use calculators for checking.

Encourage the students to work to improve their estimation skills. Problem solving can also be encouraged as students look for ways to make their estimations "better."

Sheet 1: These exercises will serve as a warm-up for the activities on succeeding sheets. Encourage students not to worry about exact answers but to try to make "good" estimates. The division key should not be used as a shortcut to estimating.

Sheet 2: Before having the students work with this sheet, do an example or two with them as a group. Starting with 15, how many successive multiplications will it take to get an answer in the interval (10 000, 10 500)? For example,

1. 15 × 600 = 9000
2. 9000 × 1.1 = 9900
3. 9900 × 1.1 = 10 890
4. 10 890 × 0.95 = 10 345.5

We did it in four steps. As the students work through the sheet encourage them to recognize when they need to multiply by a number greater than 1 or less than 1. Encourage them to make their estimate first, then to use the calculator. Students can make up their own problems (smaller intervals, larger numbers, and so on).

Sheet 3: Although this sheet is similar to sheet 2, you may want to work through one example with the class. Again, encourage the students to look for patterns and generalizations. They will need to focus on what happens when they divide by a number less than 1. You may want to allow the students to round off results.

In part II, students are asked to apply what they have done on sheet 2 and the first part of sheet 3. Encourage them to use both multiplication and division in the same problem. For some students, division by 2 will be more obvious than multiplication by 0.5.

Additional activities: Students can also look for other operations to use. For example,

Start	Range
5100	(50, 75)

They may decide on the following: $\sqrt{5100} = 71.414$.

As another example, consider the following:

Start	Range
3150	(100, 120)

a. 3150 ÷ 10 = 315
b. $\sqrt{315}$ = 17.75
c. 17.75 × 6 = 106.49

I. Estimate the second factor so that the product will fall within the given range. *Check* your estimate with a calculator.

Example: 12 × ___?___ Range: (5000, 5500)
12 × 400 = 4800 Too small
12 × 450 = 5400 We did it in two tries!

Try these. Indicate the number of tries in the space provided.

Start	Range	Number of Tries
1. 19 × ________	(800, 850)	________
2. 25 × ________	(2000, 2200)	________
3. 11 × ________	(550, 570)	________
4. 105 × ________	(1000, 1100)	________
5. 176 × ________	(20, 30)	________
6. 50 × ________	(4670, 4700)	________

II. Estimate the divisor so that the quotient will fall within the given range. *Check* your estimate with a calculator.

Example: 975 ÷ ___?___ Range: (10, 20)
975 ÷ 95 = 10.26 We did it in one try!!

Try these. Indicate the number of tries in the space provided.

Start	Range	Number of Tries
1. 850 ÷ ________	(90, 150)	________
2. 1050 ÷ ________	(200, 210)	________
3. 50 ÷ ________	(210, 220)	________
4. 125 ÷ ________	(4200, 4400)	________
5. 4360 ÷ ________	(150, 200)	________
6. 11 ÷ ________	(550, 600)	________

Use the start number and *successive* multiplication to obtain a product in the given range.

Example:

Start	Range
12	(5000, 5500)

1. $12 \times 400 = 4800$ — Too small
2. $4800 \times 1.2 = 5760$ — Too large
3. $5760 \times 0.95 = 5472$ — We did it in three steps.

See if you can do the following in as few steps as possible. Record each product in the spaces provided.

	Start	Range	Number of steps
1.	11 ×	(900, 1000)	☐
2.	400 ×	(175, 195)	☐
3.	1.1 ×	(14 100, 14 200)	☐
4.	15 ×	(8000, 8100)	☐
5.	14 222 ×	(7200, 7400)	☐
6.	0.3 ×	(280, 299)	☐
7.	π ×	(100, 110)	☐
8.	12 100 ×	(36, 37)	☐

I. Use the start number and *successive* division to obtain a quotient in the given range. Try to use fewer than five steps.

Example: Start 50, Range (2000, 2400)

	Start	Range
	50	(2000, 2400)
1.	50 ÷ 0.01 = 5000	Too large
2.	5000 ÷ 2.1 = 2380.95	We did it in two steps.

Try these. Use as few steps as possible. Record each quotient in the spaces provided.

	Start	Range	Number of steps
1.	975 ÷	(10, 20)	☐
2.	143 ÷	(60, 68)	☐
3.	743 ÷	(200, 220)	☐
4.	11 ÷	(900, 1000)	☐
5.	15 ÷	(10 000, 20 000)	☐

II. Use successive multiplication and/or division to obtain a number in the given range. Record results from each step in the spaces provided.

	Start	Range	Number of steps
1.	45	(100, 110)	☐
2.	3150	(100, 120)	☐
3.	16	(3333, 3400)	☐

ACTIVITIES

DEVELOPING ESTIMATION STRATEGIES

February 1985

By RHETA N. RUBENSTEIN, Renaissance High School, Detroit, MI 48235

Teacher's Guide

Introduction: Estimation is an important but frequently neglected objective of the mathematics curriculum. Recognizing the need for better number sense, together with mental arithmetic and estimation skills, in a society increasingly dominated by calculator and computer printouts, virtually every recent curriculum report has recommended changes that include an increased emphasis on mentally estimating results of calculations. In addition, teachers are encouraged to incorporate estimation activities into all areas of instruction on a regular and sustaining basis, in particular encouraging the use of estimating skills to pose and select alternatives and to assess what a reasonable answer may be. The following materials are intended to help students develop some computational estimation strategies.

Reys et al. (1980) characterized the processes that good estimators, from seventh graders to adults, used in computational estimation. Among these processes were "using compatible numbers," "averaging," and "rounding." These strategies are developed in this activity.

In "using compatible numbers," good estimators alter the numbers in a given situation to ones that are easier to compute mentally. This technique is frequently used in situations involving division and fractional parts, for example:

$$471 \div 6 \doteq 480 \div 6 = 80$$

$$\frac{2}{3} \text{ of } 250 \doteq \frac{2}{3} \text{ of } 240 = 2 \times 80 = 160$$

$$23\% \text{ of } 125 \doteq 25\% \text{ of } 120 = \frac{1}{4} \text{ of } 120 = 30$$

Worksheets 1 and 2 review easy arithmetic facts and ask students to identify compatible number estimates and to create their own compatible number expressions to estimate answers.

The "averaging" technique is most applicable to addition situations in which the addends are all "close" to one value, for example:

> My driving record for the first four months of the year was as follows:
>
> | | |
> |---|---|
> | January | 1978 miles |
> | February | 1043 miles |
> | March | 1095 miles |
> | April | 887 miles |
>
> About how many miles did I drive?

Recognizing that all four addends are about 1000 miles, then multiplying this "average" by 4 makes it easy to estimate the answer as 4000 miles. Worksheet 3 introduces averaging and provides some practice situations.

"Rounding" is a more commonly taught estimation strategy. Worksheet 3 provides practice in estimating grocery totals by rounding to the nearest half-dollar.

One of the key features of estimation is that it takes place over a short period of time. To provide ongoing opportunities to practice estimation in briefly timed periods, worksheet 4, a template for overhead transparencies, has been developed.

Grade levels: 6–9

Materials: Copies of worksheets 1–3 for each student and a transparency of worksheet 4

Objectives: To develop estimation strategies of using compatible numbers, averaging, and rounding and to provide opportunities for use of these strategies in real-life contexts

Procedures: Distribute the worksheets one at a time to each student. Work through the examples in each section with students before having them proceed individually. Success with sheet 1 depends on students' knowledge of basic facts and ability to work with rounded numbers. Success with sheet 2 requires a familiarity with fraction-percentage equivalences. Provide additional instruction on these prerequisite skills as necessary.

Make a transparency from sheet 4. Numbers have been omitted so the transparency can be reused with new numbers. For problems 1–4, present one item at a time on the overhead projector, giving students approximately fifteen seconds to read and respond. Solutions should then be discussed and strategies shared. The following are sample items, with sample estimates in parentheses:

1. 800 km, 90 km/h: (810 ÷ 90 = 9 h)
   500 km, 80 km/h: (480 ÷ 80 = 6 h)
2. $4.95, 500 g: (500 ÷ 500 = 1¢/g)
   $3.89, 195 g: (400 ÷ 200 = 2¢/g)
3. 1372 points, 7 minutes:
   (1400 ÷ 7 = 200 points/min.)
   1904 points, 5 minutes:
   (2000 ÷ 5 = 400 points/min.)
4. 881 students, 29%: (3/10 × 900 = 270)
   591 students, 35%: (1/3 × 600 = 200)

For problem 5, circle some set of items. Let students estimate the sum. Check totals on a calculator.

These materials have been developed with the expectation that they can be part of an ongoing program of integrating experiences in estimation with instruction. It is hoped that teachers and students will be continually "on the lookout" for opportunities to estimate and that the strategies developed here will be part of a growing repertoire of techniques.

Teaching estimation requires open-mindedness about answers. Many good estimates are usually possible. Be ready to use judgment in evaluating students' responses. Students should be encouraged to share orally their reasons for the estimates they choose.

Answers: Sheet 1: 1. a. 50; b. 5; c. 500; d. 50; e. 90; f. 90; g. 9; h. 900; i. 700; j. 8000; k. 8; 1. 5. 2. a. C, 9; b. A, 9; c. B, 5; d. C, 8; e. B, 80; f. C, 70. 3. Sample answers:

a. 720 ÷ 8 = 90 b. 6300 ÷ 9 = 700
c. 640 ÷ 80 = 8 d. 4800 ÷ 6 = 800
e. 420 ÷ 60 = 7 f. 3600 ÷ 4 = 900
g. 4500 ÷ 9 = 500 h. 540 ÷ 9 = 60

4. a. 360 ÷ 120 = 3; b. 750 000 ÷ 15 = 50 000 books; 2 400 000 ÷ 12 = 200 000 books; 700 000 ÷ 7 = 100 000 books

Sheet 2:

1. See following chart.

2. a. 9; b. 18; c. 21; d. 63; e. 400; f. 2000; g. 210; h. 4000; i. 120; j. 630; k. 350; 1. 800. 3. a. 30; b. 8; c. 2700; d. 600; e. 12; f. 48; g. 300; h. 60. 4. a. A, 200; b. B, 18; c. C, 16; d. A, 35; e. C, 150; f. B, 36. 5. a. 3/4 × 80 = 60; b. 4/5 × 200 = 160; c. 2/3 × 90 = 60; d. 9/10 × 110 = 99; e. 3/10 × 50 = 15; f. 2/5 × 35 = 14

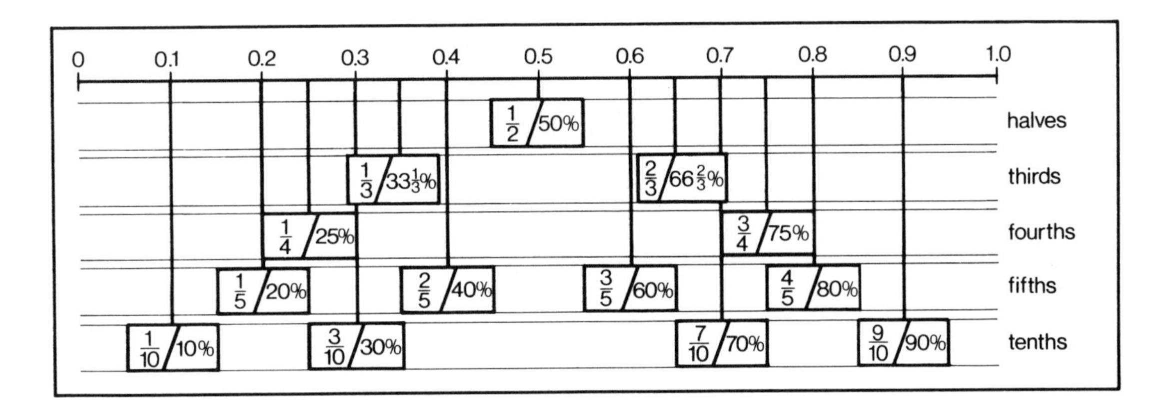

Sheet 3: 1. a. 80; b. 80 × 10 = 800. 2. a. 6000 × 8 = $48 000; b. 2000 × 12 = $24 000; c. 30 × 12 = 360 banners. 3. a. $14.50; b. $21.50–$23

REFERENCE

Reys, Robert E., Barbara J. Bestgen, James F. Rybolt, and J. Wendell Wyatt. *Identification and Characterization of Computational Estimation Processes Used by In-School Pupils and Out-of-School Adults.* Washington, D.C.: National Institute of Education, 1980.

1. Use number facts to calculate the following:

a. 450 ÷ 9 ____	e. 630 ÷ 7 ____	i. 42 000 ÷ 60 ____
b. 450 ÷ 90 ____	f. 6300 ÷ 70 ____	j. 32 000 ÷ 4 ____
c. 4500 ÷ 9 ____	g. 6300 ÷ 700 ____	k. 560 ÷ 70 ____
d. 4500 ÷ 90 ____	h. 7200 ÷ 8 ____	l. 400 ÷ 80 ____

2. To estimate the quotient 5321 ÷ 62, change the numbers to "compatible numbers" with which you can compute mentally.

   Example: 5321 ÷ 62 ≐ 5400 ÷ 60 = 90

   Estimate each answer by choosing the expression (A, B, or C) with compatible numbers and then calculating mentally.

	A	B	C	Estimate
a. 73 ÷ 8	74 ÷ 8	73 ÷ 7	72 ÷ 8	______
b. 63 ÷ 6.8	63 ÷ 7	62 ÷ 6	64 ÷ 7	______
c. 46 ÷ 8.7	46 ÷ 8	45 ÷ 9	45 ÷ 8	______
d. 57 ÷ 6.9	57 ÷ 7	58 ÷ 7	56 ÷ 7	______
e. 6504 ÷ 78	6500 ÷ 80	6400 ÷ 80	7000 ÷ 80	______
f. 291 ÷ 3.8	270 ÷ 3	300 ÷ 4	280 ÷ 4	______

3. Estimate each answer by writing a compatible number expression and calculating.

   Example: 2040 ÷ 72 ≐ 2100 ÷ 70 = 30

a. 738 ÷ 7.9 ______	e. 413 ÷ 58 ______
b. 6205 ÷ 8.7 ______	f. 3521 ÷ 3.8 ______
c. 643 ÷ 81 ______	g. 4483 ÷ 8.7 ______
d. 4729 ÷ 5.9 ______	h. 553 ÷ 9.1 ______

4. Use compatible numbers to estimate each answer:

   a. The Student Council sold 121 items at their rummage sale. The students collected $374. What was the approximate average cost of each item?

   ____

   b. In 1980 the following cities had the number of libraries and books shown. Estimate the average number of books in each library for each city.

City	Libraries	Books	Estimate
Charlotte, North Carolina	15	736 343	______
Milwaukee, Wisconsin	12	2 335 485	______
Ottawa, Ontario	7	666 404	______

COMPATIBLE NUMBERS WITH FRACTIONS AND PERCENTAGES Sheet 2

1. Complete the table with equivalent fractions and percentages. Note the relative location of each fraction and percentage on the number line.

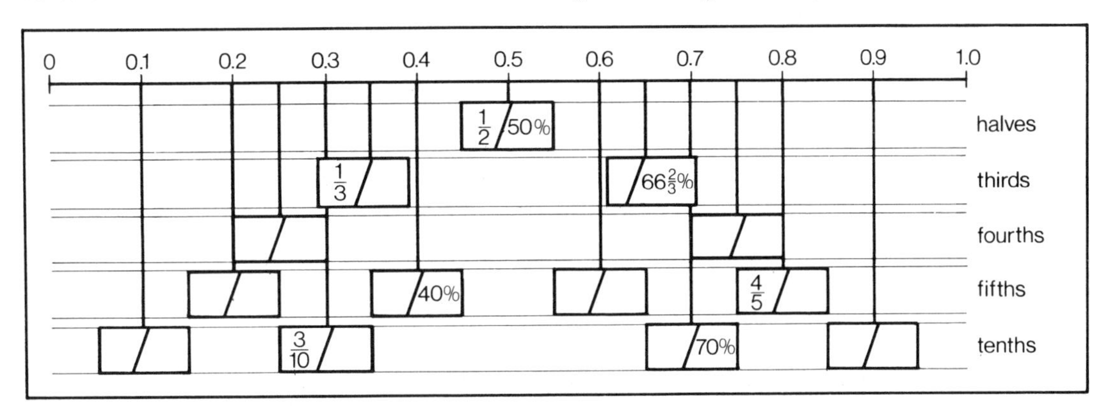

2. Use shortcuts to calculate the following problems.

Example: $4/5$ of $450 = 4/5 \times 450 = 4 \times 90 = 360$

a. 1/3 of 27 ____	e. 1/6 of 2400 ____	i. 2/3 of 180 ____
b. 2/3 of 27 ____	f. 5/6 of 2400 ____	j. 9/10 of 700 ____
c. 1/4 of 84 ____	g. 3/4 of 280 ____	k. 7/8 of 400 ____
d. 3/4 of 84 ____	h. 5/6 of 4800 ____	l. 4/5 of 1000 ____

3. Rewrite the percentages as fractions, then use shortcuts to calculate.

Example: 80% of 350 = 4/5 of 350 = 4 × 70 = 280

a. 20% of 150 ________	e. 40% of 30 ________
b. 25% of 32 ________	f. 60% of 80 ________
c. 90% of 3000 ________	g. 75% of 400 ________
d. 66 2/3% of 900 ________	h. 33 1/3% of 180 ________

4. Estimate each answer by choosing the expression (A, B, or C) with compatible numbers and then calculating.

Example: 73% of 78 ≐ 75% of 80 = 3/4 × 80 = 60

	A	B	C	Estimate
a. 26% of 800	1/4 of 800	1/5 of 800	2/6 of 800	________
b. 62% of 30	2/3 of 30	3/5 of 30	7/10 of 30	________
c. 67% of 24	4/5 of 24	1/3 of 24	2/3 of 24	________
d. 5/8 of 57	5/8 of 56	5/8 of 50	5/8 of 64	________
e. 3/5 of 257	3/5 of 200	3/5 of 300	3/5 of 250	________
f. 3/4 of 47	3/4 of 50	3/4 of 48	3/4 of 40	________

5. Estimate by using compatible numbers.

Example: 77% of 41 ≐ 75% of 40 = 3/4 × 40 = 30

a. 3/4 of 79 ____	c. 2/3 of 88 ____	e. 29% of 52 ____
b. 4/5 of 203 ____	d. 88% of 110 ____	f. 42% of 36 ____

1. The eighth graders at Middleton Middle School sold booster buttons to raise funds. Here are the numbers of buttons sold by ten homerooms:

   76 75 78 83 81 75 88 83 76 77

   a. All the homerooms sold close to the same number of buttons. What is this number? ________

   b. This number is an estimate of the average number of buttons sold by each homeroom. To estimate the total number of buttons sold, multiply the estimated average by the number of homerooms:

   ________ × ___10___ = ________

2. Use this averaging technique to estimate the totals in the following problems:

a. Profits for First Eight Months of 1984

$6459	6295
5784	6087
5924	5724
5924	6087

b. Monthly Family Earnings

$1872	1735	1709
2580	2261	1905
1932	2183	2218
2314	2017	2371

c. Numbers of Banners Sold

32	29	29
28	34	36
32	33	27
25	24	31

3. Grocery bill totals can be estimated by counting, to the nearest half, the approximate number of dollars spent and mentally keeping a running total.

Example	Rounded Amount	Running Total
$1.02	1	$1.
.80	1	2.
.34	.5	2.50
.53	.5	3.
2.14	2	5.
.49	.5	5.50
1.85	2	7.50 = Estimated total

Use the "count dollars" technique to estimate these totals:

a.

$0.47	.65	1.35
.45	1.59	.55
.34	2.92	1.39
.49	.95	2.59
.34		

Estimate = ________

b.

$5.89	.68	.75
1.30	1.09	1.88
1.19	.69	2.06
.72	1.00	2.20
2.52	.64	.24

Estimate = ________

1. Ms. Robertson is driving from Stayhere to Gothere, a distance of _____ km. She estimates she will average _____ km/h. About how long will the trip take?

2. Cleanzall laundry soap costs _____ for _____ grams. What is the approximate cost per gram?

3. Carolyn scored _____ points at the video game, Dream Machine. She played about _____ minutes. About how many points per minute did she average?

4. Kennedy Middle School has _____ students. The seventh grade makes up _____% of the school. About how many students are seventh graders?

5. Estimate the total grocery bill for the items circled:

$0.23	1.39	.49	1.49	.99	.78
.85	1.30	.72	1.09	1.88	2.46
.75	.95	.92	.55	.59	.35

ACTIVITIES

ESTIMATE AND CALCULATE

April 1985

By EARL OCKENGA and JOAN DUEA, Price Laboratory School, Cedar Falls, IA 50613

Teacher's Guide

Introduction: Recent recommendations for school mathematics strongly recommend that calculators be available for appropriate use in *all* mathematics classrooms. They futher encourage that calculators be used in imaginative ways for exploring, discovering, and developing mathematical concepts and skills. The activities that follow reflect ways that calculators can free students from computational restrictions and allow them to focus on the recognition of patterns and the application of numeration concepts in the development of their estimation skills.

Grade levels: 7–10

Materials: Calculators and a set of worksheets for each student

Objectives: To develop skills in estimating products, quotients, and percentages and in using a calculator to check estimates

Directions: Provide calculators for students who do not have their own. Distribute copies of the activity sheets one at a time to the students. After students have studied the example on a given worksheet and have made attempts at obtaining a better off-score, summarize for the class estimation strategies that you observed being used and perhaps identify other strategies that could have been used. The identification of the best total off-scores for the class and further discussion of associated estimation strategies should follow the completion of each sheet.

Sheet 1: Discuss the directions given at the top of the sheet. Have the students study the example. Then challenge them to see if they can arrange the digits 4 through 9 in the example to get an off-score that is less than 785 (9 × 7845 = 70 605, which has an off-score of 605).

As you observe students, you will find them using various strategies. Some students may use multiples of 1000 to estimate the product. For example, in round 1, they may think 6 × 8000 = 48 000, so 6 × 8745 should give a product close to the target number of 50 000. Computing the product on a calculator, they would get 52 470, resulting in an off-score of 2470. To improve on this off-score, their second estimate may be 6 × 8574, giving a product of 51 444. A third estimate may be 6 × 8457, resulting in an off-score of 742.

Other students may use the strategy of working backward to find the two factors. In round 1, they would think

50 000 ÷ 5 = 10 000, so 5 × 9876 should be close to the target number.

Sheet 2: Discuss the directions. Have the students study the example, then challenge them to arrange the digits 4 through 9 in the example to get an off-score that is less than 3 (798 ÷ 4 = 199.5, which has a rounded off-score of 0).

One strategy that students might find appropriate in this setting would entail use of multiples of 100 or 1000 to estimate the quotients. For example, in round 1 a student may think 500 ÷ 5 = 100, so 498 ÷ 5 should give a quotient close to the target number. Computing the quotient on a calculator, they would get 99.6, resulting in a rounded off-score of 0.

The working-backward strategy works well for finding dividends and divisors, too. In round 1, some students may think 100 × 9 = 900, so 897 ÷ 9 should be close to the target number.

Sheet 3: Discuss the directions. Depending on the background of your students, you may also need to discuss the use of the [%] key on their calculators. For students whose calculators do not have a [%] key, you will need to discuss the use of decimal representations of percentages. Have students study the example, then challenge them to arrange the digits in the example to get an off-score that is less than 7 (51% of 784 = 399.84, which has a rounded off-score of 0).

In estimating percentages of a number, students may use easy percentages such as 10 percent, 25 percent, or 50 percent. For example, in round 1 they may think 50 percent of 400 = 200, so 49 percent of 387 should given an answer close to the target number. Computing the percentage of the number on a calculator, they would get 189.63, resulting in a rounded off-score of 10. To improve on this off-score, a second estimate may be 51 percent of 398, giving an answer of 202.98. A third estimate may be 51 percent of 392, resulting in an off-score of 0.

Additional calculator-enhanced estimation activities can be found in Miller (1981) and in Goodman (1982).

REFERENCES

Goodman, Terry. "Calculators and Estimation." *Mathematics Teacher* 75 (February 1982):137–40, 182.

Miller, William A. "Calculator Tic-Tac-Toe: A Game of Estimation." *Mathematics Teacher* 74 (December 1981):713–16, 724.

For each round, make three attempts, as follows:

- Place the digits 4, 5, 6, 7, 8, or 9 in the boxes to get an answer close to the target number. A digit may be used only once in each try.
- Multiply.
- Subtract to find how far your answer is from the target number. Write your "off-scores" on a separate sheet of paper.
- Record your best off-score in the column at the right.

Example. Target number: 70 000

[9] × [7][8][6][5] = 70 785

70 785 − 70 000

Off-score: 785

Round 1. Target number: 50 000

[ ] × [ ][ ][ ][ ] = ______ Best off-score: ______

Round 2. Target number: 40 000

[ ] × [ ][ ][ ][ ] = ______ Best off-score: ______

Round 3. Target number: 80 000

[ ] × [ ][ ][ ][ ] = ______ Best off-score: ______

Round 4. Target number: 60 000

[ ] × [ ][ ][ ][ ] = ______ Best off-score: ______

Round 5. Target number: 40 000

[ ][ ] × [ ][ ][ ] = ______ Best off-score: ______

Round 6. Target number: 70 000

[ ][ ] × [ ][ ][ ] = ______ Best off-score: ______

Rate yourself: Total: ______

Total off-score less than 8000 _ _ _ Super estimator
8001 to 10 000 _ _ _ _ _ _ _ _ _ _ _ Excellent
10 001 to 15 000 _ _ _ _ _ _ _ _ _ _ Good
More than 15 000 _ _ _ _ _ _ _ _ _ Need practice

For each round, make three attempts, as follows:

- Place the digits 4, 5, 6, 7, 8, or 9 in the boxes to get an answer close to the target number. A digit may be used only once in each attempt.
- Divide and then round your quotient to the nearest whole number.
- Subtract to find how far your answer is from the target number. Write your off-scores on a separate sheet of paper.
- Record your best off-score in the column at the right.

Example. Target number: 200

[9][8][7] ÷ [5] ≈ 197

200 − 197

Off-score: 3

Round 1. Target number: 100

[ ][ ][ ] ÷ [ ] ≈ ______ Best off-score: ______

Round 2. Target number: 50

[ ][ ][ ] ÷ [ ] ≈ ______ Best off-score: ______

Round 3. Target number: 250

[ ][ ][ ] ÷ [ ] ≈ ______ Best off-score: ______

Round 4. Target number: 800

[ ][ ][ ][ ] ÷ [ ] ≈ ______ Best off-score: ______

Round 5. Target number: 1000

[ ][ ][ ][ ] ÷ [ ] ≈ ______ Best off-score: ______

Round 6. Target number: 200

[ ][ ][ ][ ] ÷ [ ][ ] ≈ ______ Best off-score: ______

Total: ______

Rate yourself:

Total off-score less than 40 — Super estimator
41 to 60 — Excellent
61 to 100 — Good
More than 100 — Need practice

For each round, make three attempts, as follows:

- Place the digits 1, 2, 3, 4, 5, 6, 7, 8, or 9 in the boxes to get an answer close to the target number. A digit may be used only once in each try.
- Find the percentage of the number and then round the result to the nearest whole number.
- Subtract to find how far your answer is from the target number. Write your three off-scores on a separate sheet of paper.
- Record your best off-score in the column at the right.

Example. Target number: 400

[5][1] % of [7][9][8] ≈ 407   407-400 Off-score: 7

Round 1. Target number: 200

[ ][ ] % of [ ][ ][ ] ≈ ______   Best off-score: ______

Round 2. Target number: 100

[ ][ ] % of [ ][ ][ ] ≈ ______   Best off-score: ______

Round 3. Target number: 500

[ ][ ] % of [ ][ ][ ] ≈ ______   Best off-score: ______

Round 4. Target number: 300

[ ][ ] % of [ ][ ][ ] ≈ ______   Best off-score: ______

Round 5. Target number: 100

[ ] % of [ ][ ][ ][ ] ≈ ______   Best off-score: ______

Round 6. Target number: 50

[ ][ ][ ] % of [ ][ ] ≈ ______   Best off-score: ______

Total: ______

Rate yourself:

Total off-score less than 60— —Super estimator
61 to 80— — — — — — — — — — — — —Excellent
81 to 100— — — — — — — — — — — —Good
More than 100— — — — — — — — — —Need practice

ACTIVITIES

ESTIMATION, QUALITATIVE THINKING, AND PROBLEM SOLVING

September 1987

By ZVIA MARKOVITS, RINA HERSHKOWITZ, and M. BRUCKHEIMER, Weizmann Institute of Science, Rehovot, Israel

Teacher's Guide

Introduction: Although many articles dealing with estimation have been published recently, most have been devoted to computational estimation (Rubenstein 1985; Ockenga and Duea 1985; Reys et al. 1985; Trafton et al. 1986). But computational estimation is not the only kind of estimation. The importance of estimation activities also comes from the contribution they might make to the development of qualitative thinking and to problem-solving skills (Trafton 1978; O'Daffer 1979).

In a unit on estimation written for junior high school students, we included, in addition to computational estimation, the following estimation topics.

1. *Meaningful accuracy.* The aim of this topic is to show students that they should judge the degree of accuracy required according to the situation and the circumstances. Accuracy is not always better than estimation; sometimes it is ridiculous or impossible.

2. *Estimation in "everyday life."* Activities of this type are designed to develop the ability to estimate a variety of measures in the students' environments. Also, this kind of estimation is used in problem solving and especially in judging the reasonableness of the results obtained.

3. *Estimation in measurement.* Without using measuring instruments, students should be able to appreciate the difference between absolute and relative errors, when estimating lengths and areas, for example. Examples of these activities and an accompanying game can be found in an article by Markovits, Hershkowitz, Taizi, and Bruckheimer (1986).

4. *Algorithmic estimation.* Students can develop the ability to estimate quantities by constructing a suitable algorithm and by taking into account missing data—data

needed for the computation but not appearing in the statement of the problem.

5. *Checking the reasonableness of the results.* The aim of this estimation is to show that not every problem must (or even can) be solved by using a standard algorithm and that not every problem has a unique answer. However, the answer has to be reasonable.

These additional aspects of estimation are woven through the activities presented here: Estimation in Everyday Life (sheet 1), Selecting an Estimation Strategy (sheet 2), and Using Estimation in Problem Solving (sheets 3 and 4).

Grade levels: 6–9

Materials: Copies of activity sheets 1–4 for each student

Objectives: To introduce students to a variety of estimation activities that involve qualitative thinking and to develop problem-solving skills

Procedure: The suggestions that follow are based on our experience in using the materials with sixth- and seventh-grade students and with preservice teachers. Most classes should be able to complete sheets 1–3 during one class period. Sheet 4 could then be used for homework or saved until the next class day. Distribute the activity sheets to the students, one at a time, and conduct class discussions at the points noted in this guide.

Sheet 1 introduces the student to the use of estimation strategies in everyday-life situations. The students are first asked to arrange the items in order of magnitude and only then to write the requested measure for each. In this way we tend to avoid answers given by many sixth and seventh graders, who maintained that the speed of a jet is between 0 and 100 km an hour, less than the speed of a car. Or, again, we avoid the answers of about half the students, who claimed that a 200-page book is thicker than a volume of an encyclopedia. The students should be encouraged, in both problem 1 and problem 2, to use measures known to them as reference points. For example, many students find it easier to imagine the thickness of a book with 200 pages (1–3 cm) and to calculate the thickness of one page than to estimate directly the thickness of a page. The teacher may wish to have a class discussion on the methods used in problem 1(c) before continuing with the page to emphasize the use of reference points.

Sheet 2 focuses on the use of common sense when selecting an estimation strategy. In the situations in problem 1, students frequently apply an algorithm that leads to nonsense. The intent of these questions is to have the student become familiar with the idea that common sense by itself can often lead to a reasonable answer. In problem 1(a) the standard "linear" algorithm leads to nonsense: $1.5 + 1.5$ or 2×1.5. (Be careful! We have seen students who wrote 2.10 because $1.5 + 1.5 = 2.10$.)

When solving problem 1(b), about one-third of the students we observed in grades 6–7 used the "average" algorithm and obtained an answer of five hours, which seems to be reasonable. But, by using simple reason, we realize that if one painter paints the wall in three hours, two painters together will finish in less than three hours. (Only about one-third of the sixth and seventh graders and less than half of the preservice teachers gave a reasonable estimate.)

In problem 1(c), again using the standard "linear" algorithm, we have

$$\frac{3000000}{30000} = 100,$$

$$10 \text{ million} \times 100 = 1000 \text{ million}$$

$$= 1 \text{ billion people}$$

and find that about one-fifth of the world's population lives in Argentina! Over half of the seventh graders and about one-third of the preservice teachers used the "linear" algorithm or some other algorithm. For example, consider the following line of thought:

The population of Argentina is 20 million. How did I get the answer? I multiplied the population of Belgium by 2 because the area of Argentina is bigger by 2 zeros.

The teacher may wish to lead a class discussion after students have completed problem 2 to make sure they have realized the nonsense that some of the algorithms yield in these situations.

If we have two many such problems, we may cause students to avoid algorithms altogether. So, in problem 3, we mix problems that can be solved by an algorithm with those that cannot. For each problem they must decide whether or not to use a standard algorithm and so understand that in some situations an algorithm is applicable but in others only common sense and reflection are required. For example, in problem 3(d) the algorithm 188 ÷ 6 can be used but only as a means for obtaining a reasonable answer; 31 1/3 is not reasonable!

Sheets 3 and 4 integrate estimation into problem-solving activities. The teacher might use problem 1(a) to show the procedure needed to solve this kind of problem: first, determine the missing data; next, suggest some reasonable values for each of the missing data; construct a suitable algorithm; and then solve the problem. Estimation in everyday life can help students to give reasonable values for the missing data. For problems 1 and 2 the students should be encouraged to use the estimates in one part to yield additional data for the next part. Note that because of the amount of missing data, problems 1(c) and 2(c) are significantly more difficult than the preceding problems.

In problem 3 students should be encouraged to use the algorithm of words per line × problem lines per page × pages, and not words per page × pages. From our students' answers we found that the first algorithm gives more reasonable answers than the second.

In problem 4 students can use the answer they got in 4(a) to help with their estimate in 4(b). The answers will obviously depend on the coin chosen. The connection between the number of coins and the size of the coin is a fruitful area for further investigation.

Many students tend to estimate the area of the parking lot in problem 5 and of a car and then to divide. They forget that the cars must also be able to get in and out of the parking lot, so suitable distances between cars and at the ends of the rows of cars are needed.

In problem 6 students who do not use the appropriate volume algorithm, but just imagine the bath and guess, are usually surprised by the large number of liters contained by a bathtub. Over half the students claimed that the tub contains less than fifty liters! Indeed, it may be worthwhile to ask students first to guess and then to estimate by reasonable reflection.

Solutions

Sheet 1:

1. A	B	
The fastest	kilometers/hr.	miles/hr.
jumbo jet	800–1200	500–700
passenger train	150–200	80–100
car	60–100	35–55
galloping horse	40–70	25–40
ship	30–50	20–30
bicycle	10–40	5–25
The slowest		
pedestrian	3–10	2–5

2. A	B	
The thickest	metric	English
encyclopedia	3–8 cm	1–3 in.
200-page book	1–3 cm	0.4–1.2 in.
carpet	0.5–2 cm	0.2–0.8 in.
booklet	1–5 mm	0.04–0.2 in.
cloth	1–5 mm	0.04–0.2 in.
The thinnest		
page	0.01–0.1 mm	$\frac{1}{250}$–$\frac{1}{100}$ in.

Sheet 2:

1. *a.* About 1.8–1.9 meters (not 3 meters!)
   b. About two hours. The important point here is that students realize that the answer has to be less than three hours.
   c. Cannot be determined from the data. The population density in Belgium is not the same as that in Argentina. Look up the population of Argentina and discuss population density in the two countries (and others).

3. *a.* 1200 ice-cream bars. Some students may suggest a smaller quantity because they consider a morning break or down time.
   b. About 10–12 kg (not 48 kg)
   c. Not −20. We have no way of knowing, but we can discuss possible answers, taking into account the supermarket hours and the performance times in the cinema.
   d. Many possible answers exist. For example, 31, 31, 31, 31, 32, 32; 30, 30, 30, 30, 30, 38; and so on (not 31 1/3 or 31.33)

Sheets 3 and 4

Answers are not given because they depend on the values given to the missing data.

REFERENCES

Markovits, Zvia, Naomi Taizi, Rina Hershkowitz, and M. Bruckheimer. "From the File: Estimeasure." *Arithmetic Teacher* 34 (December 1986):9.

Ockenga, Earl, and Joan Duea. "Estimate and Calculate." *Mathematics Teacher* 78 (April 1985):272–76.

O'Daffer, Phares. "A Case and Techniques for Estimation: Estimation Experience in Elementary School Mathematics—Essential, Not Extra!" *Arithmetic Teacher* 26 (February 1979):26–51.

Reys, Robert E., Barbara J. Reys, Paul R. Trafton, and Judy S. Zawojewski. "Estimating with 'Nice' Numbers." *Mathematics Teacher* 78 (November 1985): 615–25.

Rubinstein, Rheta M. "Developing Estimation Strategies." *Mathematics Teacher* 78 (February 1985):112–18.

Trafton, Paul R. "Estimation and Mental Arithmetic: Important Components of Computation." In *Developing Computational Skills,* 1978 Yearbook of the National Council of Teachers of Mathematics, edited by Marilyn N. Suydam, pp. 196–213. Reston, Va.: The Council, 1978.

Trafton, Paul R., Judy S. Zawojewski, Robert E. Reys, and Barbara J. Reys. "Estimating with 'Nice' Fractions." *Mathematics Teacher* 79 (November 1986): 629–34.

Many situations in everyday life require an estimate rather than an exact answer.

1. *a.* Using column A, order the following from the fastest to the slowest.

   galloping horse bicycle ship jumbo jet
   passenger train car pedestrian

   b. Write down the approximate speed of each object in column B.

	A	B approximate speed
fastest		
slowest		

   c. Describe the method you used to obtain your estimates in 1(*b*).

   ____

   Compare your strategy with the ones used by two of your classmates.

2. *a.* Using column A, rank the following objects from the thickest to the thinnest.

   200-page book carpet page cloth booklet
   encyclopedia

   b. Write the approximate thickness of each object in column B.

	A	B approximate thickness
thickest		
thinnest		

1. Answer each of the questions below and describe the strategy that you employed to arrive at your estimate.

   a. The height of a ten-year-old boy is 1.5 meters. What do you think his height will be when he is twenty years old?
   Estimate ________________
   Strategy ____

   b. Karen paints a wall in three hours. Dwayne paints the same wall in seven hours. The two painters work together. About how many hours will it take them to paint the wall?
   Estimate ________________
   Strategy ____

   c. The area of Belgium is about 30 000 km², and its population is about 10 million. The area of Argentina is about 3 000 000 km². What do you think the population of Argentina is?
   Estimate ________________
   Strategy ____

2. Read each part of problem 1 again together with your estimate. Are your answers reasonable? Does your common sense suggest different answers in some cases?

3. Estimate answers in each of the following problems. Make sure your answer does not contradict your common sense.

   a. The following table shows the number of ice-cream bars produced by a machine in an ice-cream factory.

Time	Number Produced by the Time Shown
8:00	300
9:00	600
10:00	900

   What do you think the number of ice-cream bars produced by the machine will be by 11:00? ____________

   b. The weight of a one-month-old baby is 4 kg. What do you think its weight will be when the baby is one year old? ____________

   c. The following table shows the number of cars in a parking lot next to a cinema and a supermarket. What do you think the number of cars will be at 5:00?

Time	Number of Cars
2:00	70
3:00	40
4:00	10

   ____

   d. One-hundred-eighty-eight books were delivered to six libraries. How many books do you think each library received? ____________

In the following problems some data are missing. Specify the missing data in column A, write reasonable values in column B, and answer the problem in column C.

1.

Problem	A Missing Data	B Estimates	C Answer
a. About 1000 children went on an outing. Each class was accompanied by 1 teacher. How many teachers went on the outing?	The average number of children in a class.	25–35	
b. About how many buses were needed for the outing?			
c. About how much did the diesel fuel cost for the outing?			

2.

Problem	A Missing Data	B Estimates	C Answer
a. Estimate the number of students in an elementary school	number of children		
b. Estimate the number of students in 300 middle schools.			
c. Estimate the total floor area of a middle school.			

3. How many words are in a 200-page book?
   Missing data ____________________
   Estimated values ____________________
   Answer ____________________

4. How many coins are needed to cover—
   a. this page?
   Missing data ____________________
   Estimated values ____________________
   Answer ____________________
   b. your desk?
   Missing data ____________________
   Estimated values ____________________
   Answer ____________________

5. A parking lot is 90 meters long and 40 meters wide.
   a. Estimate the number of cars that can park in this lot.
   Missing data ____________________
   Estimated values ____________________
   Answer ____________________
   b. Estimate the number of buses that can park there.
   Missing data ____________________
   Estimated values ____________________
   Answer ____________________

6. How many liters of water are in a full standard bathtub?
   Missing data ____________________
   Estimated values ____________________
   Answer ____________________

ACTIVITIES

EXAMINING RATES OF INFLATION AND CONSUMPTION

September 1982

By Melfried Olson, Oklahoma State University, Stillwater, OK 74078
Vincent G. Sindt, University of Wyoming, Laramie, WY 82071

Teacher's Guide

Grade level: 8–12

Materials: Student worksheets, calculators

Objectives: As students work with these problems, they will be exposed to the effects of inflation rates. Given an inflation rate, students will determine the length of time it takes for a cost to double. Similarly, when applied to consumption rates, students will examine the effect of growth rates and how quickly growth rates can deplete a finite resource.

Directions:

Sheet 1. You may want to work through the example provided before having the students complete exercise 2. It is important that the students know how to use the calculator to compute the total cost for each successive year. You may want to emphasize multiplying by 1.1 for the 10% rate, rather than multiplying by 0.1, and adding the cost increase to last year's cost to get the new cost. Also, as many calculators have an automatic constant for multiplication, you might investigate how to use it for quicker calculations. Note that the method of calculation chosen and the use of round-off or truncation can vary the amounts per year. Amounts would also change if we compounded semiannually, quarterly, monthly, or daily; so use the table as a guide to a 10% inflation rate. This is the reason for column *C* of the table on sheet 1. If we compute the 10% rate compounded at 5/6% per month, the numbers in columns *A* and *B* would be 6 and 7, respectively.

Sheet 2. Sheet 2 is similar to sheet 1 but focuses on consumption rather than inflation. The seven-year period is chosen partially for convenience but mostly because initial plans called for a 10% increase in coal production to compensate for the decline in oil supplies. Maintained at this rate per year, this means a seven-year period of time before we double consumption. Note the rules do not equate with the growth rate per year but come close enough to demonstrate clearly how a consumption rate can deplete a finite resource. The 2800 years represents the high estimate given for coal reserve supplies in the United States (Bartlett 1978).

It is informative and useful to vary the table to account for different rates of consumption, for even a 7% consumption rate will exhaust these 2800 units in under 100 years.

Sheet 3. This work is provided to help consolidate inflation and growth rates through the principle that if something doubles in cost in n years, it quadruples in cost in $2n$ years. Thus the doubling effect builds very rapidly.

Answers:

Sheet 1.

2.

R	(A)	(B)	(C)
12%	6 ($1.97)	7 ($2.21)	6
9%	8 ($1.99)	9 ($2.17)	8
8%	9 ($1.99)	10 ($2.15)	9
7%	10 ($1.97)	11 ($2.10)	10
6%	11 ($1.90)	12 ($2.01)	12

3. Note the inverse variation relationship. A general formula can be taken as $C = 70/R$ or $C = 72/R$, both of which provide a good approximation for use depending on the R values involved. Actually, the formula will vary somewhat according to the period of compounding. For a calculation of this formula, see Bartlett (1978).

Sheet 2.

2003–2009	8	56	105
2010–2016	16	112	217
2017–2023	32	224	441
2024–2030	64	448	889
2031–2037	128	896	1785
2038–2044	256	1792	3577

1. Note that in each seven-year period seven more units of coal are used than were used in total up to that period.
2. 2038–2044. Actually, in year 2041 we will have exhausted the supply.
3. This depends on the student's age now, but some students will still be alive to watch the last unit go under these rules.

Sheet 3.

1a. 5:00 P.M.

1b. Working backward from 6:00 P.M., it would be 1/64 full (1.6%)—lots of room left.

2. $23 344
3. $43 178
4. Somewhere around 43–45 years, depending on how intermediate values are rounded.

BIBLIOGRAPHY

Bartlett, A. A. "Forgotten Fundamentals of the Energy Crisis." *American Journal of Physics* 46 (September 1978):876–88.

Lange, L. H. "Some Everyday Applications of the Theory of Interest." In *Applications in School Mathematics*, 1979 Yearbook, edited by S. Sharron. Reston, Va.: National Council of Teachers of Mathematics, 1979.

Lapp, R. E. *The Logarithmic Century*. Englewood Cliffs, N.J.: Prentice-Hall, 1973.

Waits, B. K. "The Mathematics of Finance Revisited through the Hand Calculator." In *Applications in School Mathematics*, 1979 Yearbook, edited by S. Sharron. Reston, Va.: National Council of Teachers of Mathematics, 1979.

1. If one can purchase an item that costs $1 now and inflation continues at 10% per year (compounded yearly), when will the cost double? We can make a table like the following to determine the result. Show how you would compute the remaining costs.

Year	Cost
Now	$1.00
1	$1.10 = 1.00 + (0.10 × 1.00) = 1.1 (1.00)
2	$1.21 = 1.10 + (0.10 × 1.10) = 1.1 (1.10)
3	$1.33 = 1.21 + (0.10 × 1.21) = 1.1 (1.21)
4	$1.46 =
5	$1.61 =
6	$1.77 =
7	$1.95 =
8	$2.14 =

2. Complete the following table, starting with an item that now costs $1. See how many years it will take for the item to cost $2 at the various rates of inflation, compounded yearly. The 10% rate is completed for you using data from the chart above.

R Rate per Year (Compounded Yearly)	(*A*) Number of Years Just Prior to Reaching $2	(*B*) Number of Years Just After Passing $2	(*C*) Choice from (*A*) or (*B*) Closest to $2
12%			
10%	7 ($1.95)	8 ($2.14)	7
9%			
8%			
7%			
6%			

3. Graph the results above on this coordinate system.

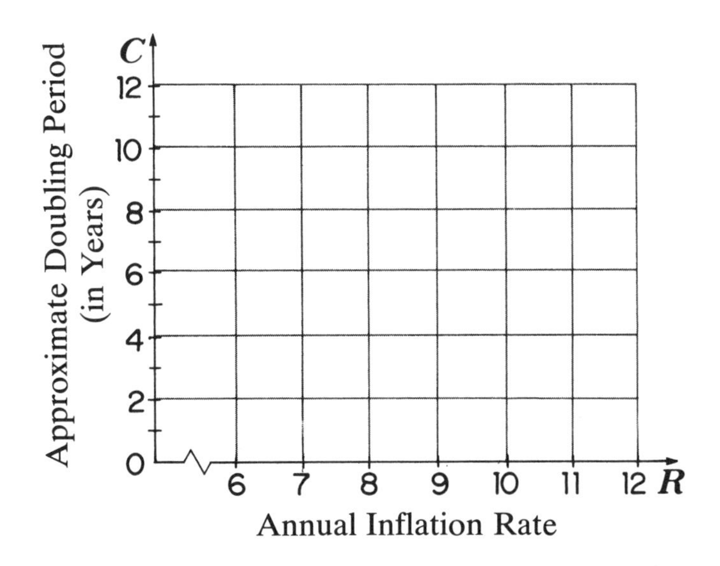

At the end of 1980, scientists in the United States made two measurements concerning coal. They determined how much coal was used in the United States during 1980 and estimated the total available. Using the 1980 amount as one unit, scientists concluded that the total supply of coal reserves in the United States is 2800 units.

However, in 1981 people found more uses for coal and proposed increasing its consumption. The debate that ensued was settled by agreeing to the following rules:

Rule 1: Use one unit of coal in the first year, 1982.

Rule 2: Use the same number of units of coal per year for a seven-year period.

Rule 3: At the end of each seven-year period, double the amount of coal that can be consumed during the next seven-year period.

Use these rules to complete the following table.

Years	Number of Units of Coal Used *Each Year*	Number of Units Used during the Seven-Year Period	Total Number of Units Used Thus Far
1982–1988	1	7	7
1989–1995	2	14	21
1996–2002	4	28	49
2003–2009			
2010–2016			
2017–2023			
2024–2030			
2031–2037			
2038–2044			

1. How does the number of units used during a seven-year period compare to the total number of units used thus far? ______________________

____

2. During which seven-year period will the last unit of coal be used under these rules? ______________________

____

3. How old will you be when the last unit of coal is used? ______________________

____

1. Bacteria grow by division so that 1 bacterium becomes 2, the 2 divide to produce 4, the 4 divide to produce 8, and so on. Scientists noticed a particular strain of bacteria, Numah, for which this division time is 1 hour. They also noted that when 1 Numah is placed in a bottle of a certain size at 6:00 A.M., the bottle is full at 6:00 P.M. the same day.

   a. At what time was the bottle 1/2 full?

   b. How full was the bottle at noon?

2. The average price for a new subcompact car is about $9000. At a 10% annual inflation rate, what will the average cost for a new car be in 10 years?

3. Suppose that you have an income of $20 000. If you receive an 8% raise yearly, what will your yearly income be in 10 years?

4. Population growth in the world is approximately 1.6% per year. If that rate exists for your town and that rate continues, how many years will it take before your town doubles its population?

Activities for

Algebra and Graphs

The study of patterns is at the center of mathematics. Searching for patterns was introduced in the first section of this volume as a useful problem-solving strategy. Activities throughout the remaining sections provide abundant opportunities for students to generalize patterns using algebra. The activities in this section promote the investigation of patterns within the system of algebra itself.

The first four activities engage students in investigating multiple representations and properties of linear equations. In "Solving Linear Equations Physically," students use a balance-scale model together with strips and small squares to represent and solve linear equations. Their experiences with concrete and pictorial representations lead to the discovery of formal methods for solving such equations. In the activity "Microcomputer Unit: Graphing Straight Lines," students use a graphing utility (a computer with function-graphing software or a graphing calculator) to explore how the values of m and b in $y = mx + b$ affect the inclination and position of the graph of the equation and to discover relationships between the slopes of parallel lines and of perpendicular lines. "Slope as Speed" provides a physical interpretation of the slope of a line. Pupils use this interpretation of slope to analyze and draw distance-time graphs and make inferences about car speeds. In the activity "Families of Lines," students explore the geometry of systems of linear equations. They use technology or by-hand methods to graph pairs of linear equations and linear combinations of these pairs and in the process discover a formal algebraic method for solving systems of linear equations. The last activity sheet requires students to complete, run, and modify a BASIC program for solving a system of equations.

The next two activities focus on a global analysis of graphs. "Interpreting Graphs" provides multiple nonlinear settings in which students produce and interpret various distance-time graphs. Focused questions illuminate the relationship between the shape of a graph and the rate of change of the quantities represented. In "Relating Graphs to Their Equations with a Microcomputer," pupils use a graphing utility to investigate how the values of a, b, and c in the equation $y = a|x + b| + c$ affect the shape and position of its graph. Included are tasks involving translation from symbolic representations to graphical representations and vice versa.

This section concludes with two activities that enable students to investigate multiple representations and properties of quadratic polynomials and equations. In the activity "Microcomputer Unit: Graphing Parabolas," students use a graphing utility to explore how the values of a, b, and c in the equation $y = ax^2 + bx + c$ affect the direction and shape of the graph of the equation. "Finding Factors Physically" uses an extension of the strip-square model of the first activity of this section to provide students opportunities to investigate concretely and pictorially the factoring of quadratic polynomials and thereby discover relationships among the coefficients. This discovery leads to a conceptual understanding of formal methods for factoring such polynomials.

ACTIVITIES

SOLVING LINEAR EQUATIONS PHYSICALLY

September 1985

By BARBARA KINACH, Lesley College, Cambridge, MA 02138

Teacher's Guide

Introduction: The "Activities" section of the *Mathematics Teacher* first appeared in 1972 as a means of providing classroom teachers ready-to-use discovery lessons and laboratory experiences in worksheet form. The importance and efficacy of these alternative instructional methodologies were reaffirmed in the mathematics education reform documents of the 1980s. It was recommended that teachers employ diverse instructional strategies, materials, and resources, including the following:

- The provision of situations that provide discovery and inquiry as well as basic practice
- The use of manipulatives, where appropriate, to develop concepts and procedures

The following materials provide opportunities for students to investigate solving linear equations using both a physical and a pictorial model and in the process discover a method that will permit them to solve such equations at the symbolic level.

Grade levels: 7–10

Materials: Scissors, cardboard, a lighter-weight tagboard, and a set of activity sheets for each student. Transparency cutouts of the strips, squares, and balance scale on sheet 4 would be useful for purposes of demonstration and for discussion of students' solutions.

Objectives: Students will (1) represent first-degree polynomials physically (with strips and squares) and pictorially; (2) solve linear equations both physically through the use of a balance scale and pictorially; and (3) discover a method for solving linear equations using algebraic manipulation.

Directions: On the day before this lesson distribute a copy of sheet 4 to each student. As indicated on sheet 1, instruct them to glue the row of strips on a piece of cardboard, glue the rows of squares on a piece of lighter-weight tagboard, and then cut out the pieces. Provide students with an envelope in which to keep their equation-solving pieces.

On the day of the lesson, distribute the worksheets one at a time. All students should be able to complete the first two sheets during a single forty-five-minute class period if the strips and squares have been precut. Depending on the time available, sheet 3 can be completed during the next class period.

Sheet 1 emphasizes the distinction between a variable and a constant. It is important that students realize that the strips were purposely constructed so that their weights were a variable quantity in relation to the weights of the squares. In addition, this first worksheet uses the strip-square diagrams to clarify the use of grouping symbols in algebraic expressions. Students thus distinguish the difference between $2S + 3$ and $2(S + 3)$ pictorially.

Sheet 2 establishes the analogy between balancing a scale and solving a linear equation of the form $ax + b = cx + d$ where a, b, c, $d \geq 0$. In this activity pupils first attempt to maintain the scale's balance while replacing each strip with the same number of squares. This experience reinforces the conditional nature of an equation. Specifically, it demonstrates that the asserted equality between two expressions need not hold for all replacements of the variable S. The second portion of this sheet provides an algorithm for physically determining the value for S that will maintain the scale's balance. Students may solve the equations in exercise 8 by making pictorial representations of their physical manipulations or by using standard symbolic methods. At this stage, either method should be accepted. You might also have students verify their solutions for exercises 7 and 8 by actually substituting the values obtained for S into the equations and then performing the arithmetic indicated.

Finally, sheet 3 introduces a pictorial method for solving linear equations as a bridge between the physical model and the ultimate goal of algebraic manipulation. It would be instructive to demonstrate how the equation $4S + (-3) = 3S + (-4)$ in the example can also be solved pictorially by adding three white squares to both sides of the configuration in step 2 and using the fact that a white square and a gray square cancel each other. The solution of exercise 9c will require students to add three gray squares to each side of the configuration. Encourage pupils to verify their solutions for exercises 9 and 10 by substituting the values obtained for S into the equations and then performing the arithmetic indicated.

After all students have completed sheet 3, carefully establish the algebraic methods for solving linear equations in terms of the corresponding pictorial manipulations. In the example at the top of sheet 3, this correspondence can be illustrated as follows.

Step 1: Represent the equation. See (1).

Step 2: Subtract (remove) 3 gray squares. See (2).

Step 3: Subtract (remove) 3 strips. See (3).

$$\begin{array}{lrcr}
(1) & 4S + (-3) & = & 3S + (-4) \\
(2) & \underline{\quad - (-3)} & & \underline{\quad - (-3)} \\
 & 4S & = & 3S + (-1) \\
(3) & \underline{-3S \quad} & & \underline{-3S \quad} \\
 & S & = & -1
\end{array}$$

Students should note the pictorial distinction made between a negative number and the operation of subtraction. A negative number is represented as a gray square. The operation of subtraction is indicated by crossing out with an "X" the strip or square to be removed. This distinction of notation should remind pupils that the symbol $(-)$ has different meanings.

Supplementary activities: Introduce notation for subtracting a constant from a variable. For example, to indicate the subtraction $3S - 4$, place (or draw) four *white* squares on top of the strips,

Note the distinction with the representation of $3S + (-4)$,

Challenge students to solve each of the following equations first pictorially and then using the corresponding algebraic manipulations.

a. $4S - 3 = S + 3$

b. $5(S - 1) = 2S + 7$

c. $-3 + 2S = 3(S - 2)$

Later in the year you may wish to use a modification of this strip-square model as described by Hirsch (1982) to factor quadratic polynomials physically.

Answers:

Sheet 1: 2. b. 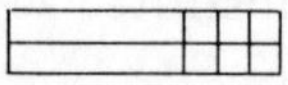 ; c. The rectangular region formed consists of two rows of identical shapes. Since the weight of one row is $S + 3$, the weight of the region is $2(S + 3)$.

3. a. 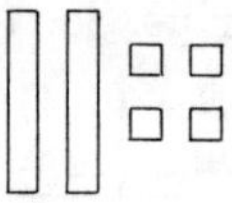 ; b. 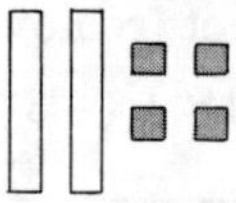 ;

c. 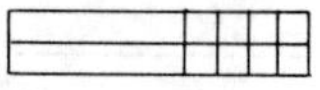 ; d. 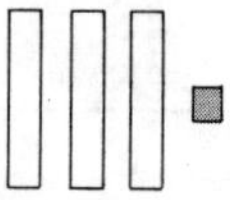 ;

e. 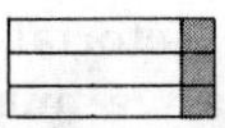 ; f. 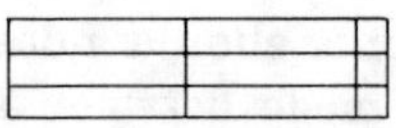 .

4. a. $3S + 6$; b. $4(S + 1)$ or $4S + 4$; c. $3S + (-9)$.

Sheet 2: 5. b. No; f. No. 6. three. 7. a. $S = 1$; b. $S = 7$; c. $S = 5$; d. $S = 2$; e. $S = 3$. 8. a. $S = 4$; b. $S = 7$; c. $S = 2$.

Sheet 3: 9. a.

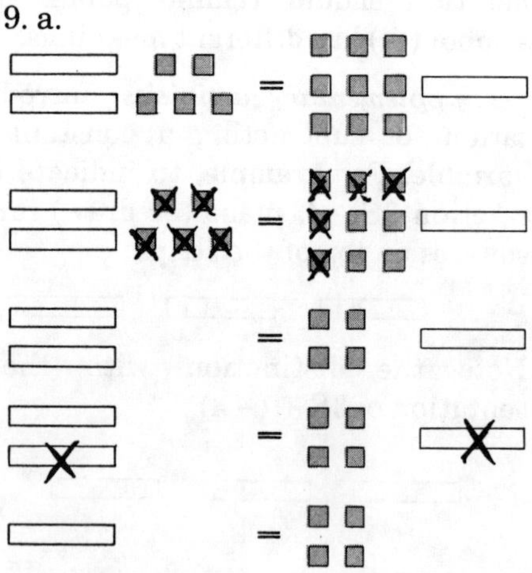

=

=

=

=

=

The solution is $S = -4$.

b.

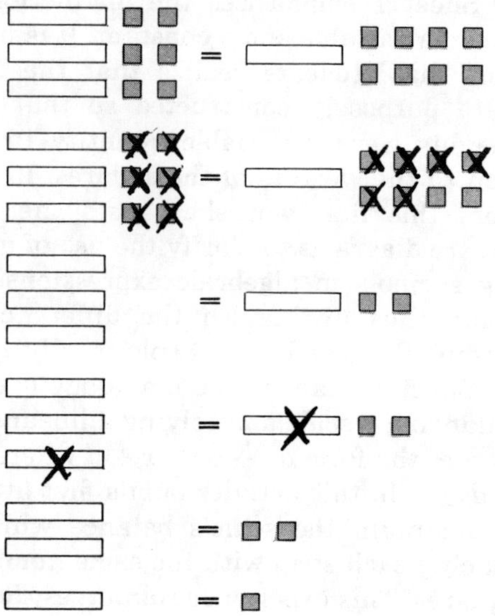

=

=

=

=

=

=

The solution is $S = -1$.

c.

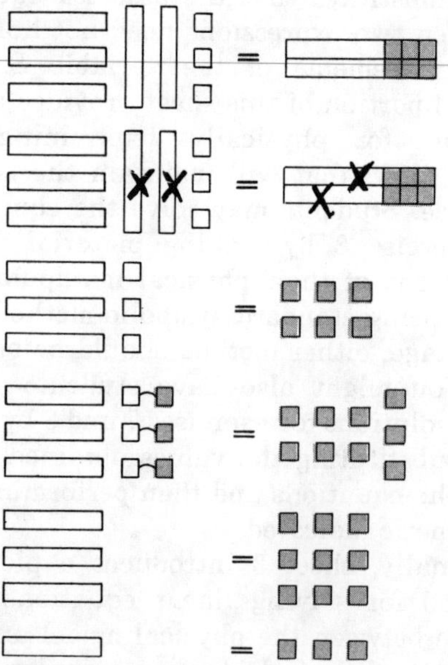

=

=

=

=

=

=

The solution is $S = -3$.

10. a. $S = -4$; b. $S = 14$

REFERENCE

Hirsch, Christian R. "Finding Factors Physically." *Mathematics Teacher* 75 (May 1982):388–93, 419.

1. Sheet 4 of this activity consists of a series of strips, white and gray squares, and a diagram of a balance scale.
   a. Cut the sheet along the two dashed lines.
   b. Glue the series of strips onto a piece of a cardboard and then carefully cut out the strips.
   c. Glue the series of squares onto a piece of light-weight tagboard, such as a file folder, and then carefully cut out the white and the gray squares.

   For this activity, assume that the weight of—
   a. a *white square* is a *positive one* $(+1)$ unit,
   b. a *gray square* is a *negative one* (-1) unit,
   c. a *strip* is a *variable* quantity S (depending on the cardboard backing used).

2. a. The expression $2S + 6$ can be represented physically by two strips and six white squares. Place these eight pieces on your desk.
   b. Rearrange the pieces to form a rectangle.
   c. Explain why the weight of these eight pieces can also be expressed as $2(S + 3)$. ______________________________

3. Represent each algebraic expression with strips and squares. Draw diagrams of your solutions in the spaces provided.

   a. $2S + 4$ b. $2S + (-4)$ c. $2(S + 4)$

   d. $3S + (-1)$ e. $3[S + (-1)]$ f. $3(2S + 1)$

4. Write an algebraic expression for each diagram.

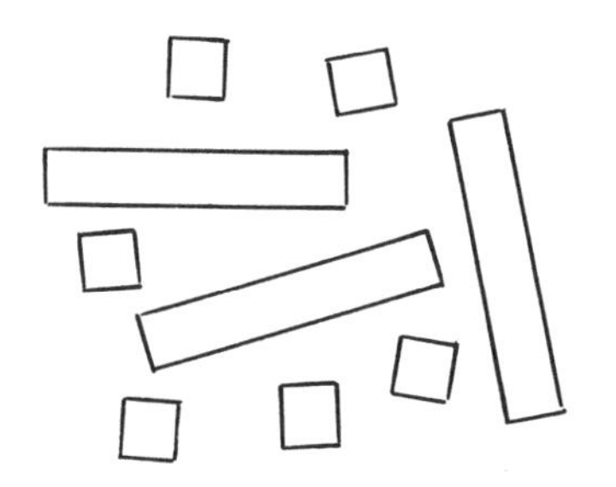
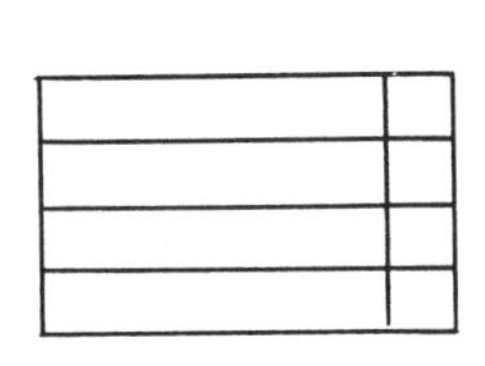
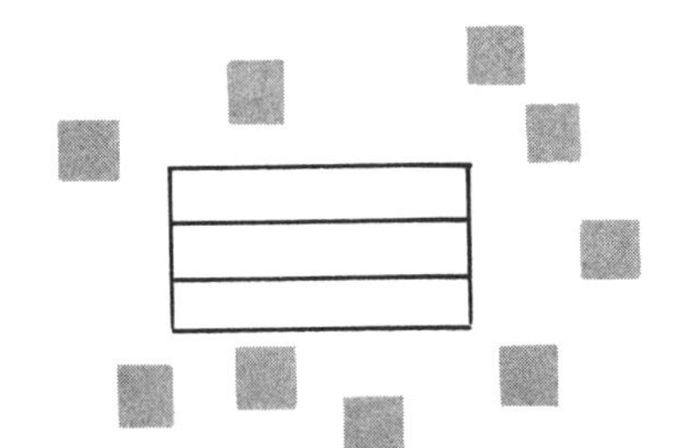

a.  b. ______________ c. ______________

5. To represent the linear equation $3S + 2 = 2S + 5$ physically, place the strip-square representation of $3S + 2$ on the left pan in your diagram of a balance scale and the corresponding representation of $2S + 5$ on the right pan.
   a. Replace each strip with one white square (that is, assume $S = 1$).
   b. Would your scale remain balanced? ______
   c. Restore the original pieces to the scale.
   d. Now replace each strip with two gray squares (that is, assume $S = -2$).
   e. Simplify by using the fact that a gray square and a white square cancel each other out, since $(-1) + (+1) = 0$.
   f. Describe the contents of the left pan: ______________________
   The right pan: ______________________ Does the scale balance? ______

6. To solve the linear equation $3S + 2 = 2S + 5$ physically, we must determine the number of squares that can be used to replace each strip and keep the scale balanced. This can be accomplished by removing *equally weighted* pieces from each side of the scale until you have only one strip remaining on the scale.
   a. Represent the equation $3S + 2 = 2S + 5$ on your balance scale.
   b. Remove two squares from each side.
   c. Now remove two strips from each side.
   d. The weight of one strip equals the weight of ______ white squares. Thus, the solution of the equation is $S = 3$.

7. Use the method in exercise 6 to solve each of the following linear equations. Write your solutions in the spaces provided.
   a. $2S + 4 = S + 5$ $S =$ ____________
   b. $4S = 3S + 7$ ____________
   c. $5S + 3 = 4S + 8$ ____________
   d. $3S = 6$ ____________
   e. $4S + 1 = 2S + 7$ ____________

8. Try solving each of the following equations without using the materials from sheet 4. Physically check your solutions by using the method in exercise 6.
   a. $S + 5 = 9$ $S =$ ____________
   b. $2S + 6 = S + 13$ ____________
   c. $5S + 2 = 3S + 6$ ____________

Since we do not usually think of putting a weight of -1 on a scale, it is helpful also to look at pictorial methods for solving linear equations. For example, to solve the equation $4S + (-3) = 3S + (-4)$ pictorially, we again use the process of removing *equally valued* pieces from (or adding *equally valued* pieces to) each side of the configuration.

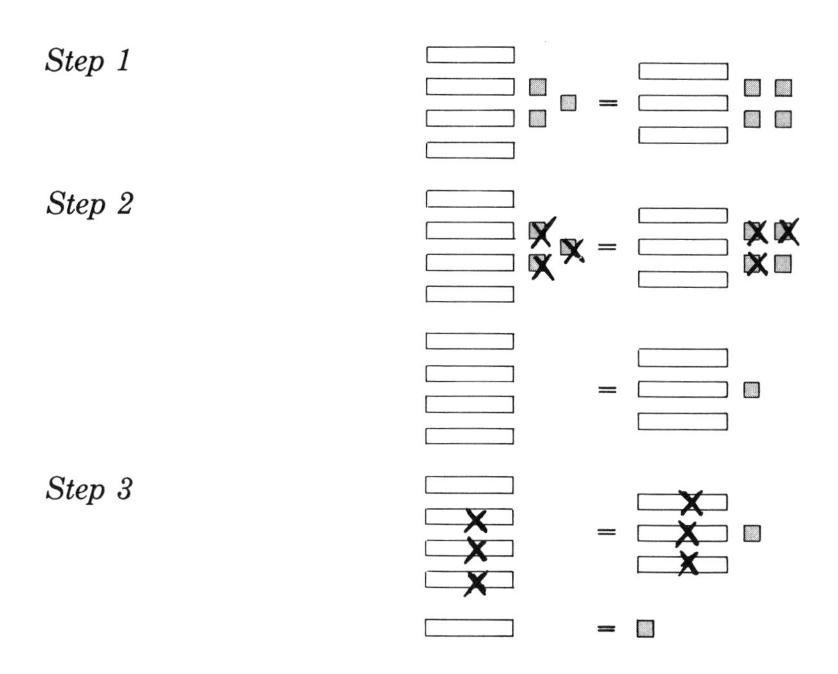

The solution is $S = -1$.

Note that to remove (or subtract) a square or strip, we cross it out with an "X." In Step 2 we could have added three white squares to each side instead of removing three gray ones.

9. Use the method described above to solve each of the following linear equations.

   a. $2S + (-5) = -9 + S$ $S =$ ____________

   b. $3S + (-6) = S + (-8)$ ____________

   c. $5S + 3 = 2[S + (-3)]$ ____________

10. Try solving each of the following equations without using pictorial representations. Check your solutions by using the method shown at the top of this sheet.

    a. $4S + (-2) = 2S + (-10)$ $S =$ ____________

    b. $5S + (-6) = 4S + 8$ ____________

EQUATION-SOLVING KIT SHEET 4

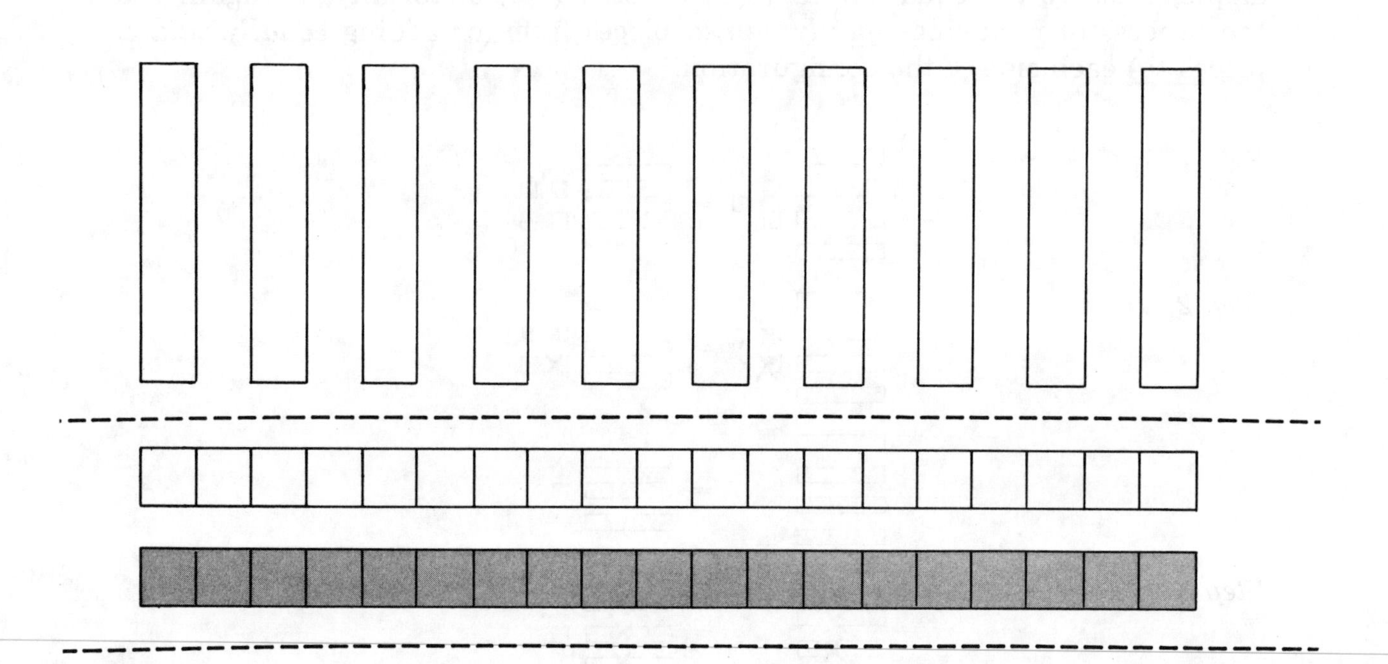

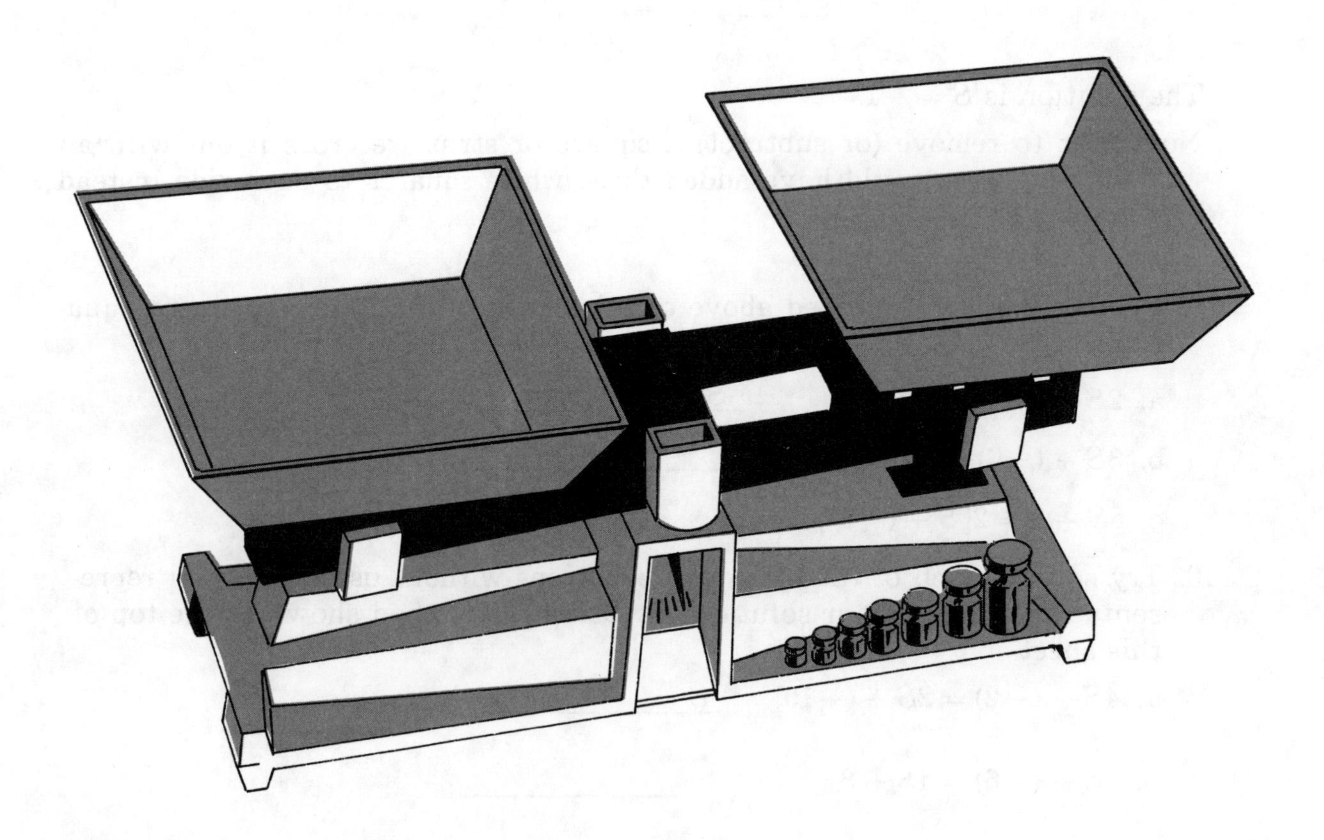

ACTIVITIES

MICROCOMPUTER UNIT: GRAPHING STRAIGHT LINES

March 1983

By *Ellen H. Hastings, Minnehaha Academy, Minneapolis, MN 55409*
Daniel S. Yates, Mathematics and Science Center, Richmond, VA 23223

Teacher's Guide

Grade level: 8–10

Completion time: One fifty-minute period

Objectives:

1. To investigate how the value of m in the equation $y = mx + b$ affects the inclination of the graph of the equation and, specifically, to arrive at these generalizations:
   a) The larger the value of m, the steeper the slope becomes.
   b) With a positive value of m, the graph is inclined up and to the right.
   c) With a negative value of m, the graph is inclined up and to the left.
2. To investigate the role of b in the equation $y = mx + b$—
   a) b is the y-intercept of the line $y = mx + b$, and
   b) the graph of $y = b$ is a horizontal line.
3. To discover the characterizing property of parallel lines and of perpendicular lines—
   a) parallel lines have the same slope, and,
   b) for perpendicular lines, the product of their slopes is $^-1$.
4. To give students an opportunity to use technology as a tool for learning.

Equipment: Microcomputers with function-graphing software or graphing calculators

Prerequisite experience: Students should have some experience with—

1. graphing ordered pairs (x, y) on a coordinate system;
2. the linear equation; for example, they should know that $y = mx + b$ is the equation for a straight line, and they should know what it means to "plot a point;" and
3. substituting values of x to find values of y in a linear equation.

Directions: How you implement this activity will depend on the number and nature of graphing calculators or computers you have available. If you have only one computer, you may want to have the students work through the activity in pairs or in small groups on successive days. A larger group can participate if the screen is large enough for easy viewing. The ideal situation is to have multiple computers or calculators so that small groups can work simultaneously. It is important to give as many students as possible the experience of keying in the equations. The activity is sequential, so each sheet should be completed before going on to the next one.

TABLE 1

```
10  TEXT : HOME
20  VTAB 12
30  PRINT "THIS PROGRAM GRAPHS STRAIGHT LINES"
40  PRINT : PRINT "IN THE GENERAL FORM
      Y = M*X + B."
50  FOR N = 1 TO 5000: NEXT : REM     PAUSE LOOP
60  HGR
65  HCOLOR = 3
70  PRINT : POKE 37,20: PRINT
80  PRINT : PRINT "GIVEN THE GENERAL FORM
      Y = M*X + B"
90  PRINT : PRINT "INPUT M,B     ";
100 INPUT M,B
110 PRINT : PRINT : PRINT
120 REM  DRAW AXES
130 HPLOT 140,0 TO 140,159
150 HPLOT 0,80 TO 279,80
170 FOR H = 0 TO 270 STEP 10
180 HPLOT H,77 TO H,83
190 NEXT H
210 FOR K = 0 TO 160 STEP 10
220 HPLOT 137,K to 143,K
230 NEXT K
240 HPLOT 271,65 TO 277,71
250 HPLOT 277,65 TO 271,71
260 HPLOT 147,0 to 151,6
270 HPLOT 155,0 TO 147,12
280 REM     PLOT THE STRAIGHT LINE
290 FOR X = - 14 TO 13.9 STEP .1
300 LET Y = M * X + B
310 IF Y > 8 THEN 340
320 IF Y < - 7.9 THEN 340
330 HPLOT 140 + 10 * X,80 - 10 * Y
340 NEXT X
345 REM     PRINT LINEAR EQUATION
350 PRINT : POKE 37,20: PRINT
360 PRINT "THIS IS THE GRAPH OF Y = "M" * X + "B
370 PRINT
380 PRINT "1=ANOTHER GRAPH  0=QUIT";
390 INPUT Z
400 PRINT : PRINT
410 IF Z = 1 THEN 60
420 TEXT : HOME
```

If graphing software is not available, the program that appears in table 1 in this teacher's guide can be entered into an Apple IIe or II+ computer and then saved on a disk (type SAVE SLOPE) ahead of time. This program, SLOPE, can be used for all exercises on sheets 1 and 2. Before beginning sheet 3, students need to type a one-line modification to the program: 410 IF Z = 1 THEN 80. They should use this modified program for exercises 7 and 8. Students should be able to complete all three sheets in turn with little further teacher direction.

To begin sheet 1 using the program above, some students might need the following directions: Insert the diskette in the disk drive, close the door, and turn on the monitor and the Apple. When the red light goes out and you see the language symbol] and the flashing cursor, type RUN SLOPE and press the RETURN key. You will be asked to supply values for m and b, in this order. Type the value for m, then a comma, then the value for b, and then press the RETURN key. Fractions will have to be typed as decimals (e.g., enter 1/4 as .25). Note also that the computer uses the symbol * to represent multiplication.

Since the purpose of this activity is to learn about the slope of a line and its general orientation, the student is expected to sketch the lines freehand on the axes provided rather than trying to produce precision drawings.

You should be advised that some screens may slightly distort the spacing on the axes so that perpendicular lines do not appear to be exacly perpendicular. If your screen does not show perpendicularity well and there is no other alternative, you may want to omit exercise 8 on sheet 3.

Students who are knowledgeable in programming in Applesoft may want to go through the program SLOPE and see what each line does. An extension of the activity for these students would be to modify the program to graph the special case $x = a$. A related programming challenge might be to write a program that would determine if two lines intersect and, if so, to announce the coordinates of the point of intersection.

Selected answers:

1. (d) It becomes steeper or more vertical in a counterclockwise direction.
2. (d) It becomes more vertical in a clockwise direction.
4. (d) y-intercept
6. (d) horizontal (or parallel to the x-axis)
7. (d) parallel
   (e) . . . parallel [if] they have the same slope.
8. (b) $y = -4x - 3$
   (c) . . . product of their slopes is $^-1$.
   (d) (The relationship always holds.)

1. In the equation $y = mx + b$, let $b = 0$ so that $y = mx$. Use a computer or graphing calculator to graph the equation with the corresponding values of m and b given in part (a). After the line is produced, sketch the graph and record the equation. Do the same for (b) and (c).

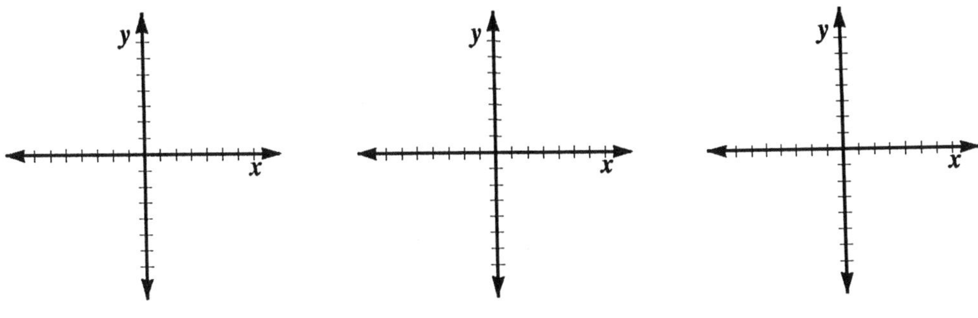

(a) $m = \frac{1}{4}$ $b = 0$ **(b)** $m = 1$ $b = 0$ **(c)** $m = 2$ $b = 0$

equation ________ equation ________ equation ________

(d) As m increases, what happens to the graph? ________________

2. Again let $b = 0$ (so $y = mx$) and use for m each of the following numbers in turn. Sketch the produced graph and record the equation.

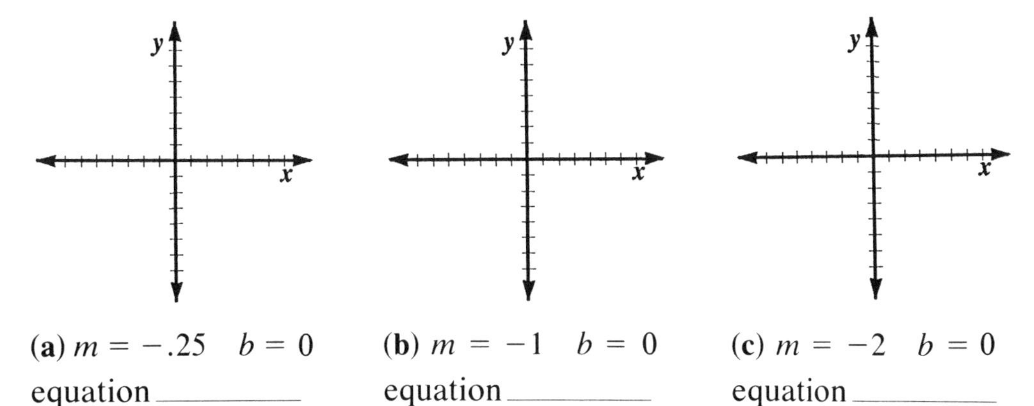

(a) $m = -.25$ $b = 0$ **(b)** $m = -1$ $b = 0$ **(c)** $m = -2$ $b = 0$

equation ________ equation ________ equation ________

(d) As m decreases (i.e., gets larger negatively), what happens to the graph?

____

3. Without using a computer or calculator, sketch what you think each of the following equations looks like. Will the graph of the equation fall or rise from left to right?

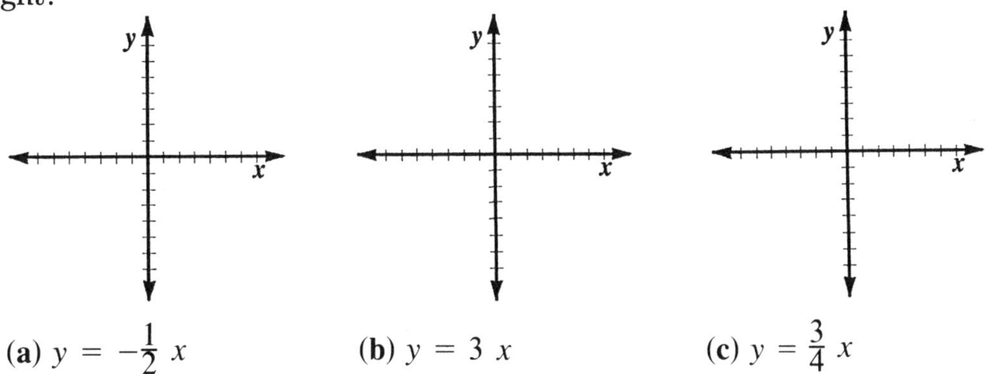

(a) $y = -\frac{1}{2}x$ **(b)** $y = 3x$ **(c)** $y = \frac{3}{4}x$

(d) Now use technology to graph each equation in turn to see if you were right. Correct your sketches if necessary.

4. Use a computer or calculator to graph each of the three lines below. For each line, sketch the graph and record the *y-intercept* (the y-value where the line crosses the y-axis).

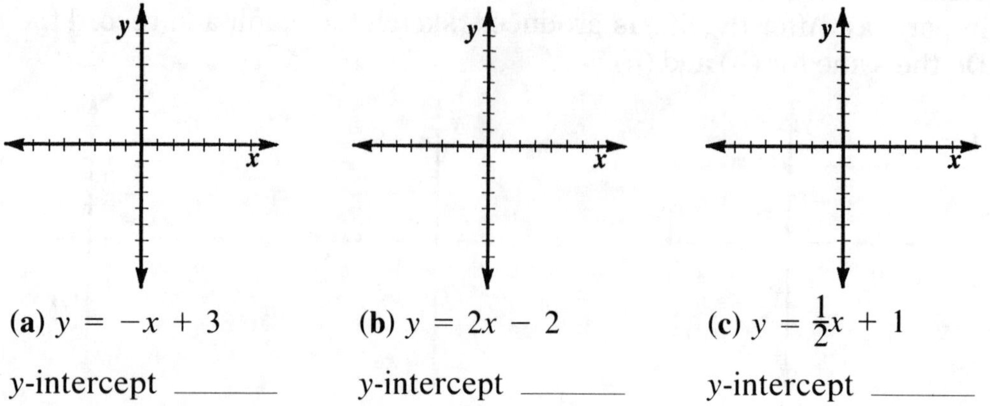

(a) $y = -x + 3$ (b) $y = 2x - 2$ (c) $y = \frac{1}{2}x + 1$

y-intercept ______ y-intercept ______ y-intercept ______

(d) In the equation $y = m\,x + b$, b is the ________________.

5. In the equation $y = mx + b$, the coefficent of x (i.e., m) is called the *slope* of the line. Study the equations below and, *without using a computer or calculator*, determine the slope and the y-intercept; then sketch the line.

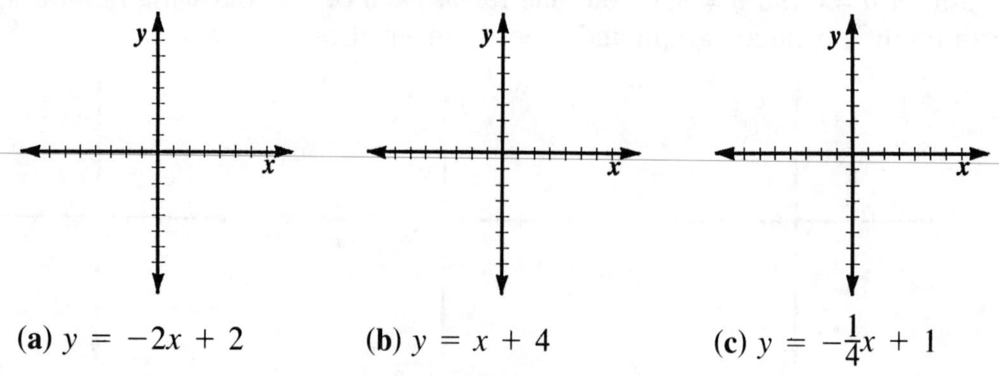

(a) $y = -2x + 2$ (b) $y = x + 4$ (c) $y = -\frac{1}{4}x + 1$

(d) Now use technology to graph each equation to see if you were right. Correct your sketches if necessary.

6. In the equation $y = mx + b$, let $m = 0$ so that $y = b$. Use a computer or calculator to graph each equation below. Sketch the graphs.

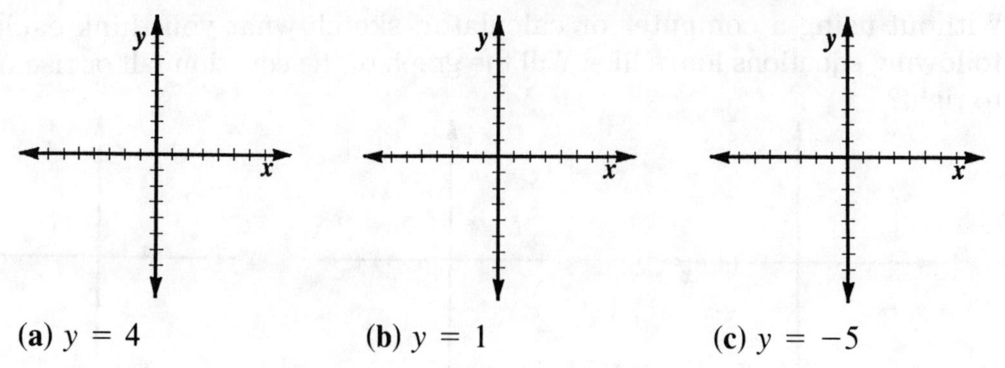

(a) $y = 4$ (b) $y = 1$ (c) $y = -5$

(d) Generalize: When the slope is 0, how are the lines positioned?______

____

7. Use a computer or calculator to graph each pair of lines on the same axes. After each pair, clear the screen and graph the next pair. Sketch the results below.

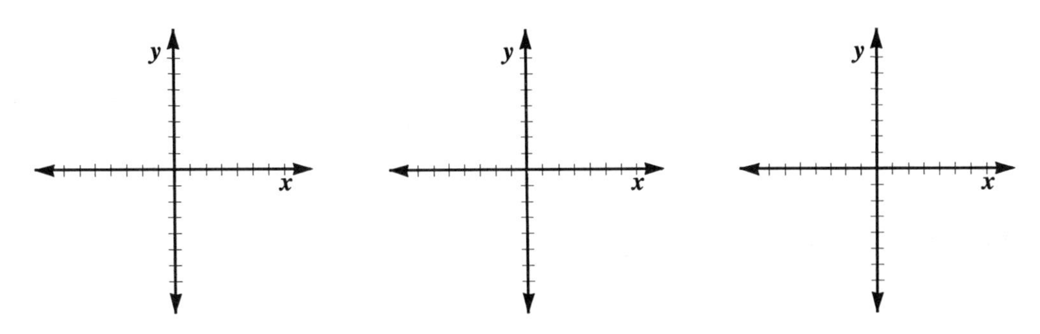

(**a**) $y = x + 3$
$y = x + 5$

(**b**) $y = -.5x + 2$
$y = -.5x + 4$

(**c**) $y = 1.6x - 3$
$y = 1.6x + 2$

(**d**) The two lines in each pair are (perpendicular) (parallel) (not related). [Cross out wrong answers.]

(**e**) Generalize: Two lines are ______________________ if ____________

____.

8. (**a**) Use technology to graph the line $y = \frac{1}{4}x - 1$.

(**b**) Then experiment with different values of m in the equation $y = mx - 3$ to find the line through (0, $^{-}3$) that is perpendicular to the first line.

(**c**) The lines $y = \frac{1}{4}x - 1$ and $y = \square x - 3$ are perpendicular. The product of their slopes is ____.

(**d**) Does the relationship in (c) *always* hold for perpendicular lines? Test the conjecture by graphing pairs of lines of your choice.

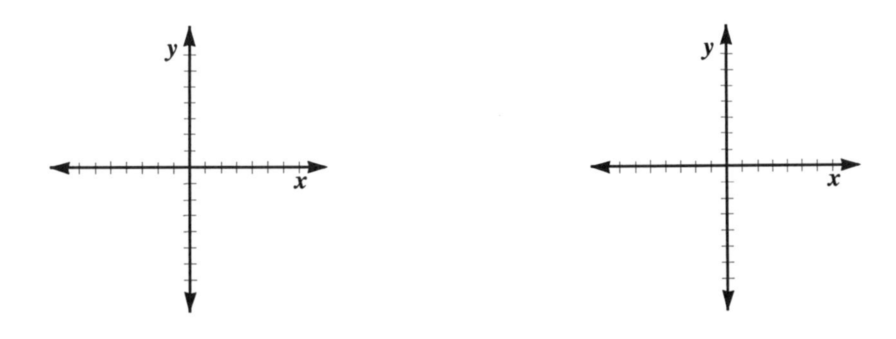

ACTIVITIES

SLOPE AS SPEED

April 1988

By JAMES ROBERT METZ, Damien High School, Honolulu, HI 96817

Teacher's Guide

Grade levels: 9–12

Materials: One set of activity sheets and a ruler for each student; a set of transparencies for class discussion

Objectives: Students will read and draw distance-time graphs and use the slopes of these graphs to draw conclusions about car speeds.

Directions: This activity assumes that students have been introduced to the concept of the slope of a line and that they can determine from a graph the y-intercept and the point of intersection of two lines. The activity sheets should be distributed one at a time. The transparencies can be used to discuss answers for each sheet before progressing to the next activity.

On sheet 1, question 2, students should determine car speed by using the familiar distance formula, $d = rt$. You may wish to use a specific pair of points to review the formula for the slope of a line, depending on the background of your students. Showing a transparency of sheet 1 would be helpful here. Check that all students have computed the slopes correctly and have discovered that the slope of a distance-time graph indicates the speed before distributing the second sheet.

Sheet 2 assumes that a constant speed is maintained over the given intervals. Note that the vertical scale for the graphs in questions 7 and 8 differs from that used in question 6. Students should observe that a visual inspection of the steepness of a distance-time graph could be used to determine the relative speed of two or more cars. This connection is explored further in sheet 3.

You might ask for the equation of each distance-time graph, depending on the level of your students. Questions 7 and 8 will require different equations for the intervals $[0.0, 0.5]$ and $(0.5, 1.0]$.

For sheet 3 it is again assumed that a constant speed is maintained over the given intervals. Note also that the vertical scale for the graph in question 10 differs from that used in questions 9 and 11. After completing this sheet, some of your students might again be encouraged to write the equations of the given lines.

Supplementary activities: You may wish to follow up this activity with a consideration of speed-time graphs. For example, suppose a car is driven at a constant speed of 55 km/h for 3 hours; draw a graph showing the relationship between speed and time. The graph is a horizontal line with an intercept of 55 on the vertical axis representing speed. If a perpendicular is drawn from this graph to the point on the time axis that corresponds with 3 hours, a rectangle will be formed. The area of the rectangle gives the distance traveled. If speed is permitted to increase at a steady rate, the foregoing process will produce a triangle, the area of which gives the distance traveled.

Answers

Sheet 1. 1. See the figure.

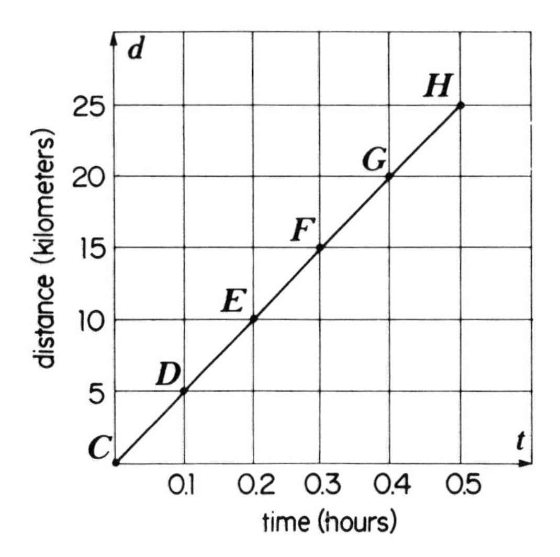

2. 50 km/h; 3. *a*. see the figure, *b*. 50, *c*. 50; 4. same; 5. speed

Sheet 2. 6. *a*. 5 km/h, *b*. 15 km/h, *c*. 25 km/h, *d*. 50 km/h; 7. *a*. 40 km/h, *b*. 80 km/h, *c*. 60 km/h; 8. *a*. 60 km/h, *b*. 0 km/h, *c*. 30 km/h

Sheet 3. 9. *a*. car *A*, *b*. 10 km/h; 10. *a*. 50 km/h, *b*. 50 km/h, *c*. cars are traveling at the same speed, *d*. (0, 20)—at time $t = 0$, car *A* was 20 km further than car *B*; 11. *a*. car *A*, *b*. (0.50, 15)—the cars meet after 0.5 hours, 15 km from the starting point for car *A*.

A navigator in a sports car rally reset her chronometer (a special instrument for measuring time) as the car passed checkpoint C. The table below gives the times and distances traveled to the next five checkpoints. Assume the car speed was constant over this portion of the course.

1. Graph the points and then connect them with line segments.

Checkpoint	Time (hours)	Distance (kilometers)
C	0.0	0
D	0.1	5
E	0.2	10
F	0.3	15
G	0.4	20
H	0.5	25

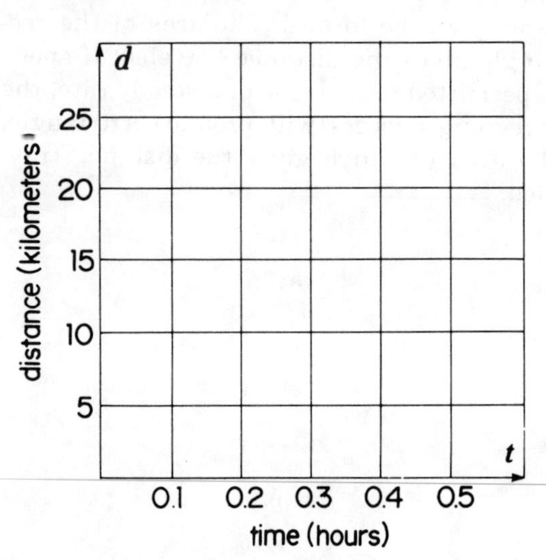

2. At what speed was the car traveling as it passed checkpoint D?__________ Checkpoint G?__________

Recall that for any two points (x_1, y_1) and (x_2, y_2) on a nonvertical line, the *slope* of the line is

$$\frac{\text{vertical change}}{\text{horizontal change}} = \frac{y_2 - y_1}{x_2 - x_1}.$$

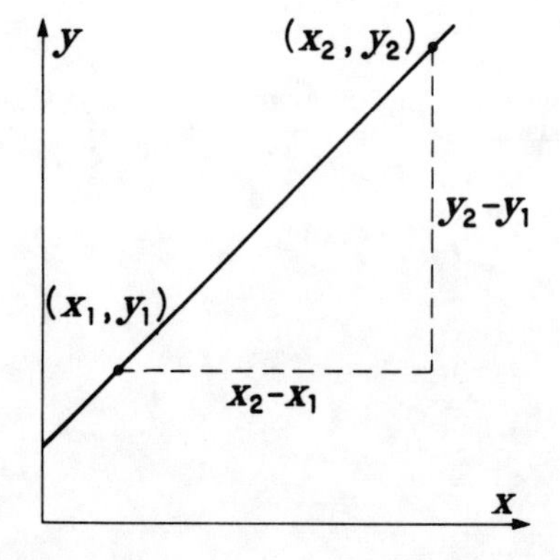

3. *a.* On the graph for exercise 1, label each point with the letter of the checkpoint to which it corresponds.

   b. The slope of the line segment joining points E and F is ________ km/h.

   c. The slope of the line segment joining points D and G is ________ km/h.

4. How do your answers for 3*b* and 3*c* compare with your answers to question 2?

____

5. The slope of a distance-time graph is the ____________________.

6. Use the slope to find the car speed for each of the following distance-time graphs.

a.

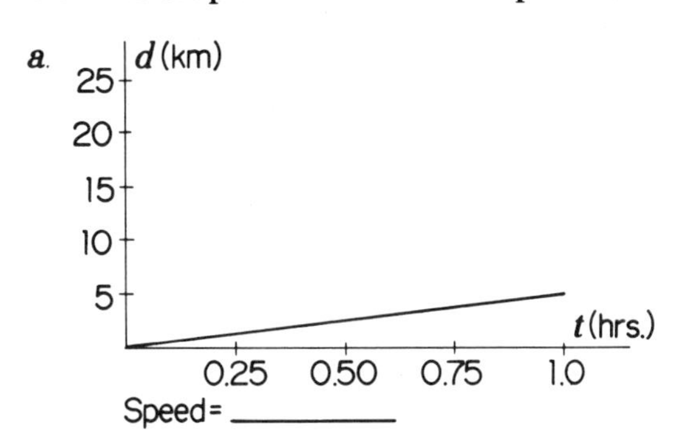

Speed = __________

b.

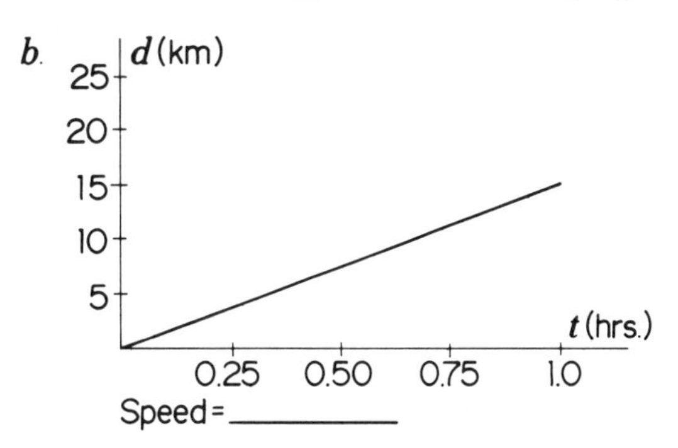

Speed = __________

c.

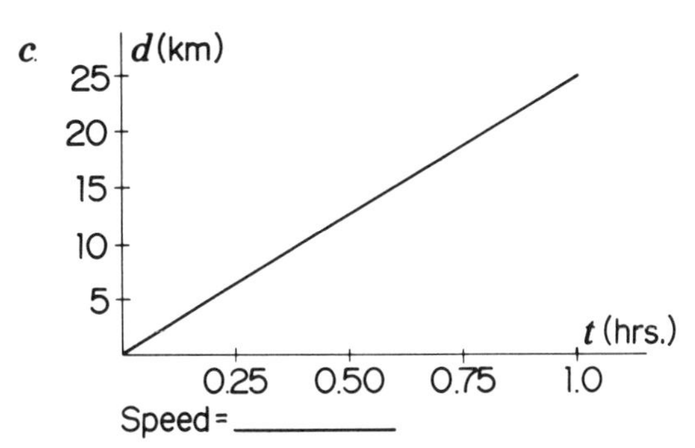

Speed = __________

d.

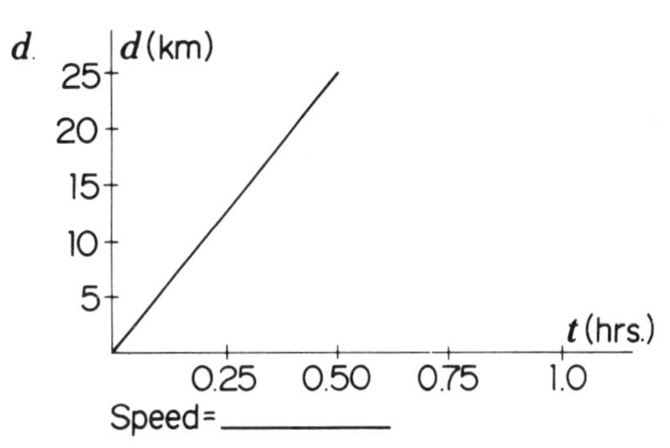

Speed = __________

7. Study this distance-time graph and answer the questions that follow.
   a. How fast was the car going during the first 0.5 hours? __________
   b. How fast was the car going during the second 0.5 hours? __________
   c. What was the average speed of the car during the 1-hour period?

   ____

   d. Draw a graph for the average speed on the same coordinate system.

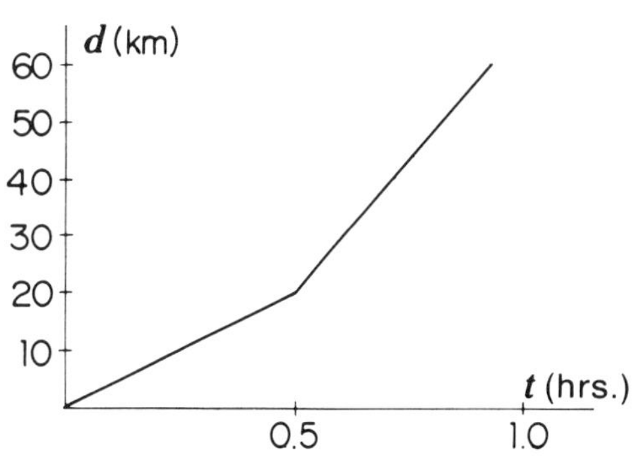

8. Use the distance-time graph at the right to answer each question.
   a. How fast was this car going during the first 0.5 hours? __________
   b. How fast was this car going during the second 0.5 hours? __________
   c. What was the average speed of the car during the 1-hour period?

   ____

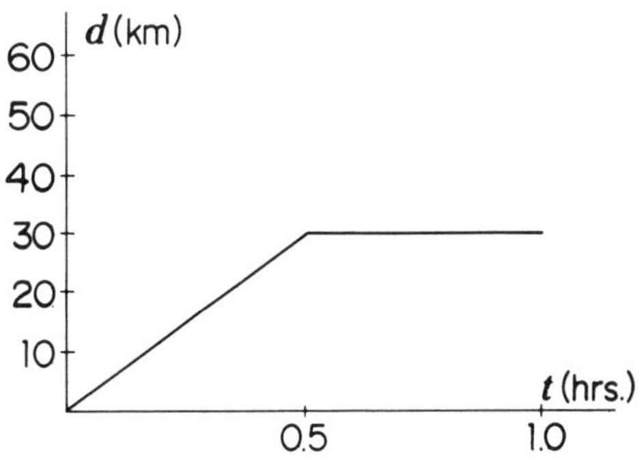

9. *a*. Which car is going faster?

____

b. How much faster?

____

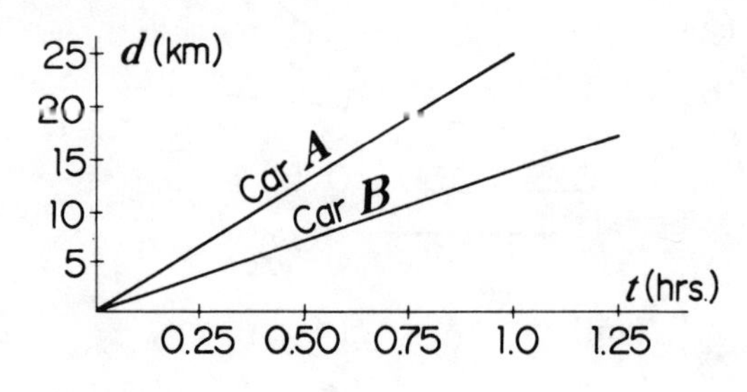

10. Study the graph at right and answer the questions that follow.

a. How fast is car *A* going?

____

b. How fast is car *B* going?

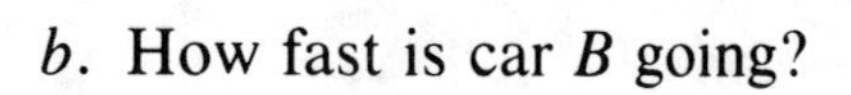

c. What do parallel lines indicate about speed on a distance-time graph?

____

d. What is the *t*-intercept of the graph for car *A*? ______________

What does this intercept indicate about the two cars?

____

11. Study the graph below and answer the questions that follow.

a. Which car is going faster?

____

b. At what point do the two graphs intersect?

____

What is the physical significance of this point?

____

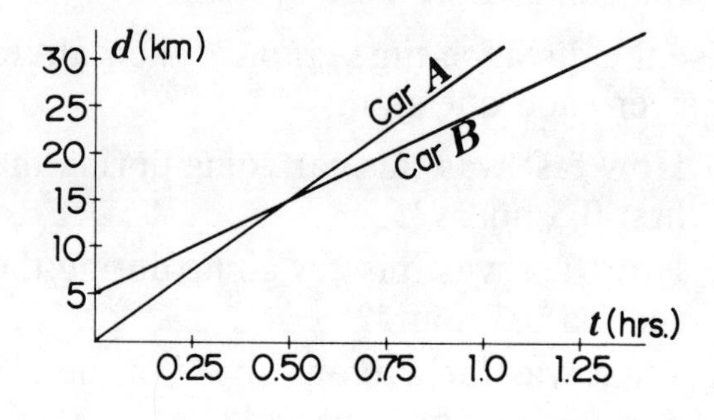

ACTIVITIES

FAMILIES OF LINES

November 1983

By **Christian R. Hirsch,** *Western Michigan University, Kalamazoo, MI 49008*

Teacher's Guide

Grade Level: 8–11

Materials: A set of activity sheets for each student. Access to a microcomputer is necessary for part of the last worksheet.

Objectives: Students will (1) graph pairs of linear equations, linear combinations of these pairs, and use the results to discover an algebraic technique for solving a system of equations; and (2) complete, run, and modify a BASIC program for solving a system of linear equations.

Prerequisites: Students should have some facility in graphing linear equations of the form $ax + by = c$ as well as of the special forms for horizontal and vertical lines. The third worksheet assumes that students are familiar with the essential elements of computer programming in the BASIC language—specifically, the form and use of INPUT, LET, IF-THEN, and PRINT statements.

Directions: This activity is designed for use in a unit on systems of linear equations. The first two worksheets should be used in conjunction with or immediately following a lesson on solving systems of equations graphically. These two sheets provide exploratory graphing exercises through which students can discover the addition-with-multiplication method for solving a system of equations together with the geometric reasons behind the method. The third worksheet can be used toward the end of the unit as a means to clarify, reinforce, and extend the algebraic concepts and methods developed in the unit.

Sheets 1 and 2: Encourage pupils to draw accurate graphs. Labeling the graphs with the corresponding equations is also helpful. Depending on the level and ability of the class, on completion of sheet 1 you might show that, in general, if (x_1, y_1) is a point of intersection of the graphs of the system

$$(1) \qquad \begin{aligned} ax + by &= c \\ dx + ey &= f, \end{aligned}$$

then $ax_1 + by_1 = c$, $dx_1 + ey_1 = f$, and thus, by the addition property of equality, $(ax_1 + by_1) + (dx_1 + ey_1) = c + f$ or $(a + d)x_1 + (b + e)y_1 = c + f$. Hence, the graph of the sum of the equations in (1) also contains the point (x_1, y_1). In a similar manner, on completion of sheet 2 you might demonstrate that if (x_1, y_1) belongs to the graphs of the equations in (1), then it also belongs to the graph of the sum of $m(ax + by = c)$

and $n(dx + ey = f)$, where m and n are any two nonzero integers.

Since the primary intent of sheets 1 and 2 is geometrically to motivate an algebraic method, students with access to a microcomputer could also complete these worksheets by using computer-generated graphs of the given and derived equations. The Apple II$^+$ BASIC program given by Hastings and Yates (1983) can be used in this manner if the following modifications and additions are made:

```
40    PRINT : PRINT "IN THE FORM
         AX + BY = C."
65    HCOLOR = 3
80    PRINT : PRINT "GIVEN THE FORM
         AX + BY = C"
90    PRINT : PRINT "INPUT A,B,C";
100   INPUT A,B,C
285   IF B = 0 THEN 343
300   LET Y = -A / B * X + C / B
342   GOTO 345
343   HPLOT 140 + 10 * C / A,
         0 TO 140 + 10 * C / A,159
360   PRINT "THIS IS THE GRAPH OF
         "A"X + "B"Y = "C
410   IF Z = 1 THEN 80
```

Sheet 3: This worksheet should be introduced by considering the solution of the general system of equations in (1). Solve the system for x by multiplying both sides of the first equation by e; then multiplying both sides of the second equation by $-b$; and finally adding the resulting two equations together and simplifying. Relate the form of the solution to lines 190 and 210 of the program given. The completion of the program will require students to solve the general system for y, translate the algebraic expression into a BASIC expression, and then supply the missing keywords and statement numbers.

Answers: Sheet 1: 1.b. (0, 4); e. They all pass through the point (0, 4). 2.b. (−2, 4); e. They all pass through the point (−2, 4). 3.b. (−3, −1); e. They all pass through the point (−3, −1). 4. The graph of E_3 passes through the point of intersection of the graphs of E_1 and E_2.

Sheet 2: 5.d. The graph is the same as that of $2x + 3y = 6$. 6.e. Each line passes through (3, −2), the point of intersection of the graphs of E_1 and E_2. f. $3x = 9$ (or $x = 3$), and $3y = -6$ (or $y = -2$). 7.e. $7x = 14$ (or $x = 2$), and $-7y = -28$ (or $y = 4$). 8. Each member of the family of lines $m \cdot E_1 + n \cdot E_2$ passes through the point (a, b).

Sheet 3: 9.a. (−4.5, 5); b. (5, −2); 10. line 200, 260; line 220, (A*F − C*D)/P; line 240, PRINT; line 310, INPUT; line 320, 110. 12. One possible modification is as follows:

```
260   LET Q = B * F - C * E
265   IF Q = 0 THEN 275
270   PRINT "PARALLEL LINES, NO
         SOLUTION."
272   GOTO 280
275   PRINT "SAME LINE, INFINITELY
         MANY SOLUTIONS."
```

BIBLIOGRAPHY

Hastings, Ellen H., and Daniel S. Yates. "Microcomputer Unit: Graphing Straight Lines." *Mathematics Teacher* 76 (March 1983):181–86.

Moser, James M. "A Geometric Approach to the Algebra of Solutions of Pairs of Equations." *School Science and Mathematics* 67 (March 1967):217–20.

1. a) Graph the following pair of equations:

$$x + 2y = 8$$
$$x + y = 4$$

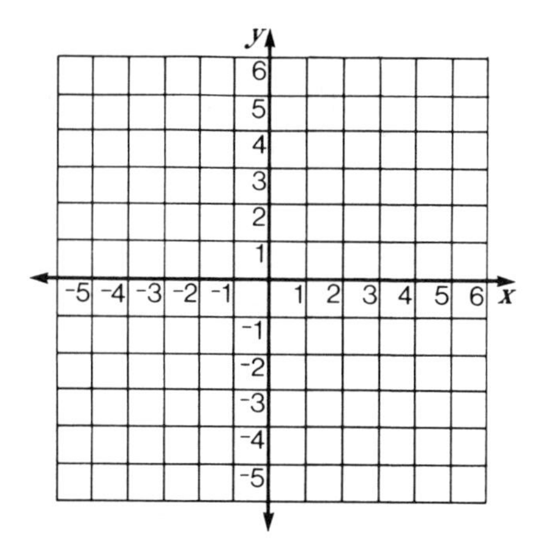

b) What are the coordinates of the point of intersection? ____________

c) Verify that these coordinates satisfy both equations.

d) Add together the two equations above. Graph the new equation on the same set of axes.

e) Do you notice anything special about these three lines? ____________

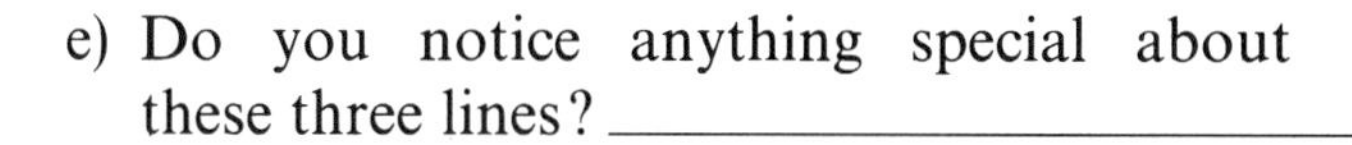

2. a) Graph the following pair of equations:

$$2x - y = -8$$
$$x + y = 2$$

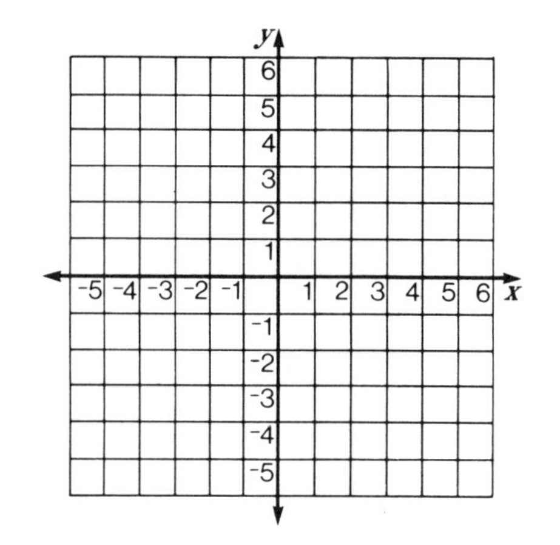

b) What are the coordinates of the point of intersection? ____________

c) Verify that these coordinates satisfy both equations.

d) Add these two equations together and graph the resulting equation on the same set of axes.

e) What appears to be true about the graphs of these three equations? ____________

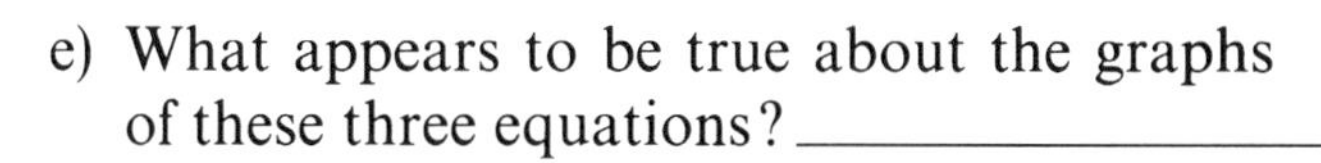

3. a) Graph the following pair of equations:

$$x - 4y = 1$$
$$-x - y = 4$$

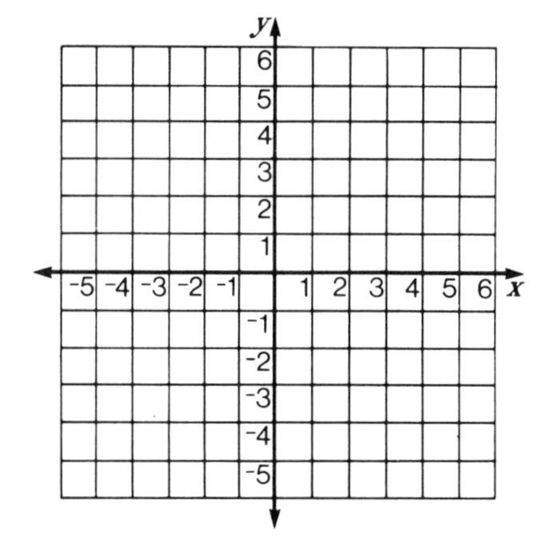

b) What are the coordinates of the point of intersection? ____________

c) Add these two equations together and graph the resulting equation on the same set of axes.

d) What appears to be true about the graphs of these three equations? ____________

____

4. If E_1 and E_2 are equations of two intersecting lines and $E_3 = E_1 + E_2$, what do you think is true about the graphs of E_1, E_2, and E_3? ____________

____

5. a) Graph the equation $2x + 3y = 6$.

   b) Multiply each term of $2x + 3y = 6$ by 5. Graph the new equation on the same set of axes.

   c) Multiply each term of $2x + 3y = 6$ by -4. Graph the new equation on this same set of axes.

   d) What do you think is true about the graph of

$$k(2x + 3y = 6)$$

   for each nonzero integer k? ____________________

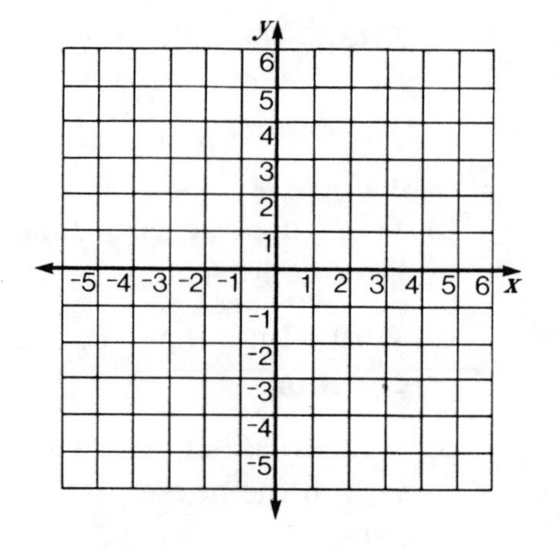

6. Let equation E_1 be $2x + y = 4$
   and equation E_2 be $x - y = 5$.
   Use this set of axes for each part.

   a) Graph E_1 and E_2.

   b) Graph $E_1 + E_2$.

   c) Graph $E_1 + 2 \cdot E_2$.

   d) Graph $E_1 + (-2) \cdot E_2$.

   e) What appears to be true about each of these lines? ____________________

   ____

   f) Which lines, other than the original two, are the most important of this family of lines? ____________________

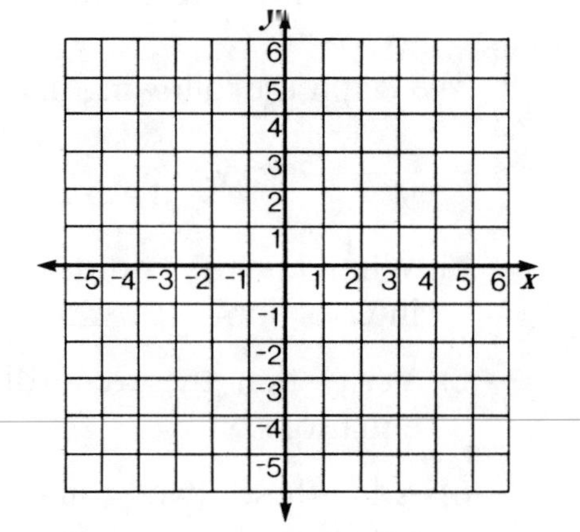

7. Let equation E_1 be $3x - y = 2$
   and equation E_2 be $x + 2y = 10$.

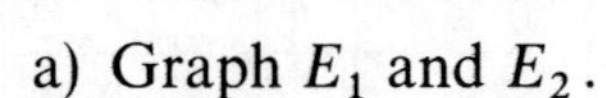

   a) Graph E_1 and E_2.

   b) Graph $E_1 + E_2$.

   c) Graph $2 \cdot E_1 + E_2$.

   d) Graph $E_1 + (-3) \cdot E_2$.

   e) Which lines, other than the original two, are the most important of this family of lines? ____________________

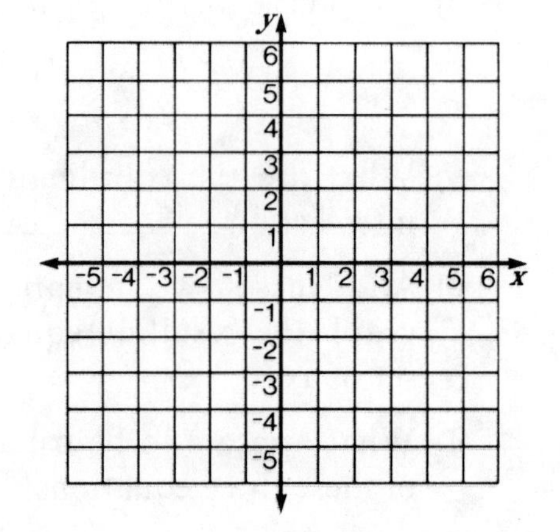

8. If E_1 and E_2 are equations of lines intersecting at a point (a, b), what do you think is true about the family of lines with equations $m \cdot E_1 + n \cdot E_2$, where m and n are any two nonzero integers? ____________________

   ____

9. Without graphing, try to find the coordinates of the point of intersection of the pairs of lines with these equations:

a) $2x + 3y = 6$
$2x + y = -4$

b) $x + 3y = -1$
$2x - y = 12$

10. The BASIC program below is designed to solve a system of linear equations. Complete the program by supplying the missing keywords, algebraic expressions, and statement numbers.

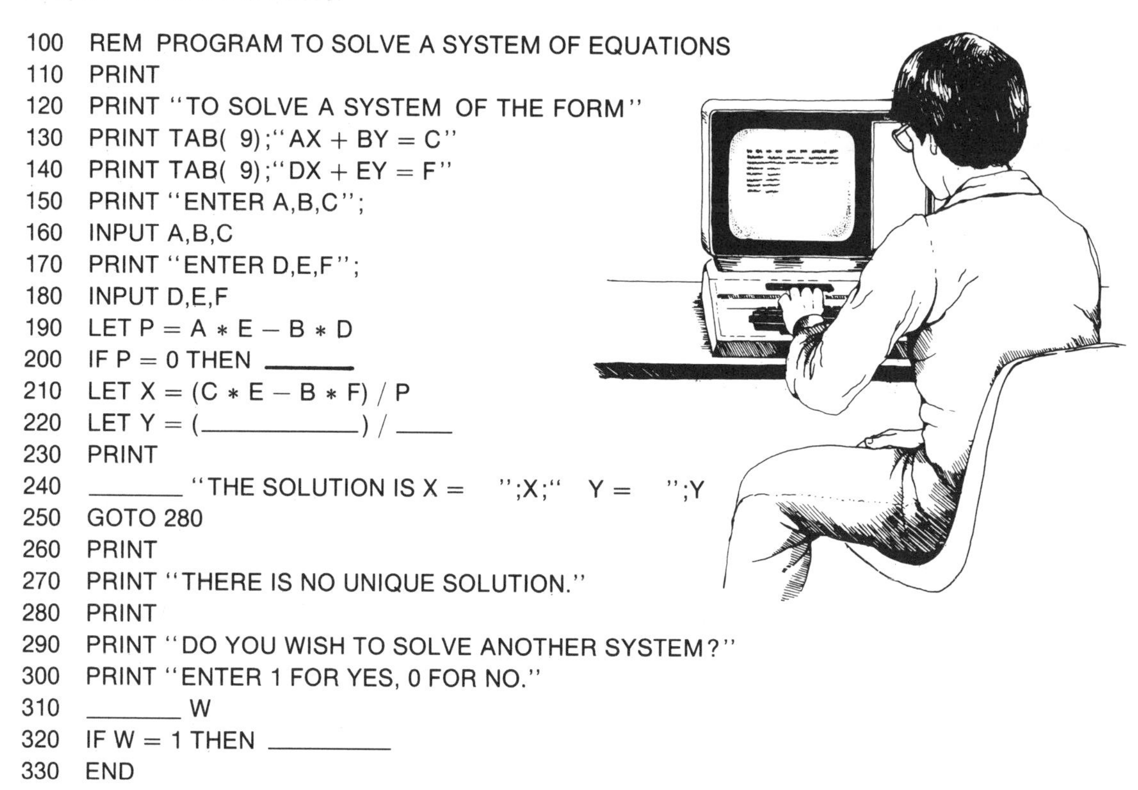

```
100  REM PROGRAM TO SOLVE A SYSTEM OF EQUATIONS
110  PRINT
120  PRINT "TO SOLVE A SYSTEM OF THE FORM"
130  PRINT TAB( 9);"AX + BY = C"
140  PRINT TAB( 9);"DX + EY = F"
150  PRINT "ENTER A,B,C";
160  INPUT A,B,C
170  PRINT "ENTER D,E,F";
180  INPUT D,E,F
190  LET P = A * E - B * D
200  IF P = 0 THEN ______
210  LET X = (C * E - B * F) / P
220  LET Y = (____________) / ____
230  PRINT
240  _______ "THE SOLUTION IS X =   ";X;"  Y =   ";Y
250  GOTO 280
260  PRINT
270  PRINT "THERE IS NO UNIQUE SOLUTION."
280  PRINT
290  PRINT "DO YOU WISH TO SOLVE ANOTHER SYSTEM?"
300  PRINT "ENTER 1 FOR YES, 0 FOR NO."
310  _______ W
320  IF W = 1 THEN _________
330  END
```

11. Copy and run the completed program for the following systems of equations:

a) $2x + 5y = 8$
$3x - y = -5$

b) $2x + 3y = 6$
$x - 5y = 1$

c) $4x + y = 8$
$8x + 2y = 5$

d) $2x + 6y = 5$
$5x - 2y = 6$

e) $3x - 6y = 6$
$4x - 8y = 8$

12. a) Modify the program so that it checks whether the given equations are for the same line or parallel lines. For parallel lines, have the program print "**NO SOLUTION**."

For the same line, have the program print "**SAME LINE, INFINITELY MANY SOLUTIONS**."

b) Run the modified program for the systems of equations in part 11.

ACTIVITIES

INTERPRETING GRAPHS

May 1989

By CHARLENE E. BECKMANN, Grand Valley State University, Allendale, MI 49401

Teacher's Guide

Introduction: Studies concerning students' understanding of Cartesian coordinate graphs indicate that students ages twelve to seventeen experience the following difficulties: (*a*) They often do not display competence with Cartesian graphs beyond plotting points and connecting these points to sketch graphs of simple functions, (*b*) they often cannot read graphs representing nonintegral data, and (*c*) they are often unable to interpret graphs in terms of the situations they represent. Students also appear to have difficulty observing global properties. This inadequacy should be expected from the emphasis of the pointwise construction of graphs that appears in most secondary school mathematics textbooks.

Recent recommendations for school mathematics suggest that students explore common phenomena that illustrate how quantities vary with respect to one another. In addition, it is recommended that students model real-world phenomena with a variety of functions in a variety of representations and that they be able to translate among tabular, symbolic, and graphical representations of functions.

The activities in this article are designed to enhance students' understanding and interpretation of graphical information by alerting students to the relationship between the shape of a graph and the rate of change of quantities represented. The activities encourage a global, rather than pointwise, interpretation of graphs.

Grade levels: 7–10

Materials: Transparencies of "spider" pictures (figs. 1 and 2); the construction-paper overlay described in the following; three full-sized transparencies depicting the functions presented in figures 3, 4, and 5; and copies of the worksheets for individual students

Objective: To enhance students' graphical interpretation skills

Directions: This activity assumes that students have been introduced to plotting points on the Cartesian coordinate plane. If they have not had such experiences, a few games of Battleship adapted to the conventions of the Cartesian coordinate-labeling system will furnish an adequate background for this activity.

TRANSPARENCY MASTER

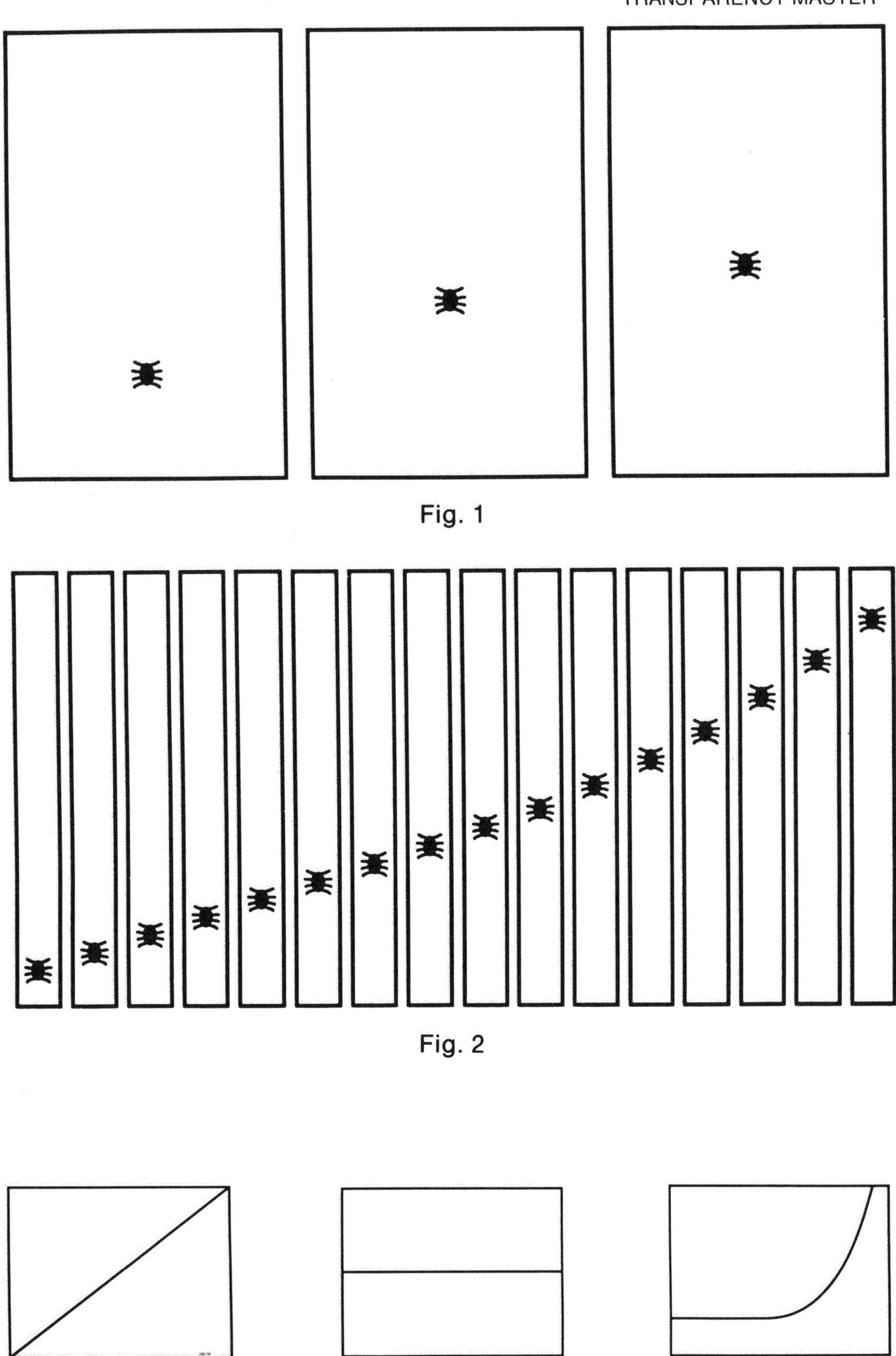

Fig. 1

Fig. 2

Fig. 3

Fig. 4

Fig. 5

As an introduction to these activities, discuss the following scenario inspired by Sawyer (1961) and display the appropriate figures on the overhead projector.

> Imagine that we are taking pictures of a spider as it climbs a wall (display fig. 1). Suppose that we take pictures of it every ten seconds, cut a narrow vertical strip containing the spider's image from each of these pictures, and line up the strips in the order that they were taken (display fig. 2).

A class discussion of the following questions sets the stage for the activities:

- How is the spider moving in the first fifty seconds?
- Does it seem to slow down slightly at any time?
- If so, when does it slow down? How can you tell from the pictures?
- Does it seem to speed up at any time? If so, when does it speed up? How can you tell from the pictures?
- Describe the spider's movement in the last fifty seconds.
- Describe the spider's movement overall. When does it appear to be moving the slowest? Fastest? Describe the appearance of the pictures when it is moving slowest and when it is moving fastest.
- If we think of the spider's images as points on a graph, how should we label the horizontal axis? How should we label the vertical axis?

The spider's progress as it climbs the wall can be simulated on the overhead projector with the instructor pulling various continuous graphs under a construction-paper overlay. Before distributing sheet 1, the instructor will need to prepare three transparencies with a full-sized, single graph on each (see figs. 3, 4, and 5) and an 8 1/2-inch-by-11-inch construction-paper overlay with a 9-inch-by-1/6-inch vertical slit down the center. The slit must be narrow so that only a small part of the graph is visible as it is pulled to the left under the overlay. It is useful to tape a sheet of paper to the left-hand edge of the transparency to facilitate pulling the transparency under the overlay.

The program Spider, written in true BASIC and listed in program 1, can also be used to perform this simulation. The program draws a function $f(x)$ and a "spider" that climbs at the same rate as the function.

Distribute sheet 1. Pull the transparency of figure 3 under the overlay. Ask students to answer questions 1a and 1b. Discuss students' responses. It might be necessary to pull the graph under the overlay a second time. After you are satisfied with students' responses, display figure 3. Repeat the procedure for figures 4 and 5. Rotate figure 5 180 degrees and repeat the procedures for this new orientation.

Students can complete sheets 2, 3, and 4 individually or in small groups. Sheet 2 encourages further discussion of the shape of a position-versus-time graph for an object traveling vertically. Sheet 3 requires students to interpret horizontal movement graphically. The change in orientation might cause initial difficulties for younger students. Use of the Spider program might be especially helpful here. Sheet 4 allows experiences in interpreting graphs in which the perceptual cues might be misleading. With careful guidance, students will begin to ignore such misleading features.

PROGRAM 1

```
def f(x) = x + sin(x)
plot 0,0
print "Enter coordinates for viewing rectangle"
input prompt "a, b, c, d" :a, b, c, d
set window a, b, c, d
let m = (b−a)/15
let n = (d−c)/10
set color 2
box circle x, x+m, y, y+n
box keep x, x+m, y, y+n in ball$
for x = a to b step abs ( (a−b)/500)
        set color 1
        plot x, f(x)
        set color 2
        box show ball$ at b−m−1/30, f(x)
next x
end
```

Answers:

Sheet 1. 1a. The spider is climbing at a constant rate, neither slowing down nor speeding up; 1b. See figure 3. 2a. The spider is not moving; it is staying in the same place on the wall; 2b. See figure 4. 3a. The spider moves first slowly then quickly; 3b. See figure 5. 4a. The spider moves quite rapidly, then begins to slow down until it is moving very slowly; 4b. See figure 5 after a half-turn rotation.

Sheet 2. 5a. The spider's graph is very shallow; the slope of the graph is nearly level; 5b. The graph is very steep; 5c. The graph is level, horizontal. 6a. 10 seconds; 6b. The elevator is stopped to let passengers off or on; 6c. 20 seconds; 6d. No, it speeds up as it leaves one floor, then travels at a constant speed, then slows down as it approaches the next floor; 6e. See the figure shown; 6f. The graph of the Empire State Building elevator would be much steeper than the graph shown.

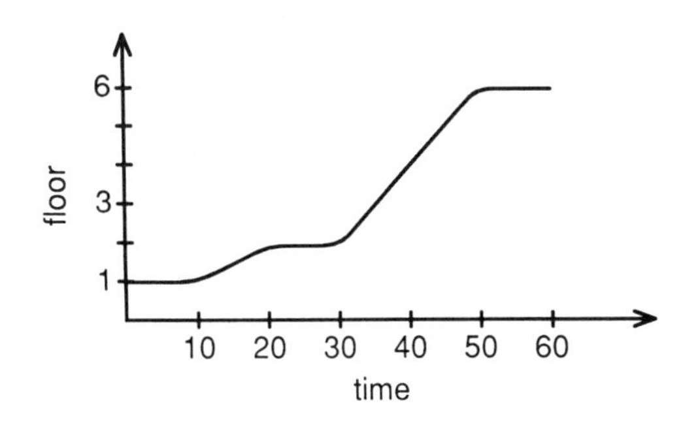

Sheet 3. 7(i) B; 7(ii) A; 7(iii) C; 7a (i). Graph B is the most level graph, showing the most slowly traveling student; 7a (ii). Graph A is the steepest of the three graphs, showing someone traveling more quickly than the others. The person on the bicycle would travel faster than the others; 7a (iii). Graph C begins after the others in time and is steeper than B and less steep than A, showing a person running rather than walking or riding a bike; 7b. Ben might have dropped his books, stopped to talk to a friend, had to tie things to his bike, and so on; 7c. The runner's pace slowed as she tired and had to walk for a while; 7d. See the figure shown.

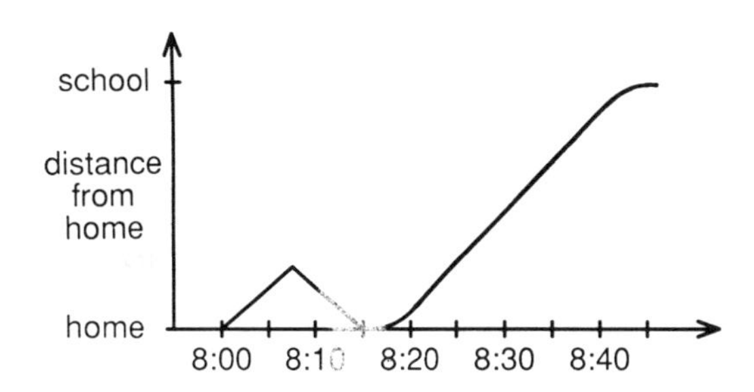

Sheet 4. 8. See the figure shown.

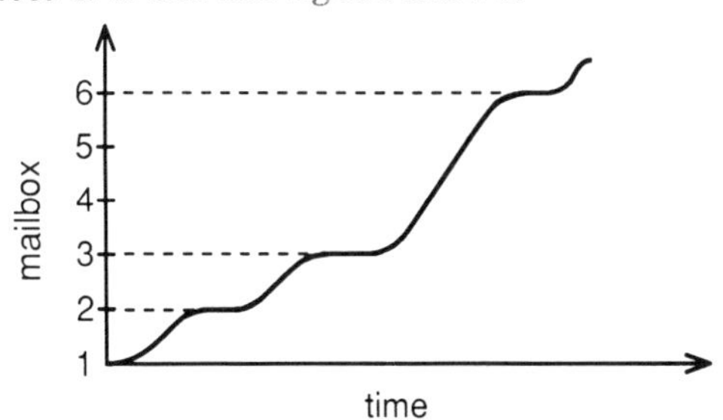

9. The middle graph is correct; it shows a cyclist traveling more slowly as he climbs the hill and then quite rapidly as he travels down the other side. 10. The first graph is correct; it shows the car moving away from, then toward, the observer as it moves around the track.

BIBLIOGRAPHY

Brown, Catherine A., Thomas P. Carpenter, Vicky L. Kouba, Mary M. Lindquist, Edward A. Silver, and Jane O. Swafford. "Secondary School Results for the Fourth NAEP Mathematics Assessment: Algebra, Geometry, Mathematical Methods, and Attitudes." *Mathematics Teacher* 81 (May 1988) 337–47.

Commission on Standards for School Mathematics of the National Council of Teachers of Mathematics. *Curriculum and Evaluation Standards for School Mathematics.* Working Draft. Reston, Va.: The Council, 1987.

Dossey, John A., Ina V. S. Mullis, Mary M. Lindquist, and Donald L. Chambers. *The Mathematics Report Card: Are We Measuring Up? Trends and Achievement Based on the 1986 National Assessment.* Princeton, N.J.: Educational Testing Service, 1988.

Janvier, C. "The Interpretation of Complex Cartesian Graphs Representing Situations—Studies and Teaching Experiments." Ph.D. diss., University of Nottingham, 1978.

Sawyer, W. W. *What Is Calculus About?* New York: Random House, 1961.

Thompson, Charles S., and Edward C. Rathmell. "NCTM's Standards for School Mathematics, K–12." *Mathematics Teacher* 81 (May 1988):348–51.

Wagner, Sigrid, Sidney L. Rachlin, and Robert J. Jensen. *Algebra Learning Project: Final Report.* Athens, Georgia: University of Georgia, Department of Mathematics Education, 1984. ■

In this activity, we investigate the relationship between the shape of a spider's position-versus-time graph and the spider's velocity.

Watch carefully as your teacher makes the spider "climb the wall."

1. a. How does the spider move as it climbs the wall? ____________

   ____

   b. Draw the spider's position-versus-time graph on the given axes.

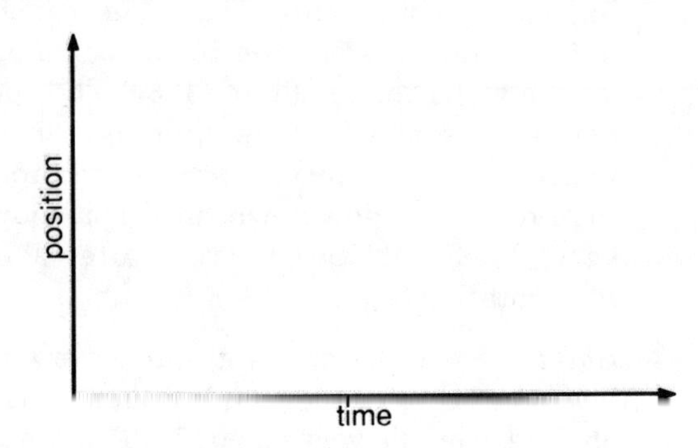

2. a. How does the spider move as it climbs the wall? ____________

   ____

   b. Draw the spider's position-versus-time graph on the given axes.

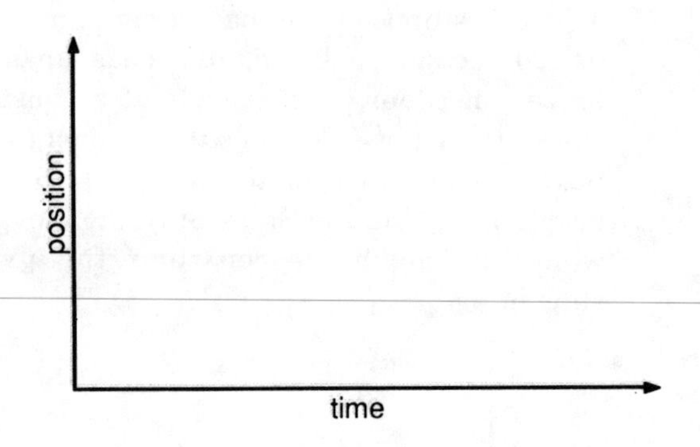

3. a. How does the spider move as it climbs the wall? ____________

   ____

   b. Draw the spider's position-versus-time graph on the given axes.

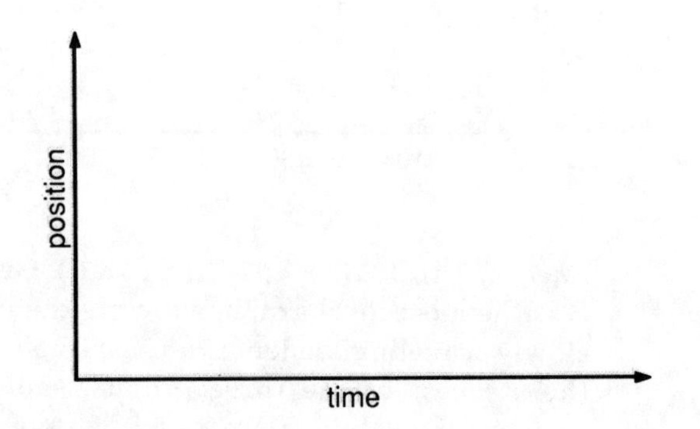

4. a. How does the spider move as it climbs the wall? ____________

   ____

   b. Draw the spider's position-versus-time graph on the given axes.

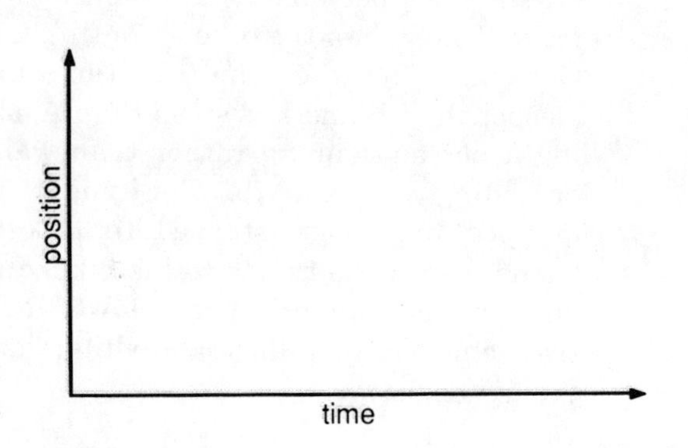

5. Describe the graph when the spider—

   a. moves slowly; ____________________

   b. moves quickly; ____________________

   c. is not moving. ____________________

Now let's consider another object that "climbs."

6. The graph at the right depicts the position of an elevator with respect to time as it travels from one floor to another in an office building.

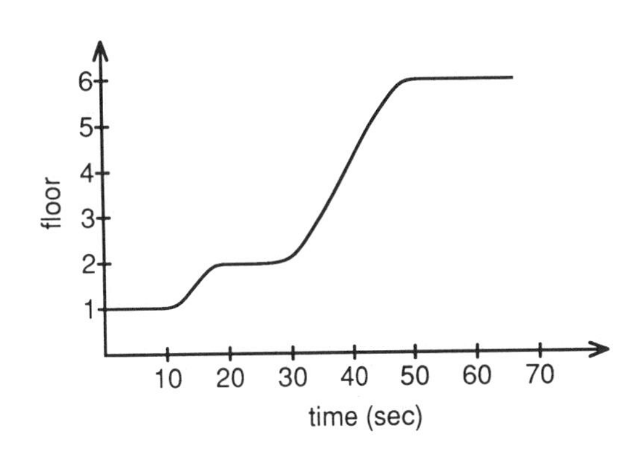

   a. How long does it take the elevator to travel from the first floor to the second floor? ____________________

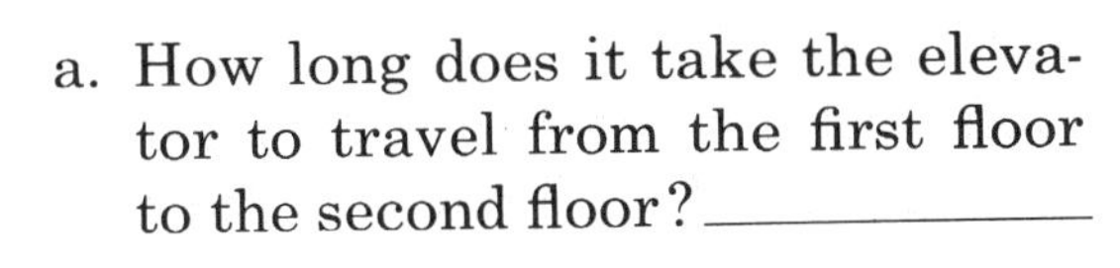

   ____

   b. The graph is level on the interval from twenty to thirty seconds. What could account for this? ____________________

   ____

   c. How long does it take the elevator to travel from the second floor to the sixth floor? ____________________

   d. Does the elevator travel at a constant rate from the moment it leaves one floor until it arrives at the next? Explain. ____________________

   ____

   e. On the axes shown, sketch the graph of an elevator that makes the same stops as the elevator depicted in the graph but travels at a constant rate between floors.

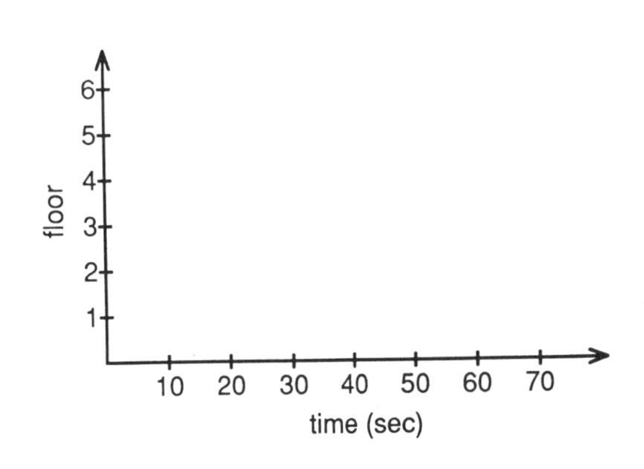

   f. The elevator in the Empire State Building in New York City ascends from the first floor to the eighty-sixth floor in 88 seconds. If a position-versus-time graph of the Empire State Building elevator is sketched on the axes shown, how do these graphs compare?

In this activity, we will explore distance-versus-time graphs modeling other familiar situations.

7. Three graphs are drawn on the axes shown. Match each graph with the situation best depicted by it.

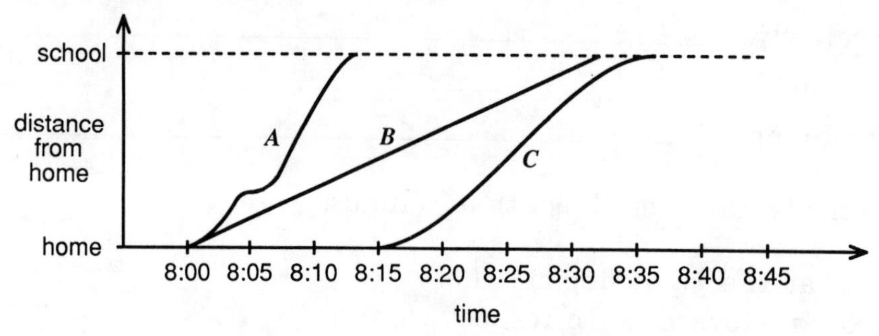

Graph

__________ (i) Colleen walked to school.

__________ (ii) Benjamin rode his bike to school.

__________ (iii) Melanie left late and had to run part of the way to school.

a. Explain why you chose each graph. ____________________

b. What could have been the cause of graph A's being level for the few minutes preceding 8:05? ____________________

c. Graph C is first steep, then more level. Describe the rate of travel of the student whose graph is depicted by graph C. ____________________

d. Suppose Colleen had walked for 7 1/2 minutes toward school and then discovered that she had forgotten her homework, returned home, and then walked to school. Assuming her pace was the same as that shown on the graph above, sketch her position-versus-time graph on the axes below.

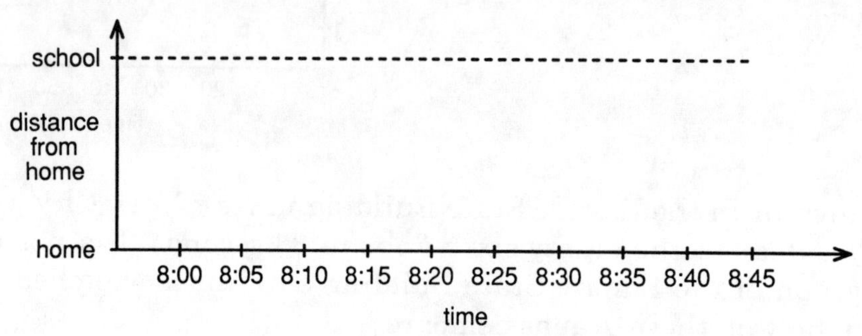

8. A mail carrier making deliveries on a rural motor route must slow down as she approaches each box, stop momentarily to drop off the mail, then proceed to the next mailbox. On the axes below, sketch the mail carrier's position-versus-time graph if she delivers mail to three customers, has no mail for the next two patrons and then delivers mail to two or more persons on her route. Assume that all the mailboxes are equal distances apart.

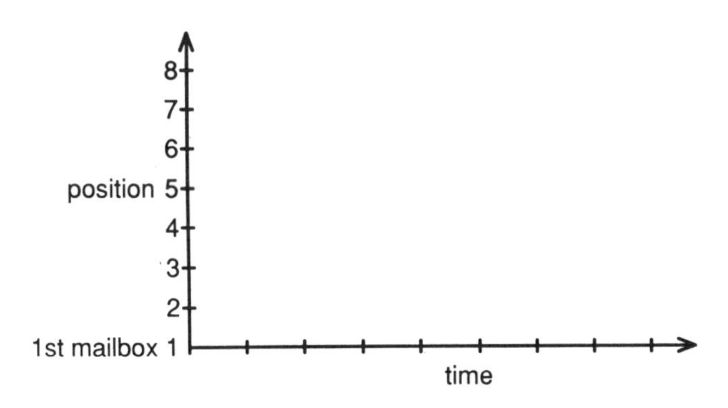

In the next two problems, some of the features of the graph might tempt us to choose the wrong one. Be careful!

9. Which of the following graphs best depicts the distance of a cyclist from the starting line of a race over time if he must travel over a large hill at the beginning of the race? Explain your choice.

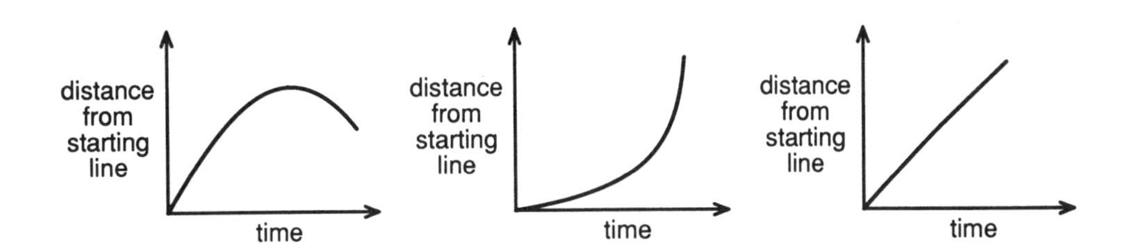

10. An observer is standing at the intersection of a figure-eight racetrack. A single car is traveling on the track. Which graph best depicts the distance of the car from the observer at any time?

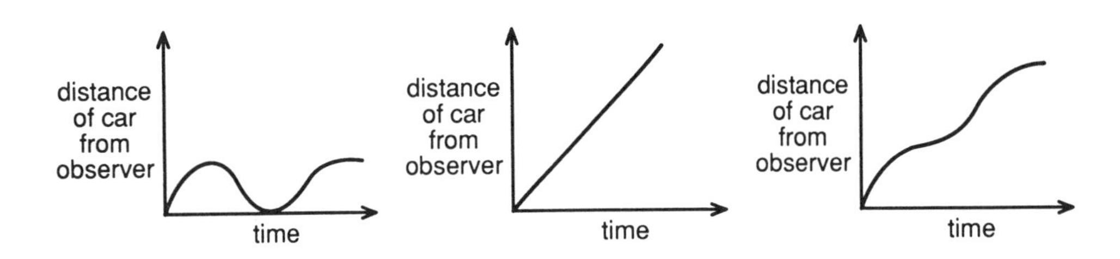

ACTIVITIES

RELATING GRAPHS TO THEIR EQUATIONS WITH A MICROCOMPUTER

March 1986

By JOHN C. BURRILL, Whitnall High School, Greenfield, WI 53228
HENRY S. KEPNER, JR., University of Wisconsin—Milwaukee, Milwaukee, WI 53201

Teacher's Guide

Introduction: The study of absolute-value equations is often slighted in the algebra sequence because of its apparent difficulty due to students' confusion about rules. The activity presented here stresses a conceptual approach to absolute value, which is based on the visualization of the graphs of absolute-value equations. Students are encouraged to compare absolute-value equations that have similar constants by picturing their graphs. An important by-product is the experience students gain in predicting a graph on the basis of the choice of constants in the generalized equation $y = A|x + B| + C$. The relationships explored here are a prelude to all later work in the visualization of functions involving reflection, dilation, and translations. With well-chosen sets of constants, students can observe how the basic V-shaped graph closes up or flattens out, opens upward or downward, or changes the position of the vertex. The simple form of the graph of an absolute-value equation allows the student to make conjectures more easily than with more complex functions, such as the hyperbola or the tangent.

This topic provides an ideal setting for integrating computing technology into the core mathematics curriculum. The activity itself is illustrative of recent recommendations that computers and calculators be used in imaginative ways for exploring, discovering, and developing mathematical concepts.

Program 1 is designed to give students a tool with which they can produce instant graphs for absolute-value equations, thereby allowing them to make and confirm conjectures about the form and position of the graph of a given equation.

Grade levels: 9–11

Materials: Microcomputers with graphing software or graphing calculators; graph paper; and a set of worksheets for each student

Objectives: To allow students to investigate how the values of A, B, and C in the equation $y = A|x + B| + C$ affect the shape and position of its graphs and, in particular, to predict (a) the changes in slope of the rays when the value of A is varied, (b)

the shifts or translations of the graph in terms of its vertex when B and C are varied, (*c*) the shape and position of the graph for an equation in which reflection, dilation, and translation(s) are involved, and (*d*) an equation for a given graph

Directions: This activity can be used in any of three forms in both introductory and intermediate algebra courses. It can be used as a demonstration that is controlled by the teacher using a computer with a large-screen monitor or an overhead graphing calculator. The teacher's questions and the opportunity for students to sketch their conjectures before seeing the computer's graph are critical features of this presentation.

The activity can also be used during class when a computer laboratory or classroom set of graphing calculators is available. Students can work individually or in small groups. Periodically, the class can stop for a general discussion or summary of their observations and conjectures in the laboratory.

Finally, students can use the worksheets as several guided-discovery assignments to be completed using technology between classes. In this format, it is important that students follow directions that call for their conjectures before verifying their work with computer-generated graphs.

The BASIC program (program 1) can be entered into an Apple IIe or II+ computer and saved on a disk ahead of time. Then make additional copies of the program as needed. If you enter this program on an Apple II+, the absolute-value symbol in the REMARK and PRINT statements should be replaced by an exclamation point (!).

Before beginning this activity, students should graph the absolute-value equation $y = |x|$ using the standard approach with ordered pairs and graph paper. Class discussion should focus on the basic characteristics of the graph and include a comparison with linear equations and their corresponding graphs.

Distribute copies of the worksheets one at a time to each student. The computer program prints the values of A, B, and C in a standard form. Thus, negative constants may require some interpretation. For example, if $A = -2$, $B = -3$, and $C = -4$, the display will look like this:

$$Y = -2|X + -3| + -4$$

This form is not unlike the notation for signed numbers often given in mathematics texts for junior high school. Remind students that fractions must be entered as decimals.

The focus of this activity is not the students' mastery of graphing absolute-value equations. A more important goal is the development of students' understanding of the relationships between an equation and its graph. The ability to read an equation and form a visual image of its graph and the ability to write an equation that approximately describes a given graph are critical tools for serious students of mathematics. This skill is another form of "equivalence" in mathematics. Students must recognize several fractional and decimal forms of the same number; they should be able to recognize equivalent equations. It is also important that they have a visual image of standard equations. This activity is valuable for students' understanding of this equivalency.

Additional computer-enhanced graphing activities can be found in Hastings and Yates (1983) and in Hirsch (1983).

Answers:

Worksheet 1: 1. *a.* $y = 3|x + 0| + 0$, $y = 3|x|$; *b.* $y = -3|x + 0| + 0$, $y = -3|x|$; *c.* $y = 0.5|x + 0| + 0$, $y = 0.5|x|$. 2. *a.* $y = 2|x|$; *b.* $y = -6|x|$; *c.* $y = -\frac{1}{3}|x|$; *d.* $y = 5|x|$; *e.* $y = -2|x|$; *f.* $y = -4|x|$. 3. If A is positive, the graph will open upward and the slope of the right ray will always be A, whereas the slope of the left ray will be $-A$. If A is negative, the graph will open downward.

Worksheet 2: 1. *a.* $y = 1|x + 0| + 3$, $y = |x| + 3$; *b.* $y = 1|x + 0| + (-6)$, $y = |x| + (-6)$; *c.* $y = 1|x + 0| + 0.6$, $y = |x| + 0.6$. 2. *a.* $y = |x| + 2$; *b.* $y = |x| + (-4)$; *c.* $y = |x| + 4.5$; *d.* $y = -3|x| + 5$; *e.* $y = 2|x| + (-6)$; *f.* $y = 0.5|x| + (-2)$. 3. *a.* The value of C affects the vertical shift (translation) of the vertex. If C is positive, the vertex is C units above the origin. If C is negative, the vertex is $|C|$ units below the origin. *b.* (0, −7), (0, *C*). 4. The values of *A* and *C* each affect the graph in their own way. The value of *A* determines the slope of the two rays of the V shape. The value of *C* places the vertex of the graph $|C|$ units above or below the origin.

Worksheet 3: 1. *a.* $y = 1|x + 4| + 0$, $y = |x + 4|$; *b.* $y = 1|x + (-5)| + 0$, $y = |x + (-5)|$; *c.* $y = 1|x + (-6)| + 0$, $y = |x + (-6)|$. 2. *a.* $y = |x + (-2)|$; *b.* $y = |x + 3|$; *c.* $y = |x + (-4)|$. 3. *a.* $y = |x + 6|$; *b.* $y = |x + (-3)|$; *c.* $y = |x + (-7)|$. 4. The value of *B* affects the horizontal shift (translation) of the vertex. If *B* is positive, the vertex is located *B* units to the left of the origin; if *B* is negative, the vertex is $|B|$ units to the right of the origin. Thus, the vertex has coordinates $(-B, 0)$.

PROGRAM 1

```
10   REM    TRANSLATION AND DILATION OF
       ABSOLUTE VALUE GRAPHS
20   HOME
30   GOSUB 480
40   GOSUB 700
50   PRINT "A MORE GENERAL EQUATION IS
       Y=A|X+B|+C"
60   FOR J = 1 TO 3000: NEXT J
70   FOR J = 1 TO 1000: NEXT J
80   PRINT : PRINT : PRINT
90   PRINT " FOR THIS GRAPH A=1, B=0 AND
       C=0"
100  FOR J = 1 TO 3000: NEXT J
110  PRINT : PRINT : PRINT
120  PRINT "WOULD YOU LIKE TO CHANGE
       THE VALUE OF A, B OR C? (Y/N)";
130  INPUT Y$
140  PRINT : PRINT : PRINT : PRINT
150  IF Y$ = "Y" THEN 180
160  IF Y$ = "N" THEN 470
170  GOTO 110
180  PRINT : PRINT : PRINT : PRINT : PRINT :
       PRINT : PRINT
190  INPUT "A MUST BE CHOSEN BETWEEN
       -0.1 AND -20 OR +0.1 AND +20 A= ";A
200  PRINT : PRINT : PRINT : PRINT
210  INPUT "B SHOULD BE CHOSEN
       BETWEEN -9 AND 9. B= ";B
220  PRINT : PRINT : PRINT : PRINT
230  INPUT "C MUST BE CHOSEN BETWEEN
       -7 AND +7. C= ";C
240  B1 = B * 12
250  PRINT : PRINT : PRINT : PRINT
260  PRINT "WE WILL NOW PLOT
       Y="A"|X+"B"|+"C
270  FOR J = 1 TO 2000: NEXT J
280  PRINT : PRINT : PRINT : PRINT :
       HCOLOR= 3
290  FOR T = 0 TO 80 STEP .5
300  X = 120 - B1 + T:X1 = 120 - B1 - T
310  Y = 80 - 8 * C - (A * T)
320  IF (A * T) > 80 THEN 370
330  ONERR GOTO 370
340  IF Y < 0 GOTO 370
350  HPLOT X,Y: HPLOT X1,Y
360  NEXT T
370  PRINT : PRINT : PRINT : PRINT
380  PRINT "Y="A"|X+"B"|+"C
390  FOR J = 1 TO 4000: NEXT J
400  PRINT : PRINT : PRINT : PRINT
410  PRINT "WOULD YOU LIKE TO CHANGE
       A, B OR C AGAIN? (Y/N)"
420  INPUT Y$
430  IF Y$ = "Y" THEN 450
440  IF Y$ = "N" THEN 470
450  GOSUB 480
460  GOSUB 700
470  END
480  HGR : HCOLOR= 3: PRINT : PRINT :
       PRINT : VTAB (21)
490  PRINT : PRINT : PRINT : PRINT
500  REM  PLOTS AXES
510  HPLOT 120,0 TO 120,160
520  HPLOT 0,80 TO 240,80
530  REM     INCREMENTS
540  FOR A = - 120 TO 120 STEP 12
550  FOR B = - 2 TO 2
560  X = 120 + A:Y = 80 + B
570  HPLOT X,Y
580  NEXT B: NEXT A
590  FOR A = 0 TO  - 80 STEP  - 8
600  FOR B = - 2 TO 2
610  X1 = 120 + B:Y1 = 80 + A
620  HPLOT X1,Y1
630  NEXT B: NEXT A
640  FOR A = 0 TO 80 STEP 8
650  FOR B = - 2 TO 2
660  X1 = 120 + B:Y1 = 80 + A
670  HPLOT X1,Y1
680  NEXT B: NEXT A
690  RETURN
700  REM       PLOTS     |X|
710  HCOLOR= 6
720  FOR T = 0 TO 120
730  X = 120 + T:X1 = 120 - T
740  Y = 80 - T
750  IF T > 80 THEN 790
760  HPLOT X,Y
770  HPLOT X1,Y
780  NEXT T
790  PRINT : PRINT
800  PRINT " THIS GRAPH REPRESENTS
       Y=|X|"
810  FOR J = 1 TO 3500: NEXT J: PRINT
820  IF Y$ = "Y" THEN 180
830  GOTO 50
```

Worksheet 4: 1. *a.* $y = 2|x + 2| + (-6)$; *b.* $y = 2|x + (-4)| + (-7)$; *c.* $y = -0.5|x + 2| + 2$; *d.* $y = -3|x + (-3)| + 5$; *e.* $y = 0.5|x + 1| + (-4)$; *f.* $y = 3|x + (-6)| + 3$. 2. *a.* vertex $(-6, -4)$, slope 2, opens upward; *b.* vertex (2, 0), slope -3, opens downward; *c.* vertex (6, 1), slope 0.5, opens upward; *d.* vertex (4, 5), slope -5, opens downward; *e.* vertex $(-1, 3)$, slope 0.25; opens upward; *f.* vertex $(-4, -2)$, slope -0.25, opens downward. 3. *a.* $y = |x + (-5)|$; *b.* $y = |x + 3|$; *c.* $y = |x + (-2)| + 2$; *d.* $y = -2|x + (-4)| + 4$; *e.* $y = -7|x + (-1)| + 7$; *f.* $y = 0.5|x + 4| + 1$.

Note: For individuals with access to the Minnesota Educational Computing Consortium's utility program *Small Characters,* we have an enhanced version of program 1 that calls *Small Characters* and places numerical values on the axes for easier use by students. We also have another version of program 1 that permits graphing equations of the form $y - D = A|B(x - (-C/B))|$ and thereby facilitates experimentation with horizontal translations through $|(-C/B)|$ units. Copies of one or both of these programs can be obtained by writing either of the authors.

REFERENCES

Hastings, Ellen H., and Daniel S. Yates. "Microcomputer Unit: Graphing Straight Lines." *Mathematics Teacher* 76 (March 1983):181–86.

Hirsch, Christian R. "Families of Lines." *Mathematics Teacher* 76 (November 1983):590–94.

RELATING GRAPHS AND THEIR EQUATIONS SHEET 1

In this activity you will investigate relationships between equations of the form $y = A|x + B| + C$ and their graphs.

1. Consider first the case where $B = 0$ and $C = 0$ so that you will be graphing $y = A|x + 0| + 0$, which has a simplified form of $y = A|x|$. For each set of constants below, use a computer or graphing calculator to graph the corresponding equation. After the graph is generated, sketch the graph and write both the general and simplified forms of the equation.

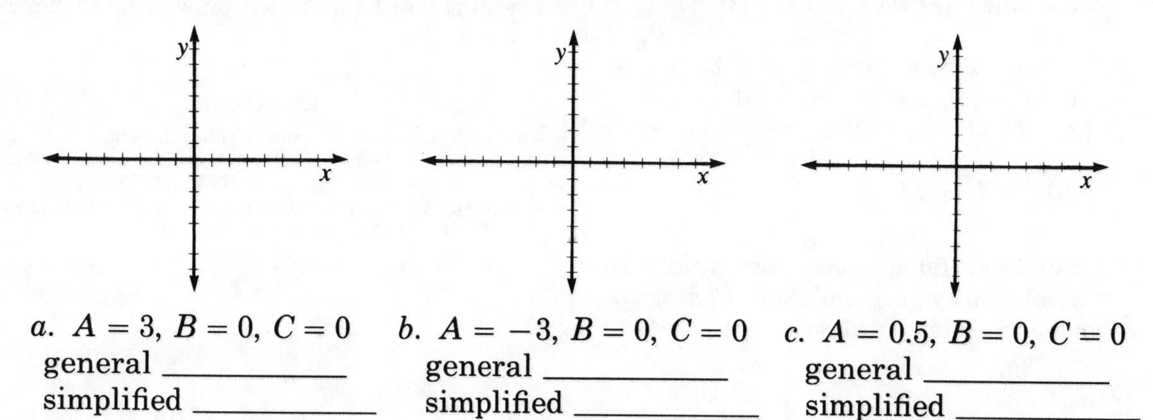

a. $A = 3$, $B = 0$, $C = 0$
general ____________
simplified ____________

b. $A = -3$, $B = 0$, $C = 0$
general ____________
simplified ____________

c. $A = 0.5$, $B = 0$, $C = 0$
general ____________
simplified ____________

2. For each part, without using a computer or calculator, sketch the graph of the equation with the following constants and write the simplified equation. Then use technology to produce the graph and compare it with your sketch. Correct your sketch if necessary.

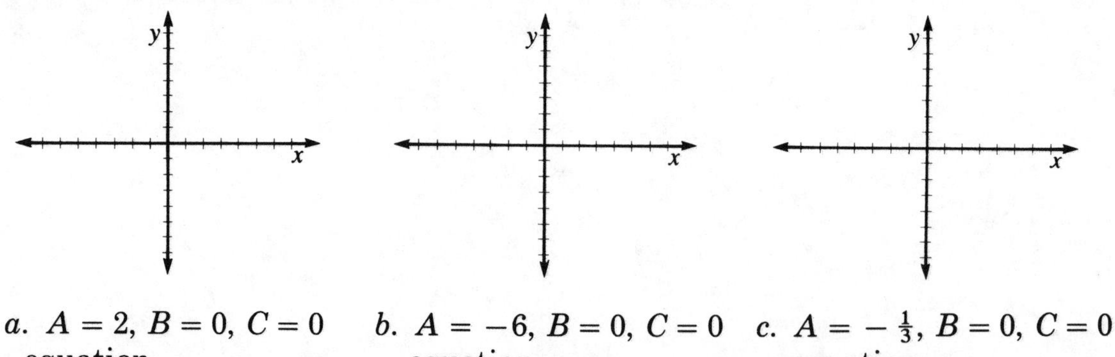

a. $A = 2$, $B = 0$, $C = 0$
equation ____________

b. $A = -6$, $B = 0$, $C = 0$
equation ____________

c. $A = -\frac{1}{3}$, $B = 0$, $C = 0$
equation ____________

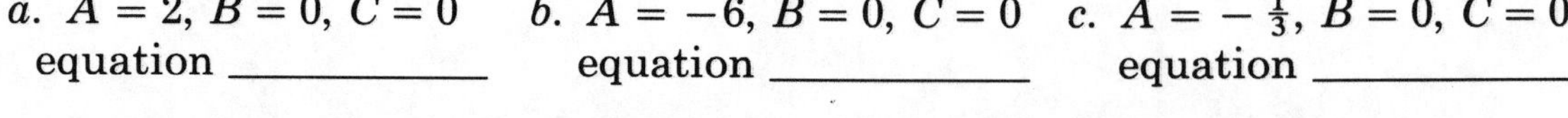

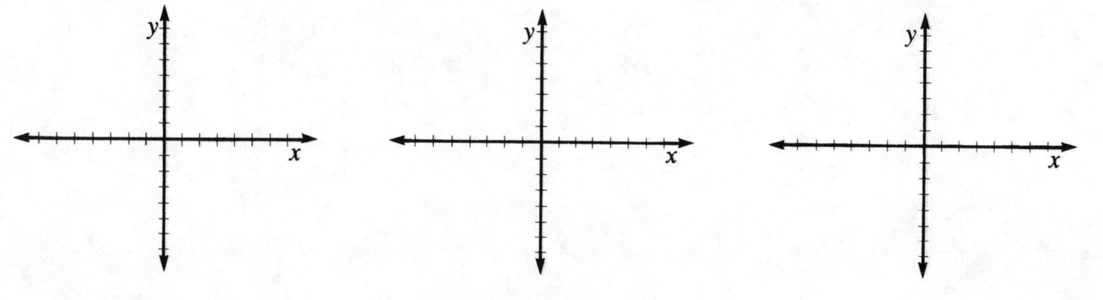

d. $A = 5$, $B = 0$, $C = 0$
equation ____________

e. $A = -2$, $B = 0$, $C = 0$
equation ____________

f. $A = -4$, $B = 0$, $C = 0$
equation ____________

3. In your own words describe how the graph of the equation $y = A|x|$ changes as the value of A changes. (Can you use your knowledge of slope to aid in your description?) ____

____

1. In the general equation $y = A|x + B| + C$, let $A = 1$ and $B = 0$ so that you will be graphing $y = 1|x + 0| + C$, or simply $y = |x| + C$. For each set of constants, use a computer or calculator to graph the corresponding equation. Copy the graph and write both the general and simplified forms of the equation.

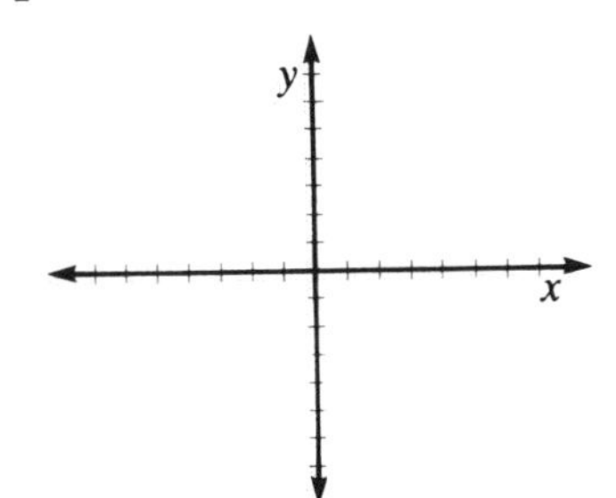

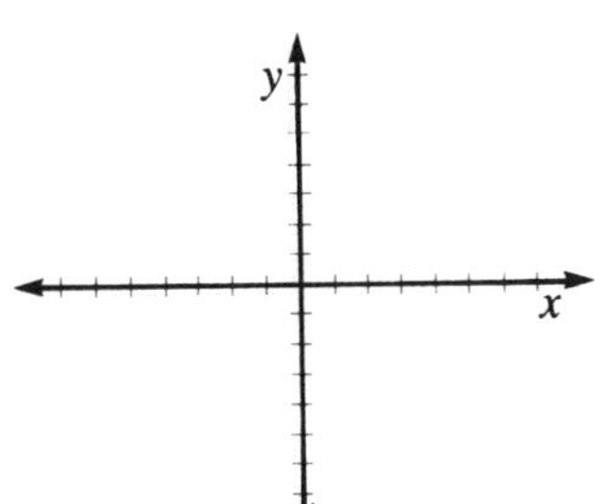

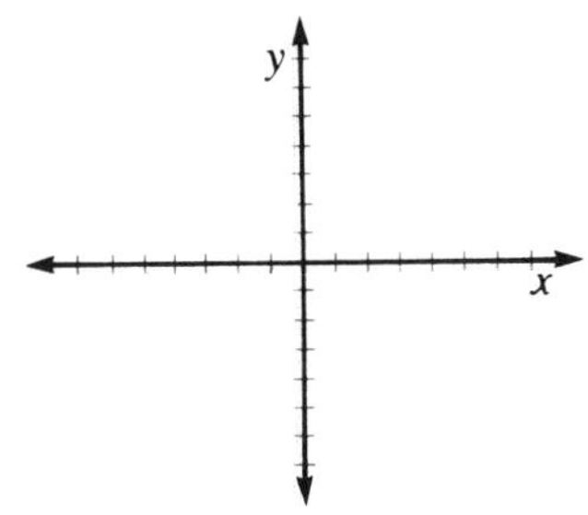

a. $A = 1$, $B = 0$, $C = 3$
general ______________
simplified ______________

b. $A = 1$, $B = 0$, $C = -6$
general ______________
simplified ______________

c. $A = 1$, $B = 0$, $C = 0.6$
general ______________
simplified ______________

2. For each part, without using technology, sketch the graph and write the simplified equation for the set of constants. Then use technology to generate the graph and compare it with your sketch. Correct your sketch if necessary.

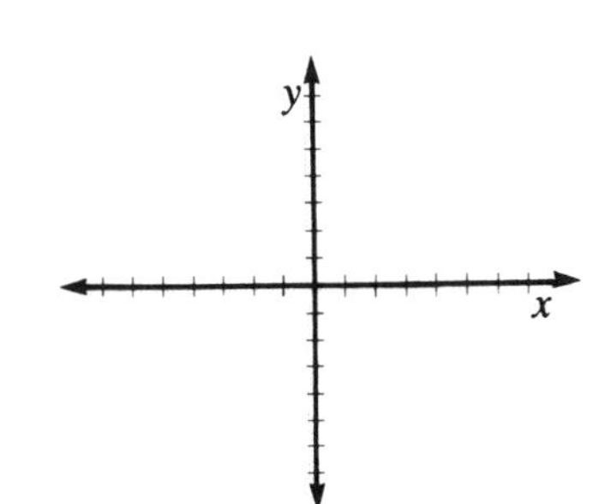

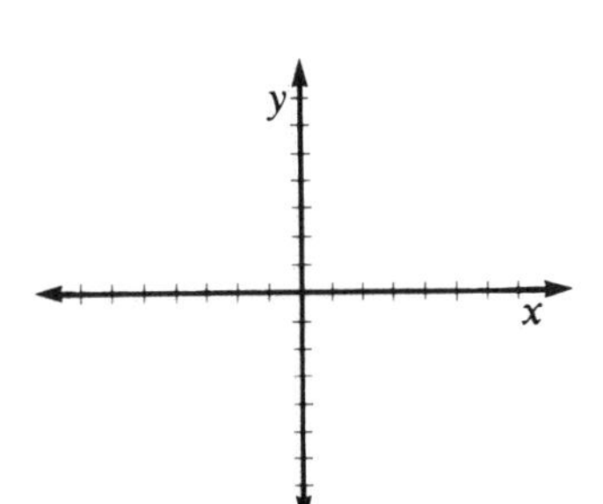

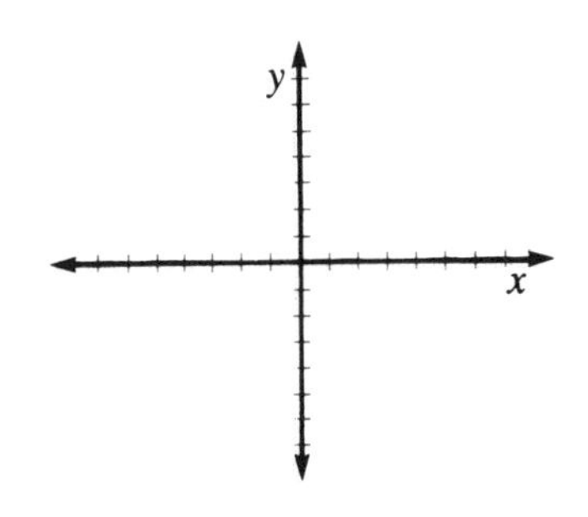

a. $A = 1$, $B = 0$, $C = 2$
equation ______________

b. $A = 1$, $B = 0$, $C = -4$
equation ______________

c. $A = 1$, $B = 0$, $C = 4.5$
equation ______________

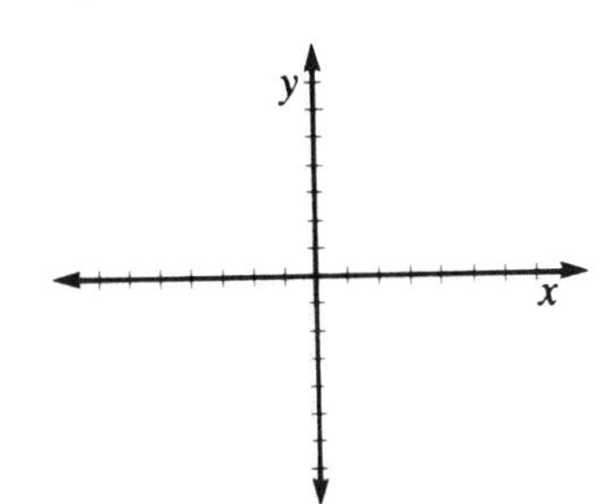

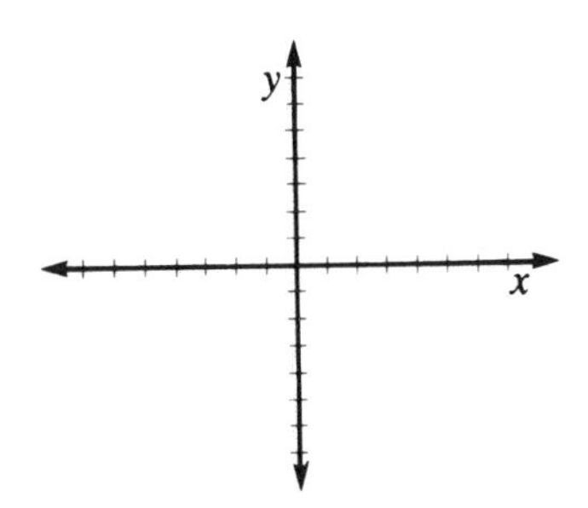

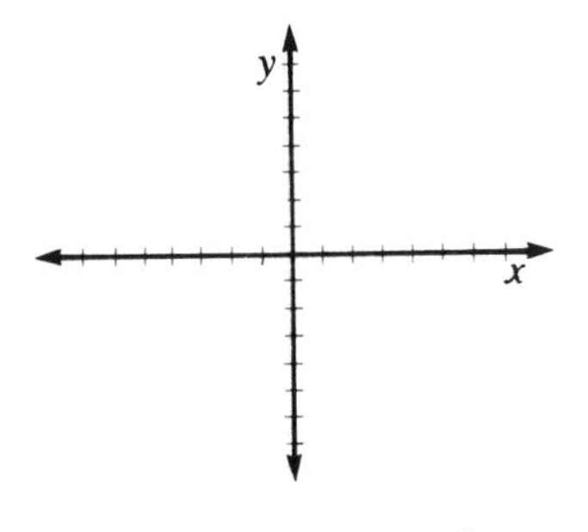

d. $A = -3$, $B = 0$, $C = 5$
equation ______________

e. $A = 2$, $B = 0$, $C = -6$
equation ______________

f. $A = 0.5$, $B = 0$, $C = -2$
equation ______________

3. *a.* Write in your own words how the graph of $y = |x| + C$ changes as the value of C is changed. ______________

b. Predict the coordinates of the vertex of the graph of $y = |x| + (-7)$ ______________ and of $y = |x| + C$ ______________.

4. Write in your own words how the graph of $y = A|x| + C$ is changed when the two values A and C are changed. Use the idea of slope and the location of the vertex. ______________

1. In the general equation $y = A|x + B| + C$, let $A = 1$ and $C = 0$ so that you will be graphing $y = 1|x + B| + 0$, or simply $y = |x + B|$. For each set of constants below, use technology to generate the graph. Then sketch the graph and write both the general and simplified forms of the equation.

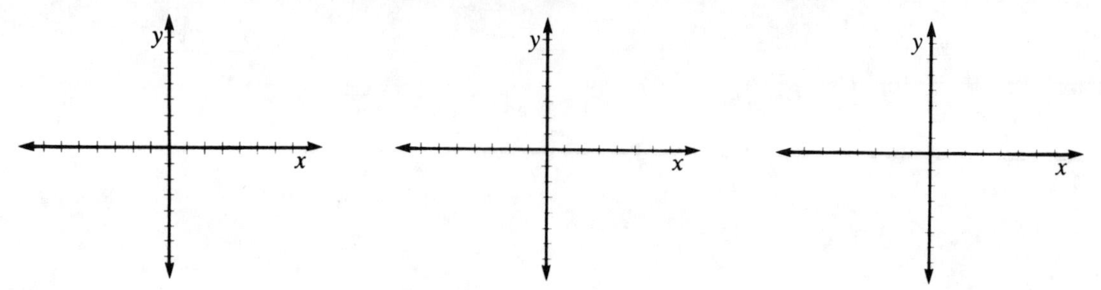

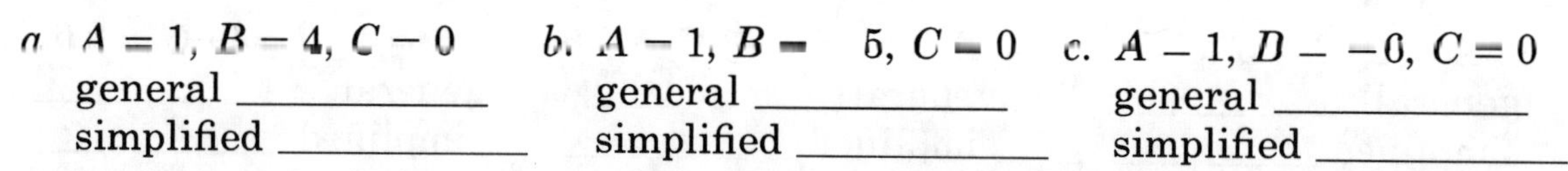

a. $A = 1$, $B = 4$, $C = 0$
general ____________
simplified ____________

b. $A = 1$, $B = 5$, $C = 0$
general ____________
simplified ____________

c. $A = 1$, $B = -6$, $C = 0$
general ____________
simplified ____________

2. For each part, without using technology, make a sketch of the graph of the equation with the given constants and write the simplified equation. Then use technology to produce the graph and compare it with your sketch. Correct your sketch if necessary.

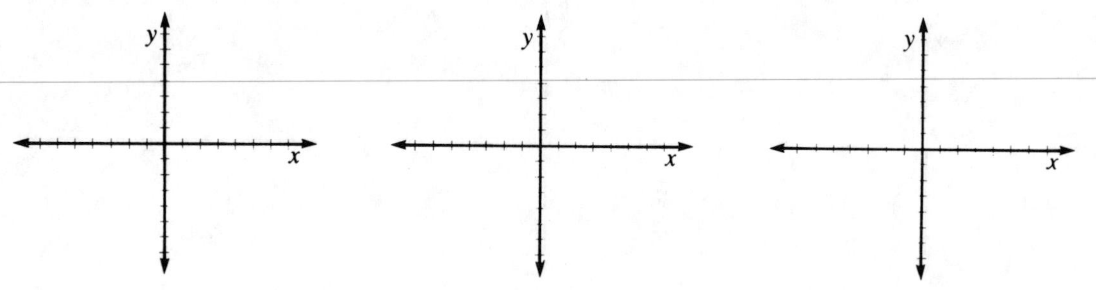

a. $A = 1$, $B = -2$, $C = 0$
equation ____________

b. $A = 1$, $B = 3$, $C = 0$
equation ____________

c. $A = 1$, $B = -4$, $C = 0$
equation ____________

3. For each part, without using technology, sketch the graph and write the simplified equation for the set of constants. Then use technology to generate the graph and compare it with your sketch. Correct your sketch where necessary.

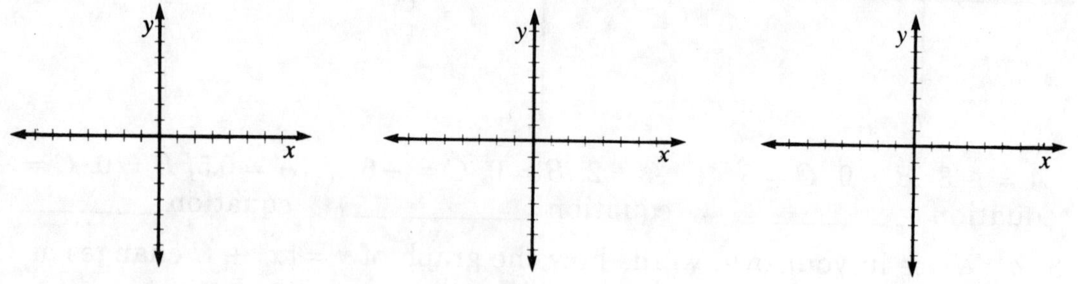

a. $A = 1$, $B = 6$, $C = 0$
equation ____________

b. $A = 1$, $B = -3$, $C = 0$
equation ____________

c. $A = 1$, $B = -7$, $C = 0$
equation ____________

4. Write in your own words how the graph of $y = |x + B|$ is changed when the value of B is changed. (*Hint:* Consider the location of the vertex.)

____

____

1. For each set of constants, write the corresponding equation below and then sketch the graph on a sheet of graph paper. Check your graph by using a computer or calculator or by comparing your result with that of other students.

	A	B	C	Equation
a.	2	2	−6	________
b.	2	−4	−7	________
c.	−0.5	2	2	________

	A	B	C	Equation
d.	−3	−3	5	________
e.	0.5	1	−4	________
f.	3	−6	3	________

2. For each equation, determine the coordinates of the vertex, the slope of the right ray, and whether the **V** opens upward or downward. Sketch the graph for each equation on a sheet of graph paper. Check your graph by using technology or by comparing your result with that of other students.

	Vertex	Slope	Upward/Downward
a. $y = 2\|x + 6\| + (-4)$	______	______	__________
b. $y = -3\|x + (-2)\|$	______	______	__________
c. $y = 0.5\|x + (-6)\| + 1$	______	______	__________
d. $y = -5\|x + (-4)\| + 5$	______	______	__________
e. $y = 0.25\|x + 1\| + 3$	______	______	__________
f. $y = -0.25\|x + 4\| + (-2)$	______	______	__________

3. For each graph, write an equation and then graph the equation using technology. If necessary, modify your equation so that the produced graph matches that shown.

a. equation__________

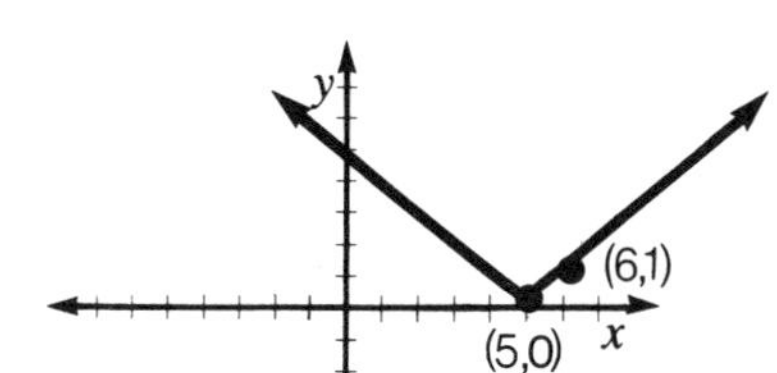

b. equation__________

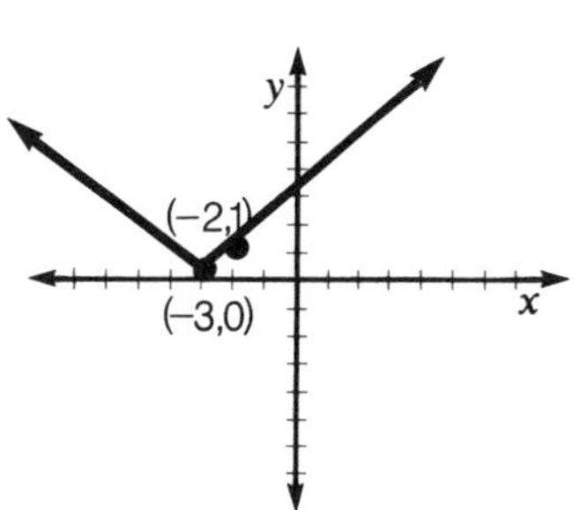

c. equation__________

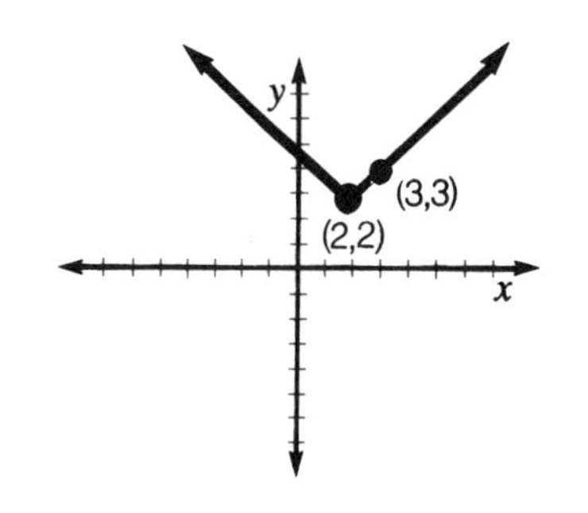

d. equation__________

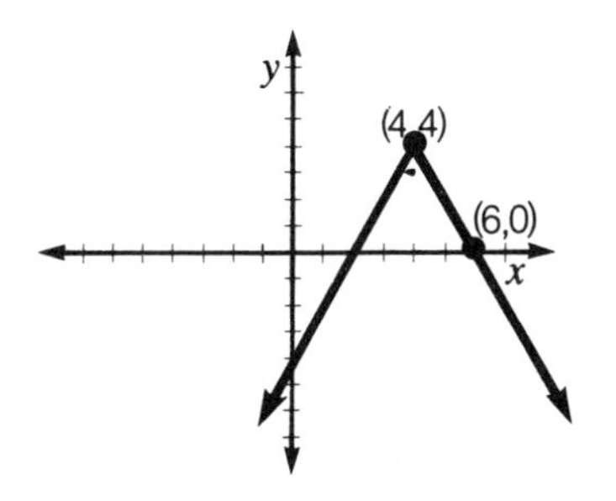

e. equation__________

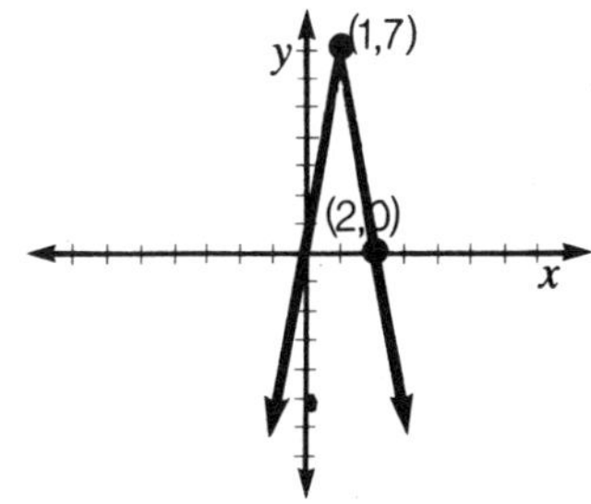

f. equation__________

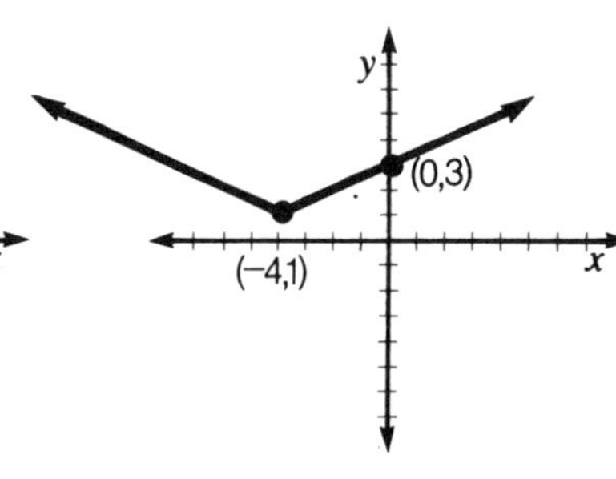

ACTIVITIES

MICROCOMPUTER UNIT: GRAPHING PARABOLAS

December 1986

By ELLEN H. HASTINGS and BRIAN PETERMAN, Minnehaha Academy, Minneapolis, MN 55406

Teacher's Guide

Grade levels: 9–11

Materials: Microcomputers with function-graphing software or graphing calculators; graph paper; and a set of activity sheets for each student

Objectives

1. To investigate how the value of A in the equation $y = Ax^2 + Bx + C$ affects the direction and shape of the graph of the equation and, specifically, to arrive at three generalizations:

a) A positive value of A indicates that the parabola opens upward.

b) A negative value of A indicates that the parabola opens downward.

c) The larger the absolute value of A, the narrower the parabola.

2. To investigate the role of C, the y-intercept of the parabola, in the equation $y = Ax^2 + Bx + C$

3. To investigate how the values of A, B, and C in the equation $y = Ax^2 + Bx + C$ determine whether the graph of the equation intersects the x-axis, specifically, to discover the following generalizations:

a) If the value of the discriminant ($B^2 - 4AC$) is zero, then the parabola has exactly one x-intercept.

b) If the value of the discriminant is greater than zero, then the parabola has two x-intercepts.

c) If the value of the discriminant is less than zero, then the parabola has no x-intercepts.

Prerequisites: Students should have had some experience in graphing relations in a coordinate system, particularly linear equations. Sheet 3 will be most beneficial to students if they have been introduced to the quadratic formula and have had opportunities to use the formula in solving quadratic equations.

Directions: The implementation of this activity will depend on the number and nature of computers or graphing calculators you have available. The activity can be done as a teacher-directed demonstration with a single microcomputer and a large-screen monitor or an overhead graphing calculator, in a laboratory setting where students can work individually or in small groups, or as the basis for individual or small-group investigations to be completed outside class.

For those with Apple IIe or II+ microcomputers, the BASIC program (program 1) could be entered into the computer, checked that it runs properly as copied, and then saved on a disk (type SAVE QUADRATIC) ahead of time. Make additional copies of the program as needed. The program gives directions about how to exit from it.

The activity is sequential, and thus it is important that each sheet be completed before going on to the next one. A class discussion summarizing students' discoveries should follow the completion of the three activity sheets.

Supplementary activities: An extension of this activity might include determining the equation of the axis of symmetry and the coordinates of the vertex of the parabola. The graphs sketched on the sheets can be used to estimate the coordinates of the vertices and the equations of the axes of

PROGRAM 1

```
10   HOME
20   PRINT "TO GRAPH AN EQUATION OF THE
        FORM"
30   PRINT
40   PRINT "          2"
50   PRINT "Y = AX + BX + C, ENTER YOUR
        VALUES FOR"
60   PRINT : PRINT
70   PRINT "A, B, AND C, SEPARATED BY
        COMMAS."
80   PRINT : PRINT
90   PRINT "IF A = 0, THE EQUATION IS NOT
        QUADRATIC,"
100  PRINT
110  PRINT "SO DO NOT ENTER ZERO FOR
        A."
120  PRINT : PRINT
130  PRINT "TO QUIT, PRESS CTRL-RESET."
140  VTAB 23: PRINT "PRESS SPACE BAR TO
        CONTINUE "; : GET A$: HOME
150  HCOLOR= 3
160  VTAB 23: FOR J = 1 TO 40 : PRINT
        " " ; : NEXT J
170  VTAB 23: INPUT "TYPE A, B, AND C:
        ";A,B,C
180  IF A = 0 THEN 160
190 AB = B:SB$ = " + ": IF B < 0 THEN SB$ =
        " - ":AB = - B
200 AC = C:SC$ = " + ": IF C < 0 THEN SC$ =
        " - ":AC = - C
210  VTAB 21: FOR J = 1 TO 40: PRINT
        " "; : NEXT J
220  VTAB 22: PRINT "Y = ";A;"X"; : VTAB 21:
        PRINT "2"; : VTAB 22: PRINT SB$;AB;"
        X";SC$;AC
230  REM   DRAW AXES
240  HGR
250  HPLOT 140,0 TO 140,159
260  HPLOT 0,80 TO 279,80
270  FOR I = 8 TO 279 STEP 12
280  HPLOT I,77 TO I,83
290  NEXT I
300  FOR I = 8 TO 159 STEP 9
310  HPLOT 137,I TO 143,I
320  NEXT I
330  HPLOT 271,65 TO 277,71
340  HPLOT 277,65 TO 271,71
350  HPLOT 147,0 TO 150,3 TO 153,0
360  HPLOT 150,3 TO 150,6
370  REM  DRAW PARABOLA
380 H = - B / (2 * A)
390 K = (4 * A * C - B * B) / (4 * A)
400  IF ABS (H) > 11.58 OR  ABS (K) > 8.77
        THEN 660
410 I = 0
420 XL = H:XR = H
430 Y1 = K
440 J = A * I * I
450 Y = K + J
460  IF Y > 8.88 OR Y < - 8.77 THEN D = 0:
        GOTO 610
470 X = H + I
480  IF X < - 11.58 OR X > 11.58 THEN R =
        0: GOTO 540
490  IF D = 0 OR R = 0 THEN 510
500  HPLOT 140 + 12 * XR,80 - 9 * Y1 TO
        140 + 12 * X,80 - 9 * Y: GOTO 530
510  HPLOT 140 + 12 * X,80 - 9 * Y
520 D = 1:D1 = 1:R = 1:R1 = 1
530 XR = X
540 X = H - I
550  IF X < - 11.58 OR X > 11.58 THEN L =
        0: GOTO 610
560  IF D = 0 OR L = 0 THEN 580
570  HPLOT 140 + 12 * XL,80 - 9 * Y1 TO
        140 + 12 * X,80 - 9 * Y: GOTO 600
580  HPLOT 140 + 12 * X,80 - 9 * Y
590 D = 1:D1 = 1:L = 1:L1 = 1
600 XL = X
610  IF D = 1 THEN Y1 = Y
620  IF D = 0 AND D1 = 1 THEN 660
630  IF R = 0 AND R1 = 1 AND L = 0 AND
        L1 = 1 THEN 660
640 I = I + .0833
650  GOTO 440
660  GOTO 160
670  END
```

symmetry. The use of the quadratic formula to find the average of the roots of $Ax^2 + Bx + C = 0$ will yield the equation $X = -B/2A$, which is the equation of the axis of symmetry. A substitution of this value of x into the general equation for a parabola will give the y-coordinate of the vertex. Students can check these values against the graphs of the parabolas they have sketched on the sheets. Rewriting the equation $y = Ax^2 + Bx + C$ in the form $y = A(x - H)^2 + K$ also yields the coordinates of the vertex, (H, K).

Students who are proficient in programming might modify the program to graph the axis of symmetry (perhaps in a different color) and give the coordinates of the vertex as output. A related programming activity would be to write a subroutine that would compute the x-intercepts (if they exist) of the given parabola and return the values to the main program for output.

For additional computer-enhanced graphing activities, see Hastings and Yates (1983), Hirsch (1983), and Burrill and Kepner (1986).

Selected answers: Sheet 2: 2*a*. vertex at the origin (and symmetric with respect to the y-axis); *b*. $A > 0$, $A < 0$; *c*. The graph gets narrower. 4*a*. The value of C affects the vertical shift of the parabola and thus determines the y-intercept of the graph. *b*. $(0, -3)$; *c*. $(0, 4)$. 5*a*. opens downward, relatively wide, y-intercept $(0, -4)$; *b*. opens upward, relatively narrow, y-intercept $(0, 3)$; *c*. opens downward, answer will vary, y-intercept $(0, 7)$; *d*. opens upward, relatively wide, y-intercept $(0, -5)$

Sheet 3: 6*a*. discriminant 0; *b*. discriminant 256; *c*. discriminant -16; *d*. discriminant 0; *e*. discriminant 25; *f*. discriminant -7. 7*a*. When $B^2 - 4AC = 0$; *b*. when $B^2 - 4AC > 0$; *c*. when $B^2 - 4AC < 0$. 8*a*. opens upward, relatively wide, y-intercept $(0, -6)$, two x-intercepts; *b*. opens upward, relatively narrow, y-intercept $(0, 1)$, no x-intercepts; *c*. opens downward, answers will vary, y-intercept $(0, -4)$, one x-intercept; *d*. opens downward, relatively narrow, y-intercept $(0, -2)$, no x-intercepts; *e*. opens downward, relatively narrow, y-intercept $(0, 1)$, two x-intercepts; *f*. opens upward, relatively narrow, y-intercept $(0, 8)$, one x-intercept

REFERENCES

Burrill, John C., and Henry S. Kepner, Jr. "Activities: Relating Graphs to Their Equations with a Microcomputer." *Mathematics Teacher* 79 (March 1986):185–88, 193–94, 196–97.

Hastings, Ellen H., and Daniel S. Yates. "Activities: Microcomputer Unit: Graphing Straight Lines." *Mathematics Teacher* 76 (March 1983):181–86.

Hirsch, Christian R. "Activities: Families of Lines." *Mathematics Teacher* 76 (November 1983):590–92, 597–98. ■

The graph of an equation of the form $y = Ax^2 + Bx + C$, where $A \neq 0$, is called a *parabola*. In this activity you will investigate how the values of A, B, and C affect the shape and position of the parabola.

1. Let us begin by considering the special case $y = Ax^2$ where $B = 0$ and $C = 0$. Use a computer or graphing calculator to graph each equation. Carefully sketch each graph displayed on the axes provided.

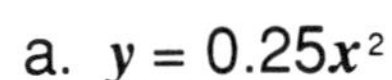
a. $y = 0.25x^2$

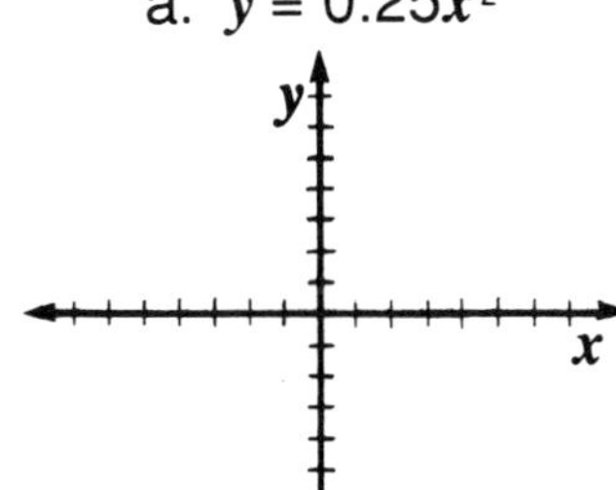

b. $y = 0.5x^2$

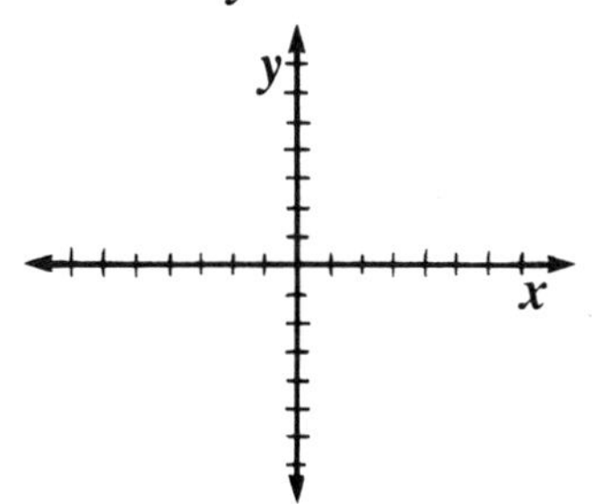

c. $y = 2x^2$

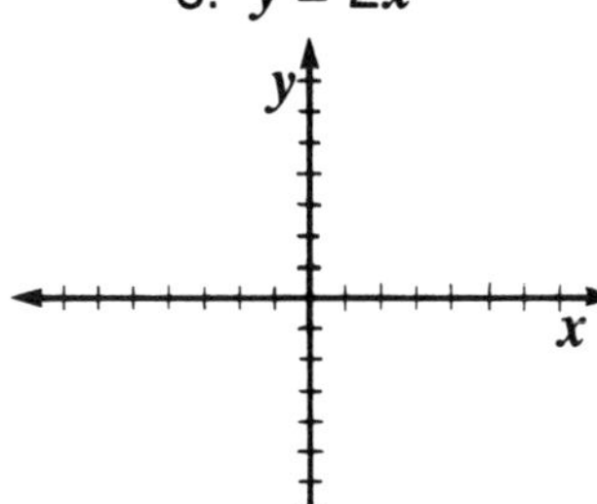

d. $y = 6x^2$

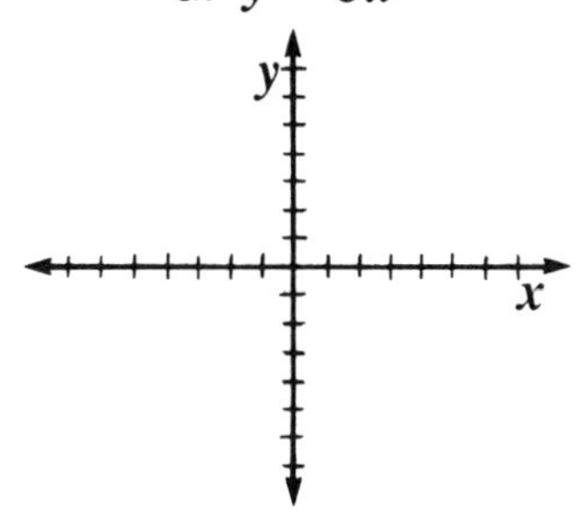

e. $y = -0.25x^2$

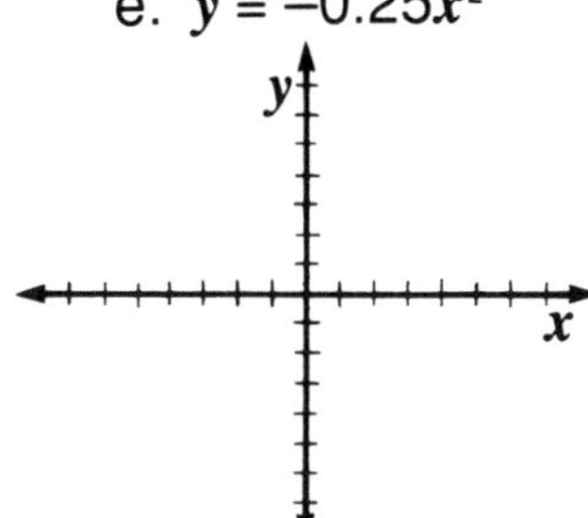

f. $y = -0.5x^2$

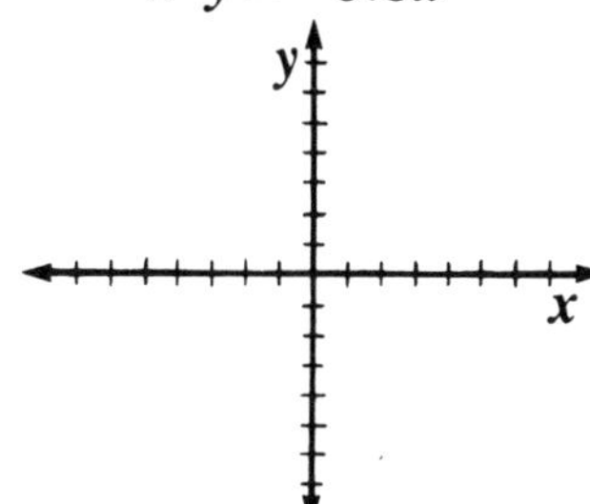

g. $y = -2x^2$

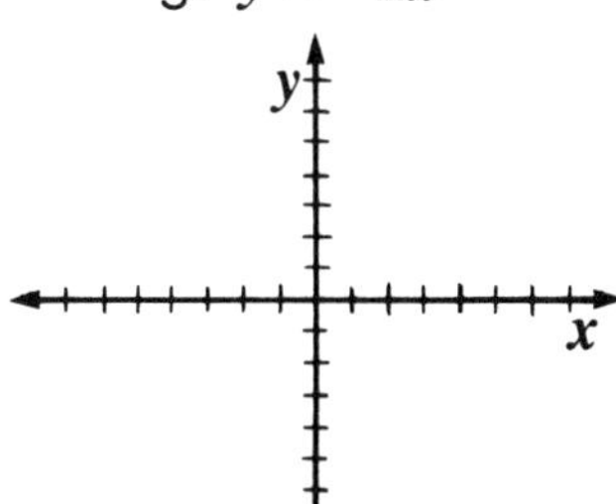

h. $y = -6x^2$

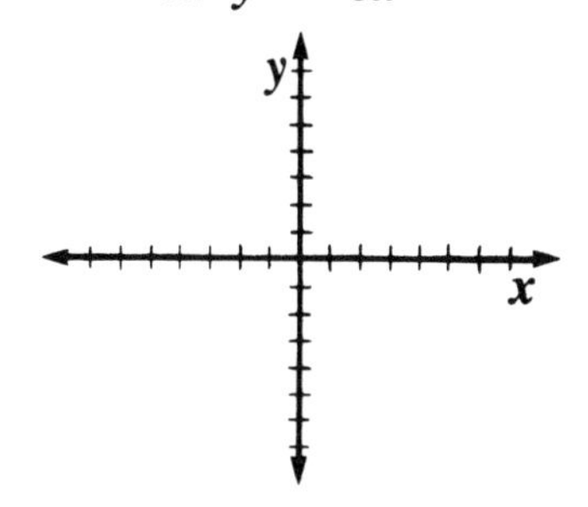

2. Use your sketches on sheet 1 to help answer the following questions:

   a. What property is common to all of the graphs? ____________

   b. Under what condition does the graph of $y = Ax^2$ open upward? ____________ Downward? ____________

   c. As A increases in absolute value, what happens to the graph of $y = Ax^2$? ____________

3. Next consider the equation $y = Ax^2 + Bx + C$ where $B = 0$. In this case the equation becomes $y = Ax^2 + C$. Use a computer or graphing calculator to produce graphs of the following equations. Sketch the graphs from your display on the corresponding axes.

a. $y = 2x^2 + 3$

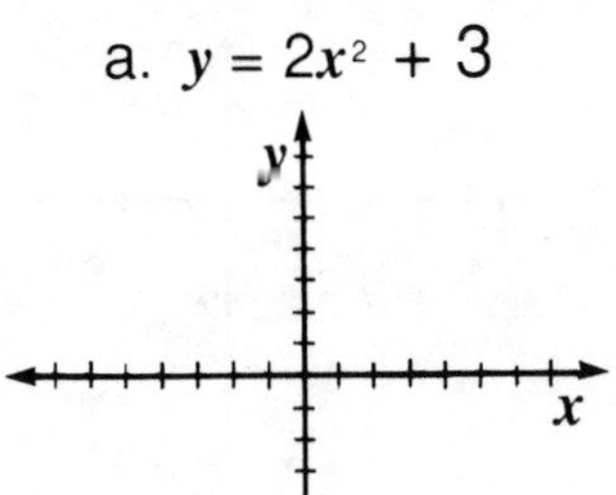

b. $y = 2x^2 - 3$

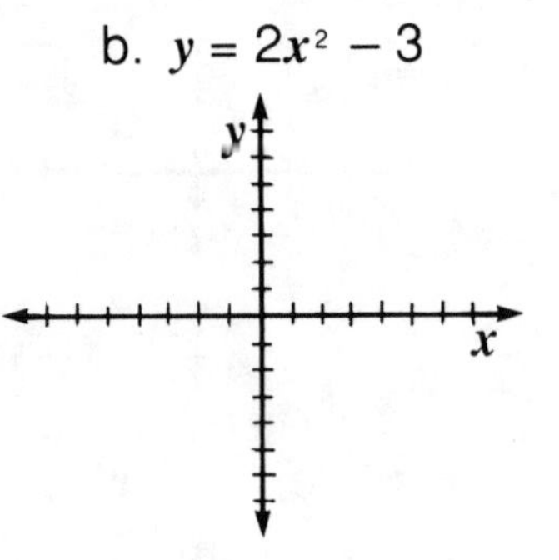

c. $y = 2x^2 - 4$

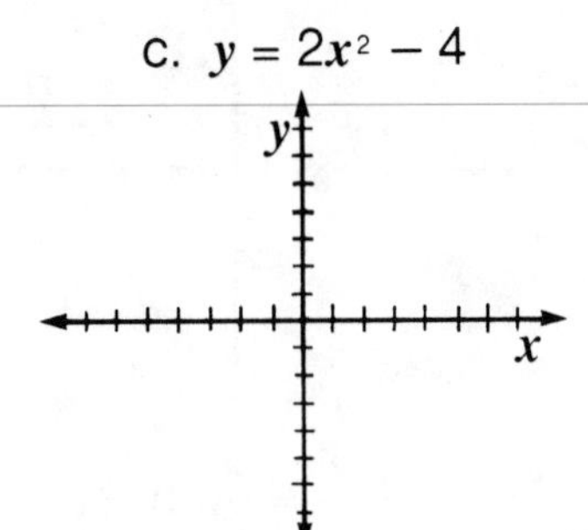

d. $y = 2x^2 + 5$

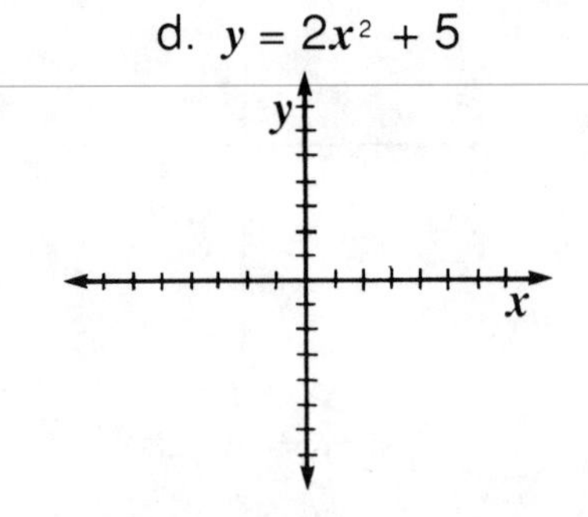

4. *a.* How does the value of C affect the graph of $y = 2x^2 + C$? ____________

   b. The graph of $y = 0.5x^2 - 3$ intersects the y-axis at the point (____, ____).

   c. The y-intercept of the graph of $y = 2x^2 + 4$ is the point (___, ___). Check your answers with a computer or calculator.

5. Without using technology, indicate whether the graph of each of the following equations will open upward or downward; decide if it will be relatively narrow or wide; and determine the y-intercept. Sketch the graph of each equation on a sheet of graph paper.

   a. $y = -0.6x^2 - 4$ ____________

   b. $y = 5x^2 + 3$ ____________

   c. $y = -x^2 + 7$ ____________

   d. $y = 0.3x^2 - 5$ ____________

   Now use a computer or calculator to check your answers and sketches.

6. For each of the following equations, use a computer or calculator to graph the equation; carefully sketch the parabola on the corresponding axes, noting the y-intercept and also the x-intercepts; and calculate the value of the discriminant ($B^2 - 4AC$) and record it in the space provided.

a. $y = x^2 + 4x + 4$
discriminant____

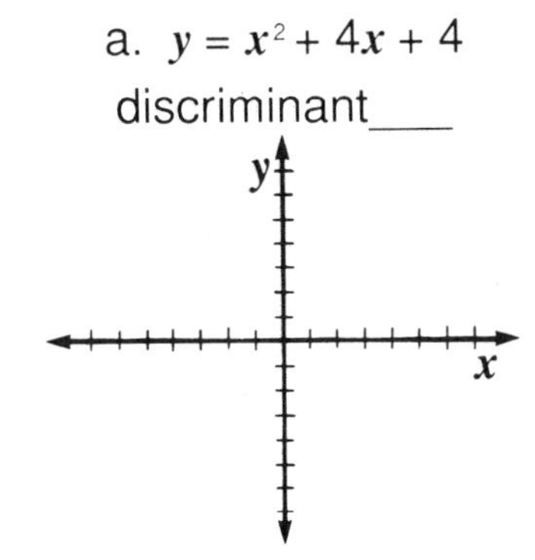

b. $y = -8x^2 + 8$
discriminant____

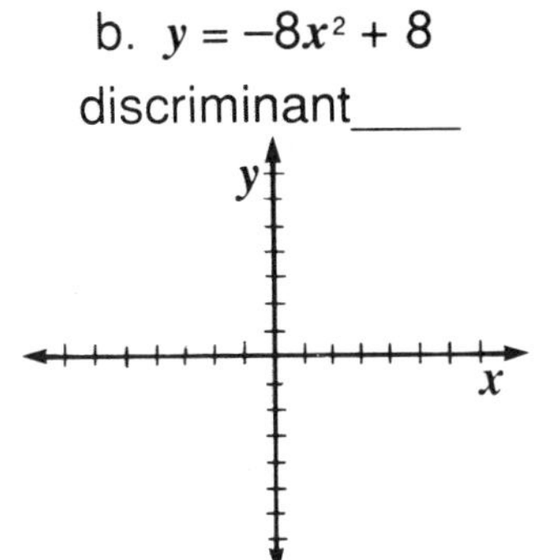

c. $y = x^2 - 2x + 5$
discriminant____

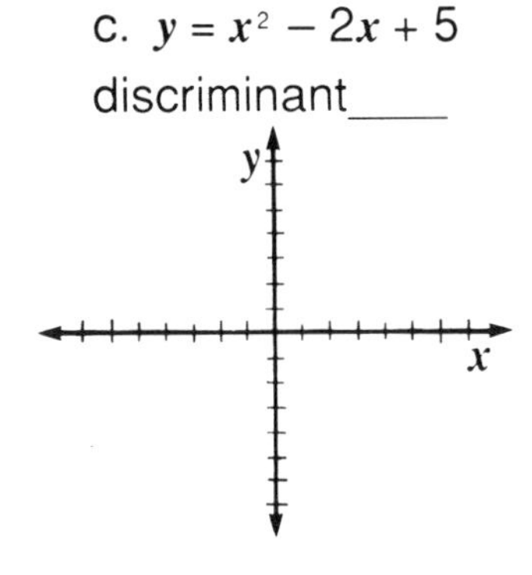

d. $y = -x^2 + 2x - 1$
discriminant____

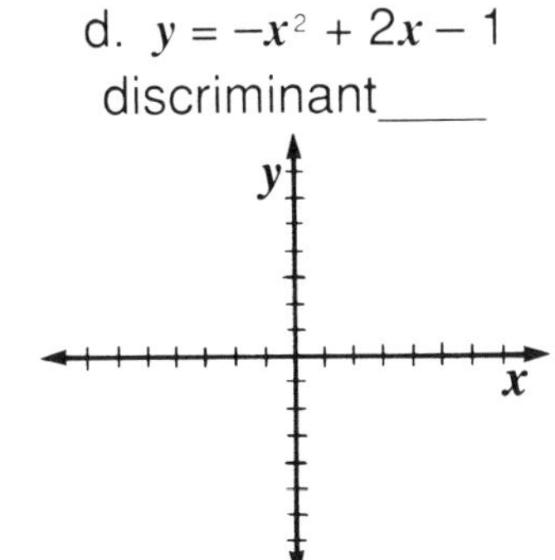

e. $y = 2x^2 - 7x + 3$
discriminant____

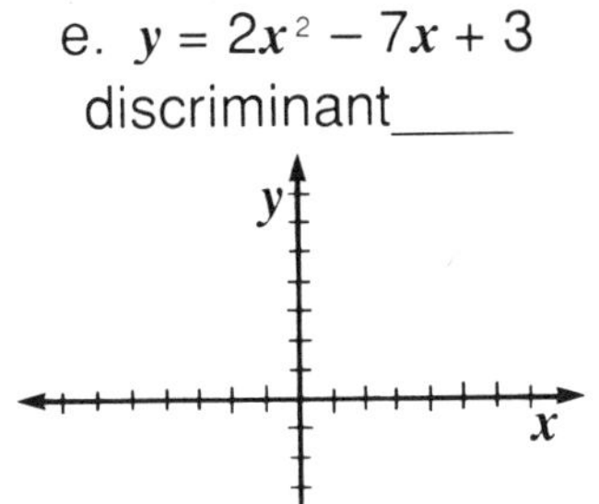

f. $y = -2x^2 + x - 1$
discriminant____

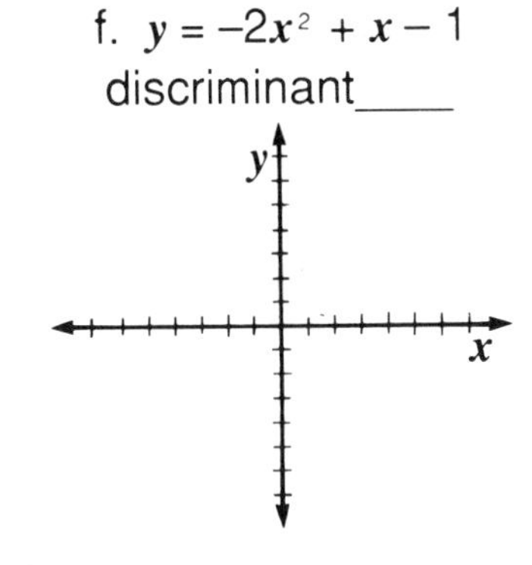

7. Use your results from exercise 6 to answer the following questions:

a. When will the graph of $y = Ax^2 + Bx + C$ have exactly one x-intercept?
b. When will the graph of $y = Ax^2 + Bx + C$ have two x-intercepts?
c. When will the graph of $y = Ax^2 + Bx + C$ have no x-intercepts?

8. Complete each row of the chart without using technology. Before continuing to the next row, graph the equation using a computer or calculator and check if the characteristics of the displayed graph agree with your predictions.

Equation	Does graph open up or down?	Is graph narrow or wide?	y-intercept	Number of x-intercepts
a. $y = 0.5x^2 - 2x - 6$				
b. $y = 6x^2 + 2x + 1$				
c. $y = -x^2 + 4x - 4$				
d. $y = -3x^2 + x - 2$				
e. $y = -4x^2 + 4x + 1$				
f. $y = 2x^2 - 8x + 8$				

ACTIVITIES

FINDING FACTORS PHYSICALLY

May 1982

By Christian R. Hirsch, Western Michigan University, Kalamazoo, MI 49008

Teacher's Guide

Grade level: 8–10.

Materials: Scissors, red pens or markers, and a set of the activity sheets for each student. Cutouts from two transparencies of sheet 4 would be helpful for discussing student solutions to the problems.

Objectives: Students will investigate factoring quadratic polynomials over the integers using a physical model and thereby discover regularities that will permit them to factor such polynomials at the symbolic level.

Background: A quadratic trinomial of the form $ax^2 + bx + c$, where $a > 0$, can be factored over the integers if and only if a rectangular shape can be formed using the corresponding pieces (with additions if necessary) from sheet 4. The dimensions of the rectangle formed are the factors of the trinomial. For example, the factors of $x^2 + 3x + 2$ can be found by arranging the pieces for x^2, $3x$ and 2 into a rectangle that measures $x + 1$ by $x + 2$. Perfect-*square* trinomials are always represented by *square* arrangements of the pieces.

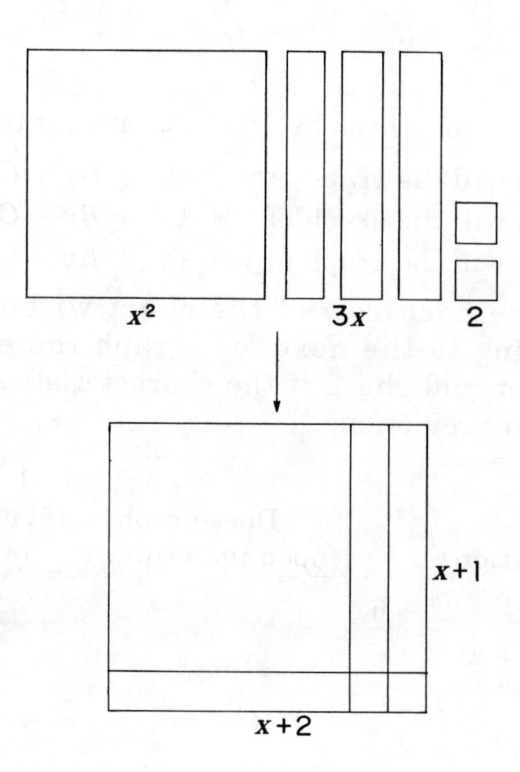

Directions: Activity sheets 1, 2, and 3 should be distributed one at a time. Students will need two copies of sheet 4—one to be distributed together with sheet 1, the other with sheet 3.

Sheets 1 and 2 introduce factoring of quadratic polynomials over the positive integers, and both can be completed during a single class session. To ensure adequate class time for the completion of these two sheets, you may wish to have pupils cut out the pieces on sheet 4 the evening prior to beginning the activity. You might suggest that students glue their copies of sheet 4 onto tagboard before cutting the pieces out. Cutouts from a transparency of sheet 4 can be used to illustrate the solution to problem 3 and thereby help ensure that all students understand how the pieces are to be manipulated to form a rectangle. You might also use the transparency cutouts to show that the x-length is not a multiple of the unit length and, in particular, the area of six of the small squares is not the same as the area of one of the rectangles. Pupils who complete sheet 2 more quickly than others might be challenged to find the factors of expressions such as $4x^2 + 20x + 25$, $6x^2 + 13x + 6$, or $9x^2 + 30x + 16$. When all students have completed sheet 2, you might have them verify their answers for problem 8 by actually multiplying the factors.

Sheet 3 focuses on factoring quadratic polynomials over the integers. The coloring of the rectangles and smaller squares can be avoided by reproducing the second copy of sheet 4 on colored paper. Similarly colored film can be used when making the second transparency of sheet 4. Use the transparency cutouts to demonstrate how the red pieces are placed on top of the white pieces to form the rectangular representation for $x^2 - 3x + 2$. Since students will find it more difficult to find the appropriate rectangles when using the red pieces, you may wish to have them complete this sheet by working in small groups. Encourage students to verify their answers for problem 12 by actually forming the corresponding rectangles (using the appropriate pieces) or by multiplying the factors. Again, some students might be challenged to find the factors of expressions such as $x^2 - 3x - 4$ or $6x^2 - 5x - 21$.

Supplementary activity: An interesting follow-up project for several students would be to investigate factoring cubic polynomials by forming rectangular solids from wooden blocks of the following sizes:

Block	Dimensions	Volume
small cube	1 by 1 by 1	1
rectangular solids	1 by 1 by x	x
	1 by x by x	x^2
larger cube	x by x by x	x^3

To ensure uniqueness of factorization, the x-length should again not be a multiple of the unit length. By using this model, students can easily see that $x^3 + 6x^2 + 12x + 8$ is the volume of a cube with edge $x + 2$, and thus $x^3 + 6x^2 + 12x + 8 = (x + 2)^3$. The article by Hendrickson (1981) illustrates nicely how Cuisenaire materials can be used for this model.

Answers:

Sheet 1: 1.b. x, c. 1; 3.c. $x + 1$ by $x + 2$; 4.a. $(x + 1)(x + 5)$, b. $(x + 2)(x + 4)$, c. $(x + 3)(x + 5)$, d. $(x + 3)(x + 4)$, e. $(x + 4)(x + 4)$, f. $x(x + 2)$, g. $(2x + 1)(x + 3)$, h. $(2x + 3)(x + 1)$, i. $(3x + 2)(x + 1)$, j. $(2x + 3)(x + 2)$.

Sheet 2: 5.a. $pq = c$, b. $p + q = b$; 6.a. $(x + 2)(x + 3)$, b. $(x + 3)(x + 3)$, c. $(x + 4)(x + 5)$, d. $(x + 1)(x + 9)$; 7.a. $pq = c$, b. $p + aq = b$; 8.a. $(2x + 3)(x + 2)$, b. $(2x + 1)(x + 4)$, c. $(2x + 3)(x + 5)$, d. $(3x + 4) \cdot (x + 2)$.

Sheet 3: 10.a. $x - 1$ by $x - 2$, b. $(x - 1)(x - 2)$; 11.a. $(x - 1)(x - 1)$, b. $(x - 2) \cdot (x - 3)$, c. $(x - 2)(x - 4)$, d. $(2x - 1)(x - 2)$, e. $(2x - 3)(x - 2)$, f. $(x - 1)(x + 2)$; 12.a. $(x - 2)(x - 6)$, b. $(x - 2)(x + 3)$, c. $(2x - 3)(x - 5)$.

BIBLIOGRAPHY

Bidwell, James K. "A Physical Model for Factoring Quadratic Polynomials." *Mathematics Teacher* 65 (March 1972):201–5.

Bruner, Jerome S. *Towards a Theory of Instruction.* New York: W. W. Norton & Co., 1968.

Flax, Rosabel. "A Squeeze Play on Quadratic Equations." *Mathematics Teacher* 75 (February 1982):132–34.

Hendrickson, A. Dean. "Discovery in Advanced Algebra with Concrete Models." *Mathematics Teacher* 74 (May 1981):353–58.

1. Sheet 4 of this activity consists of a series of large squares, rectangles, and smaller squares. The dimensions of the pieces are given below. The area of a large square is $x \cdot x = x^2$. Determine the areas of the remaining pieces.

Piece	Dimensions	Area
a. large square	x by x	x^2
b. rectangle	1 by x	______
c. smaller square	1 by 1	______

2. Carefully cut out the pieces on sheet 4.

3. a. The expression $x^2 + 3x + 2$ can be represented physically by one large square, x^2, three rectangles, $3x$, and two smaller squares, 2. Place these six pieces on your desk.

   b. Now rearrange the pieces to form a rectangle.

   c. What are the dimensions of the rectangle you formed? ______

   d. The dimensions of the rectangle formed are called the *factors* of $x^2 + 3x + 2$. We can write $x^2 + 3x + 2 = (x + 1)(x + 2)$ since both sides represent the same amount of area.

4. Form a rectangle with the pieces representing each expression given below. If a rectangle can be formed for the expression, write its dimensions in the column headed Factors.

Expression	Factors
a. $x^2 + 6x + 5$	$(x + 1)(\quad)$
b. $x^2 + 6x + 8$	______
c. $x^2 + 8x + 15$	______
d. $x^2 + 7x + 12$	______
e. $x^2 + 8x + 16$	______
f. $x^2 + 2x$	______
g. $2x^2 + 7x + 3$	______
h. $2x^2 + 5x + 3$	______
i. $3x^2 + 5x + 2$	______
j. $2x^2 + 7x + 6$	______

5. In problems 4a through 4e, you formed rectangles to represent expressions of the form $x^2 + bx + c$. The dimensions of the rectangles were of the form $(x + p)$ by $(x + q)$. That is,

$$x^2 + bx + c = (x + p)(x + q).$$

Use the results of problems 4a through 4e to answer the following questions.

a. What relationship exists between p, q and the number c?

____

b. What relationship exists between p, q and the number b?

____

6. Use your answers to 5a and 5b to find the factors of each expression below. Check your answers by actually forming the corresponding rectangles with the pieces from sheet 4.

Expression	Factors
a. $x^2 + 5x + 6$	________
b. $x^2 + 6x + 9$	________
c. $x^2 + 9x + 20$	________
d. $x^2 + 10x + 9$	________

7. In problems 4g through 4j, you formed rectangles to represent expressions of the form $ax^2 + bx + c$ where a was a prime number. The dimensions of the corresponding rectangles were of the form $(ax + p)$ by $(x + q)$; that is,

$$ax^2 + bx + c = (ax + p)(x + q).$$

Use the results of problems 4g through 4j to answer the following questions.

a. What relationship exists between p, q and the number c?

____

b. What relationship exists between a, p, q and the number b?

____

8. Use your answers to 7a and 7b to find the factors of each expression below. Check your answers by actually forming the corresponding rectangles with the pieces from sheet 4.

Expression	Factors
a. $2x^2 + 7x + 6$	________
b. $2x^2 + 9x + 4$	________
c. $2x^2 + 13x + 15$	________
d. $3x^2 + 10x + 8$	________

9. Using a red pencil or marker, color the rectangles and smaller squares on the second copy of sheet 4, and then carefully cut out these pieces.

10. The red pieces are assumed to have "negative" area. The area of a red rectangle is ^{-}x, and the area of a small red square is $^{-}1$. Thus, the areas of two congruent shapes of differing colors add to zero when placed one on top of the other. This fact can be used to find factors of an expression such as $x^2 - 3x + 2$.

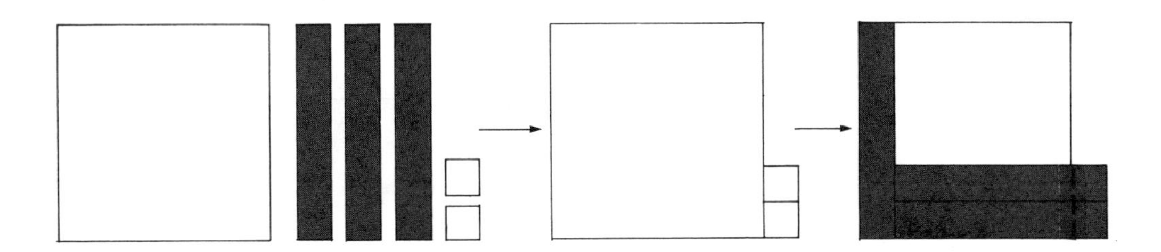

a. What are the dimensions of the white rectangle that was formed? ______

b. In factored form, $x^2 - 3x + 2 =$ (______)(______) .

11. Use your white and red pieces and the method illustrated above to form a white rectangle representing each expression below. In each case, begin by first placing the white pieces next to each other as was done above. In the column headed Factors, write the dimensions of the white rectangle that is formed.

Expression	Factors
a. $x^2 - 2x + 1$	______
b. $x^2 - 5x + 6$	______
c. $x^2 - 6x + 8$	______
d. $2x^2 - 5x + 2$	______
e. $2x^2 - 7x + 6$	______
f. $x^2 + x - 2$	______

(*Hint:* Add one white and one red rectangle to your collection of pieces representing the expression.)

12. Analyze your results for problem 11 in a manner similar to that suggested in problems 5 and 7. Now try finding the factors of each expression below without using the pieces from sheet 4.

Expression	Factors
a. $x^2 - 8x + 12$	______
b. $x^2 + x - 6$	______
c. $2x^2 - 13x + 15$	______

Activities for

Geometry and Visualization

The study of geometry helps students represent and make sense of both the world in which they live and the world of mathematics. Geometric models provide a perspective from which students can analyze and solve problems as exemplified in the first and last sections of this volume. As seen in the previous section, geometric interpretations can often help make symbolic representations more easily understood. The widespread use of computer graphics within and outside mathematics has increased the importance of visualization and visual thinking. The activities in this section are rich in opportunities for students to discover and apply geometric relationships, to visualize and work with three-dimensional shapes, and to make logical deductions.

The first three activities focus on the exploration and application of geometric relationships in the plane. In "Those Amazing Triangles," students use the basic measuring and drawing instruments of geometry to help them discover several recent and remarkable generalizations about triangles, including the theorems of Morley and Napoleon. The activity could also be completed using software tools for drawing and measuring figures. In the next activity, "Investigating Shapes, Formulas, and Properties with Logo," students work within a computer-graphics environment, created by supplied Logo procedures, to explore relationships between areas and perimeters of squares, to derive formulas for the areas of parallelograms and triangles, and to discover the relationship that the line segment connecting the midpoints of two sides of a triangle is parallel to, and one-half the length of, the third side. In "Using Calculators to Fill Your Table," students calculate, organize data in tables, and then analyze their tables to solve maximization problems involving areas of three different garden layouts.

The development of spatial skills is the focus of the remaining four activities in this section. "Spatial Visualization" engages pupils in relating solids made from small cubes and their two-dimensional representations, in making drawings of given solids, and in building solids from isometric drawings. The emphasis throughout is on moving back and forth visually from a solid to a drawing. The activity "Semiregular Polyhedra" begins with an analysis of models of the five regular polyhedra leading to the discovery of Euler's theorem relating the numbers of faces, vertices, and edges. Students are next introduced through models to a method for truncating a regular polyhedron and then use this method together with visualization and deductive counting techniques to discover patterns that enable them to predict the number of faces, vertices, and edges of a truncated polyhedron.

"Visualization, Estimation, Computation" exemplifies how the use of concrete models and computers can be effectively blended to give students opportunities to confront, in a meaningful way, an interesting and significant applied problem without a knowledge of advanced mathematics. Students construct a model of a cone, visualize how the radius, height, area, and volume change as they alter its shape, estimate the position for maximum volume, and then analyze a BASIC program and its output to check their estimates. Pupils are then encouraged to modify the program to obtain a sharper estimate of the radius for maximum volume. "Generating Solids" engages students in visualizing, identifying, and describing solids of revolution and then computing the corresponding areas and volumes.

ACTIVITIES

THOSE AMAZING TRIANGLES *September 1981*

By Christian R. Hirsch, Western Michigan University, Kalamazoo, MI 49008

Teacher's Guide

Grade Level: 7–12.

Materials: Compass, protractor, ruler, extra sheets of paper, and a set of activity sheets for each student.

Objectives: Students will practice using the basic measuring and drawing devices of geometry in the process of discovering several surprising generalizations about arbitrary triangles.

Directions: The three activity sheets are independent of one another and thus can be used individually at different times to reinforce understanding of the related topics as they are presented in your curriculum. Accurate drawings and/or constructions are a must with each of the activities. The completion of each sheet should heighten motivation for further study of geometric ideas.

Sheet 1: In this activity, pupils discover Morley's theorem:. The points of intersection of the adjacent trisectors of the angles of any triangle are the vertices of an equilateral triangle (fig. 1). This surprising result was first discovered about 1899 by Frank Morley. Some students may need help interpreting the definition of adjacent trisectors given in exercise 1b. Depending on the level of your students, you may wish to suggest that for exercise 2 they draw triangles whose angle measures are integral multiples of 3, for example, a 30°-60°-90°

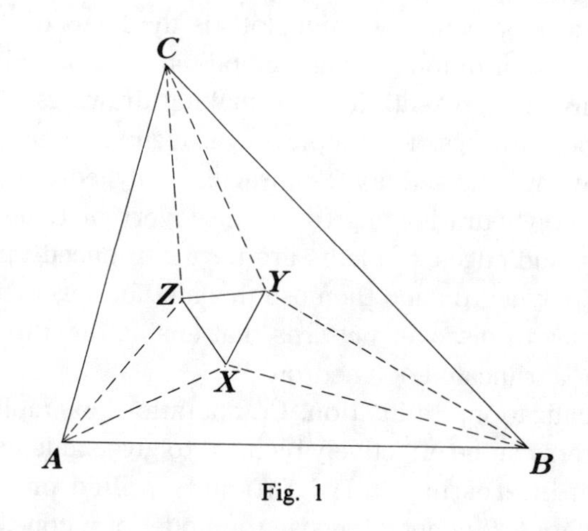

Fig. 1

triangle and a 30°-45°-105° triangle. A proof of Morley's theorem as given in Coxeter and Greitzer (1967) could be provided for better high school students. Other students might be encouraged to investigate the figures formed by the adjacent trisectors of the angles of *regular* polygons. For a discussion of this generalization of Morley's theorem see Hirsch et al. (1979).

Sheet 2: Students in high school geometry could be instructed to use a straightedge and compass to construct the parallel lines for this activity. Younger students might place a sheet of lined paper under the activity sheet to assist them in drawing the parallel lines. Pupils will be amazed to find that points X and X'' always coincide (fig. 2). Of course if X is chosen to be the midpoint of $\overline{AB}$ then X, X', and X'' will coincide. Tenth-grade geometry students might be challenged to prove these results. The proof of the first result depends only on noting that parallelograms $AXYZ$ and $X'BYZ$ have a common side $\overline{YZ}$ and thus $AX = X'B$. Using the definition of betweenness and subtraction, we can see that $AX' = XB$ and hence X and X' are equidistant from the midpoint M of $\overline{AB}$. If we repeat steps a–c starting with point X', then X' and X'' will be equidistant from M and thus $X = X''$. An entertaining account of this problem of closure may be found in Court (1953).

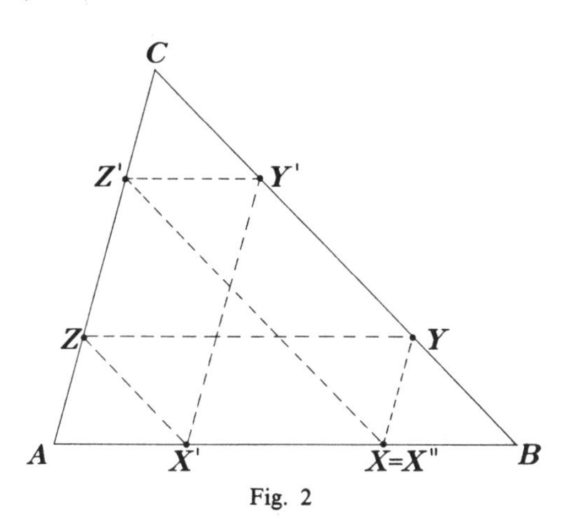

Fig. 2

Sheet 3: Depending on the background of your students, you may wish to have them construct the medians in exercise 1d. Students should compare their answers to 1g. In all cases $\triangle XYZ$ should be equilateral (fig. 3). This triangle is sometimes called the *outer Napoleon triangle* of $\triangle ABC$. Some students might be encouraged to carry out a similar investigation in which the equilateral triangles are constructed inward on the sides.

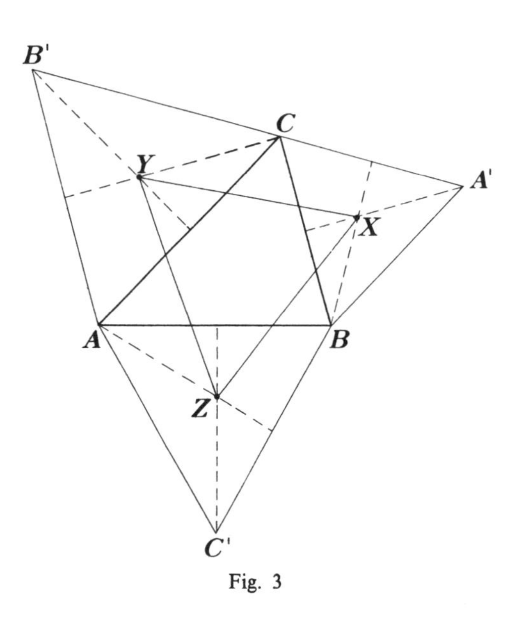

Fig. 3

In exercise 2 pupils should find that $\overline{AA'}$, $\overline{BB'}$, and $\overline{CC'}$ are concurrent and the same length. You might ask them to find the measures of the six angles around the point of concurrency to uncover yet another interesting result. If no angle of $\triangle ABC$ has a measure greater than 120°, then the point of concurrency of $\overline{AA'}$, $\overline{BB'}$, and $\overline{CC'}$ provides the solution to the following real-world problem. Suppose points A, B, and C locate consumption centers that are to be supplied materials from a single supply depot D. Where should the depot be located so that the sum of the distances from D to the three centers is as small as possible? A justification of this solution appropriate for high school students may be found in Hirsch et al. (1979).

REFERENCES

Court, Nathan A. "Geometrical Magic." *Scripta Mathematica* 19 (June–September 1953): 198–200.

Coxeter, H. S. M., and S. L. Greitzer. *Geometry Revisited.* New York: Random House, 1967.

Hirsch, Christian R., Mary Ann Roberts, Dwight O. Coblentz, Andrew Samide, and Harold L. Schoen. *Geometry.* Glenview, Ill.: Scott, Foresman & Co., 1979.

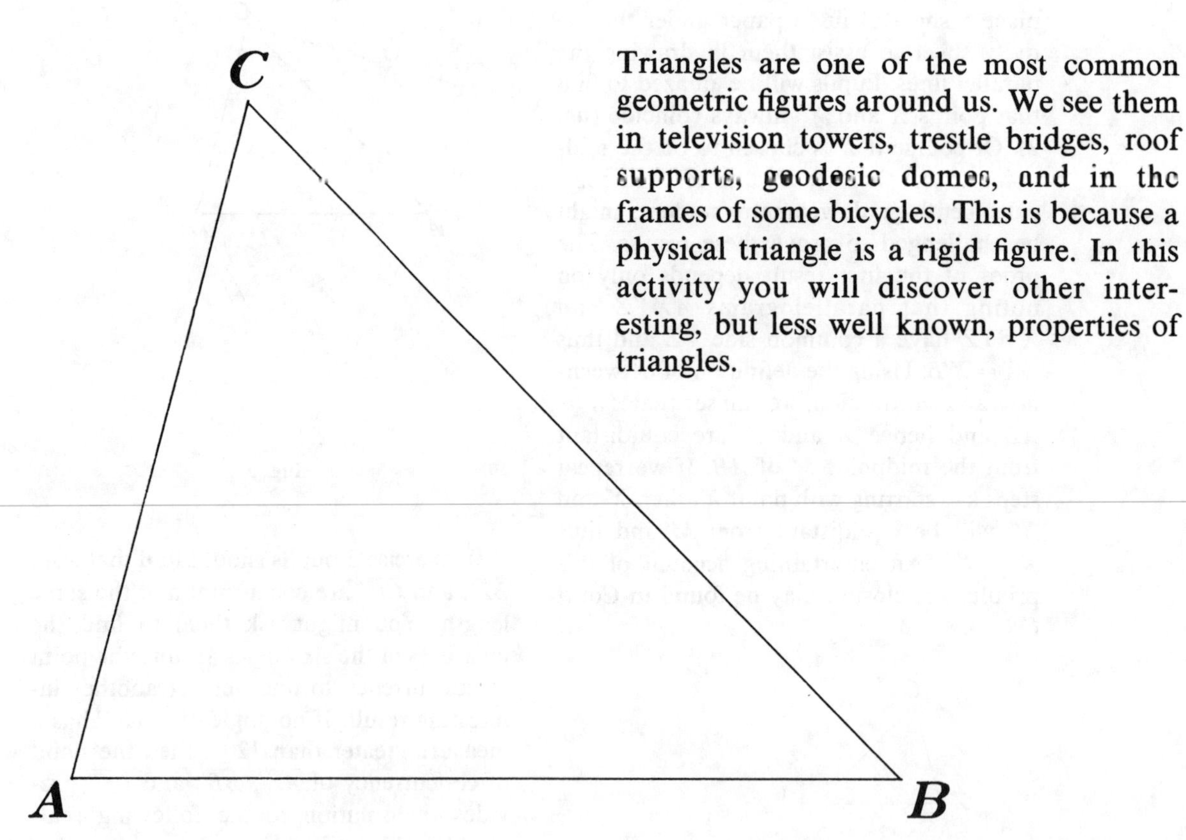

Triangles are one of the most common geometric figures around us. We see them in television towers, trestle bridges, roof supports, geodesic domes, and in the frames of some bicycles. This is because a physical triangle is a rigid figure. In this activity you will discover other interesting, but less well known, properties of triangles.

1. a. Use a protractor to trisect each angle of the triangle above. (To trisect an angle means to divide it into three congruent angles.)
   b. Label the points where the adjacent trisectors of different angles intersect X, Y, and Z. (The adjacent trisectors of two angles of a triangle are the trisectors, one at each vertex, "closest" to the common side of the angles.)
   c. Draw $\triangle XYZ$.
   d. Use a ruler or a compass to compare the lengths of the sides of $\triangle XYZ$.
   e. What kind of triangle is $\triangle XYZ$? ____________________
2. On a separate sheet of paper, draw at least two more large, differently shaped triangles and repeat steps a–e in each case.
3. It's amazing! For each triangle, the three points of intersection of the adjacent trisectors of its angles form ____________________.

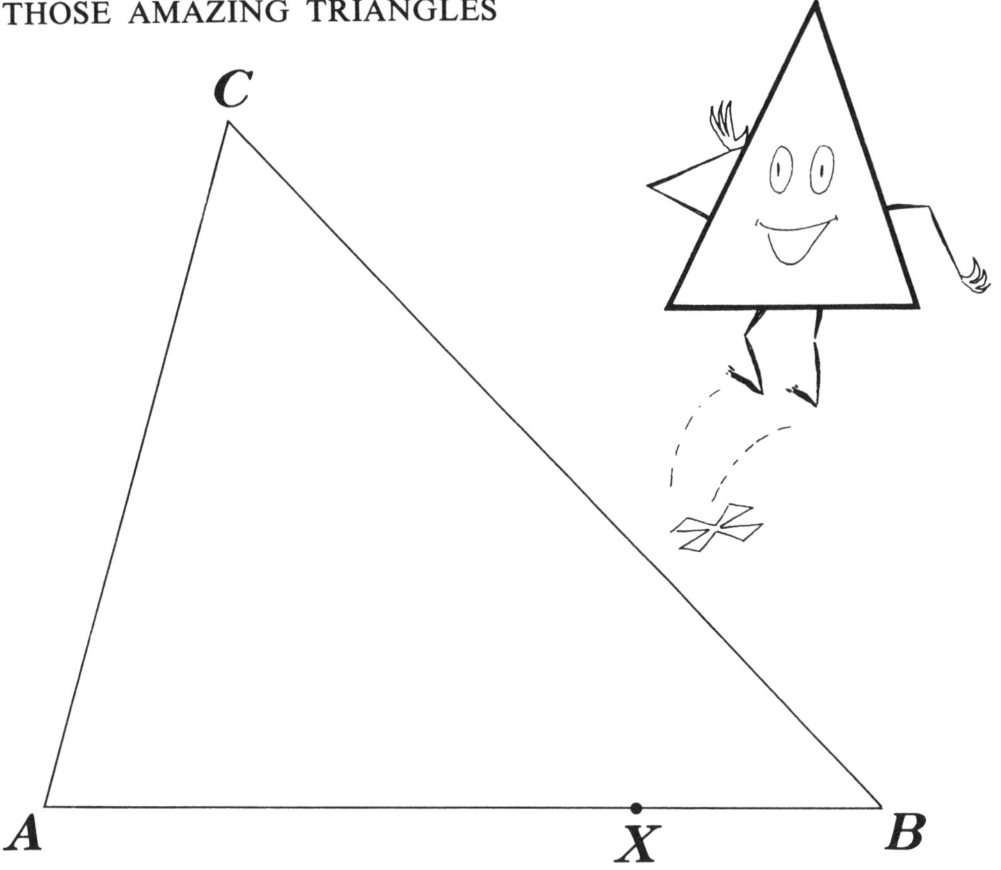

1. On the triangle above, a point X has been marked between A and B.
   a. Use a ruler to draw a line through X parallel to side $\overline{AC}$. Label the intersection with $\overline{BC}$ point Y.
   b. Through Y, draw a line parallel to side $\overline{AB}$. Label the intersection with $\overline{AC}$ point Z.
   c. Through Z, draw a line parallel to side $\overline{BC}$ and label the intersection with $\overline{AB}$ point X'.
   d. Repeat steps a–c, starting this time with point X'. Label the intersection points on the sides of the triangle Y', Z', and X'', respectively.

   What appears to be true about points X and X''? ____________________

2. Trace $\triangle ABC$ on a sheet of paper. Choose any point between A and B and label it X. Follow steps a–d. What appears to be true about points X and X'' this time? ________

   ____

3. Check to see if this amazing result holds for any triangle you choose to draw.

4. Try to find a point X on side $\overline{AB}$ of $\triangle ABC$ above so that when steps a through c are completed points X and X' coincide.

1. a. On a sheet of paper draw a large triangle. Label its vertices A, B, and C.
   b. Use a compass to construct an equilateral $\triangle A'BC$ outward on side $\overline{BC}$ as in the diagram below.
   c. Similarly construct an equilateral $\triangle AB'C$ on side $\overline{AC}$ and an equilateral $\triangle ABC'$ on side $\overline{AB}$.
   d. Draw two medians in $\triangle A'BC$, in $\triangle AB'C$, and in $\triangle ABC'$. (A median of a triangle is a segment from any vertex to the midpoint of the opposite side.)
   e. Label the points of intersection of the three pairs of medians X, Y, and Z.
   f. Use a different color pen or pencil to draw $\triangle XYZ$.
   g. What kind of triangle is $\triangle XYZ$? ______________________________

That's amazing! It is believed that this result was first discovered by Napoleon. Here are a few other surprising results he missed.

2. a. Repeat steps 1a, 1b, and 1c.
   b. Draw $\overline{AA'}$, $\overline{BB'}$, and $\overline{CC'}$. What appears to be true about these segments? __________
   c. Use a ruler or a compass to compare the lengths of $\overline{AA'}$, $\overline{BB'}$, and $\overline{CC'}$. What appears to be true? __________

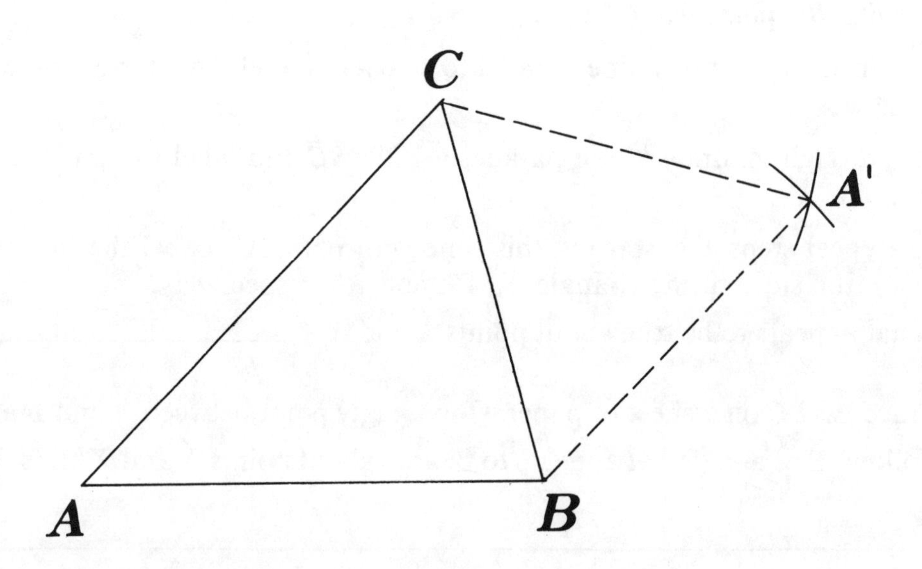

ACTIVITIES

INVESTIGATING SHAPES, FORMULAS, AND PROPERTIES WITH LOGO

May 1985

By DANIEL S. YATES, Mathematics and Science Center, Richmond, VA 23223

Teacher's Guide

Introduction: School mathematics programs have been encouraged to take full advantage of the power of computers at all grade levels. It has been recommended that—

- computers should be used in imaginative ways for exploring, discovering, and developing mathematical concepts and not merely for drill and practice;
- curriculum materials that integrate the use of the ... computer in diverse and imaginative ways should be developed and made available.

The following activities are offered in support of these curricular recommendations. The focus of each activity is on the use of the computer as an instructional tool rather than as an object of instruction.

Grade levels: 7–10

Materials: Copies of the three worksheets for each student; Apple IIe or Apple II$^+$ computers with 64K of memory (preferably no more than two or three students at each computer); Apple Logo disk (Krell or Terrapin Logo can be used with the changes noted near the end of this guide; the activities can be adapted for other versions of Logo to run on other machines, such as the IBM PCjr and Commodore 64); one initialized Logo file disk for each computer

Objectives: (1) To provide informal experiences and opportunities for discovery with squares, rectangles, parallelograms, and triangles; (2) to demonstrate the concepts of similarity and congruence of simple plane figures; (3) to explore relationships between areas and perimeters; (4) to develop formulas for finding the area of a parallelogram and the area of a triangle; (5) to demonstrate that the line segment joining the midpoints of two sides of a triangle is parallel to the third side and equal to one-half its length; and (6) to provide some initial experiences in using, but not programming in, the Logo language

Prerequisites: Students need not have any experience with Logo, but a familiarity with the Apple keyboard would be helpful. They should know that the area of a rectangle equals its base × height, and they should know the meaning of congruent and similar figures. Exercise 6d requires that the student find the dimensions of a square whose area is 200 square units; this problem will be a challenge for many students (unless they have a calculator!). They should also know that in a parallelogram, opposite sides are equal in length and are parallel, and they should know the meaning of "superimpose."

Directions: Before the class session, load the Logo language and then type in the procedures listed at the end of this teacher's guide. Simply type each line as it appears in the listing and press RETURN. After each END statement, hold down the CTRL key while typing C. After a short delay, the computer will state that the procedure has been defined. Then type the next procedure. When all procedures have been entered, make sure that one of the initialized Logo file disks is in the disk drive and type SAVE "GEOMETRY Ⓡ. Throughout this activity the symbol Ⓡ means to press the RETURN key. Be certain that the procedures are typed exactly as they appear, including spaces. Spaces are particularly important in Logo.

Work through the activities prior to the class meeting to test that the procedures work properly. If a procedure has a typographical error and will not execute properly and you are not familiar with the editing commands in Logo, then type ER "(name of procedure), for example, ER "PAR Ⓡ. Then retype the procedure correctly.

When all the procedures are checked for correctness, insert each of the file disks in turn and type SAVE "GEOMETRY Ⓡ. Then type CATALOG to verify that the file is on the disk; it should be listed as GEOMETRY.LOGO.

Special instructions for students: (1) Diagonal lines may appear "jagged" because of the way graphics images are produced on the video screen. Imagine that they are perfectly straight. (2) If you type an incorrect character by mistake, you can erase it by pressing the left arrow key (←). Make sure that each line is typed correctly before pressing RETURN. If you get an error message, ignore it and retype the line correctly. (3) The symbols [and] have separate keys on the Apple IIe and Franklin ACE. On the Apple II$^+$, [is made by pressing the SHIFT and N keys simultaneously and] is made by using the SHIFT and M keys; (4) CAPS LOC must be down on the Apple IIe and ON on the Franklin ACE for Logo to function properly.

Sheet 1: Before or at the beginning of class, boot the Logo language disk. Then insert the Logo file disk and type LOAD "GEOMETRY Ⓡ. This step loads the needed procedures into the computer's "work space." Do this step for each computer.

The student begins by typing in a procedure to draw a variable-sized square. The exercises provide experiences with areas and perimeters of squares, as well as some initial exposure to the Logo language.

Sheet 2: Starting with the formula for the area of a rectangle, the student is led to conclude that the area of a parallelogram is also its base times its height. In each of many examples, the computer draws a random parallelogram and then superimposes a rectangle with the same base. (For a related treatment see Donald Schultz's article "Using Sweeps to Find Areas" in the May 1985 *MT*.) By analyzing the displays produced in multiple examples, the student should discover that the areas of the two figures are the same. To work correctly, the REC procedure must be executed after PAR is executed.

The student must correctly answer exercise 1d on sheet 2 to deduce correctly the formula for the area of a triangle in exercise 2e.

Sheet 3: These activities lead to the discovery that the line segment joining the midpoints of two sides of a triangle is parallel to and equal to one-half the length of the third side.

The computer draws a random triangle (TRI) and then connects the midpoints (TRI2). This display reveals a variety of congruent triangles, similar triangles, and parallelograms. Using many examples and the properties of a parallelogram, the student should be able to discover the intended result.

Changes for Krell or Terrapin Logo:

1. When entering the procedures, make these changes: Change SETPOS [−100 −50] to SETXY (−100) (−50) in the procedures SETUP and PAR2.

   Change SETPC to PC in the procedures

REC, PAR2, and TRI2. ERNS can be omitted from the procedures REC and TRI2.

Change the second line of PAR2 to read:
MAKE "Z ATAN (140 – (80 * SIN :X)) (80 * COS :X)

2. Change LOAD "GEOMETRY to READ "GEOMETRY in the teacher's guide.
3. Throughout the teacher's guide and the student activity sheets, change the left arrow key (←) to the ESC key to erase the previously typed character and change CS to DRAW to clear the screen.

Procedure listings:

```
TO SETUP
HT PU SETPOS [-100 -50] PD
END

TO PAR
SETUP
MAKE "X 10 + RANDOM 70
RT :X
REPEAT 2 [FD 80 RT 90 - :X FD 140
        RT 90 + :X]
LT :X
END

TO REC
SETPC 4
REPEAT 2 [FD 80 * COS :X RT 90 FD 140
        RT 90]
SETPC 1
ERNS
END

TO PAR2
PU RT :X FD 80
MAKE "Z ARCTAN ((140 - (80 * SIN :X)) /
        (80 * (COS :X)))
RT (90 - :X) + (90 - :Z)
SETPC 2 PD
FD (80 * (SIN (90 - :X))) / SIN (90 - :Z)
PU SETPOS [-100 -50] SETH 0 PD
SETPC 1
END

TO TRI
SETUP
MAKE "A RANDOM 61
MAKE "B 110 + RANDOM 40
MAKE "AB 100 + RANDOM 50
RT :A FD :AB RT :B
FD (:AB * SIN (90 - :A)) / COS (180 -
        (:A + :B))
RT (360 - (:A + :B)) SETX -100
END

TO TRI2
MAKE "BC (:AB * SIN (90 - :A)) / COS
        (180 - (:A + :B))
RT :A FD (.5 * :AB) RT (90 - :A)
SETPC 5
FD ((.5 * :AB) * COS (90 - :A)) +
        ((.5 * :BC * SIN (180 - (:A + :B)))
RT (90 + :A) FD (.5 * :AB)
RT :B FD (.5 * :BC)
RT (180 - :B) PU BK (.5 * :AB) LT :A PD
SETPC 1
ERNS
END
```

Answers:

Sheet 1: 2. b. a small square; d. a larger square with an angle in common with the first square. 3. a. 80; 400.

4. b and c.

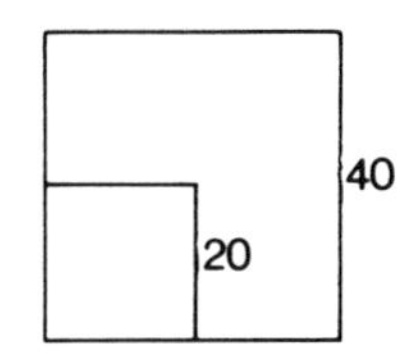

d. 160

5. SQUARE 100 Ⓡ. 6. a. SQUARE 10 Ⓡ; b. SQUARE 40 Ⓡ; c. SQUARE 10 Ⓡ; d. SQUARE 14.14 Ⓡ

Sheet 2: 1. c. equal; d. area = base × height. 2. b. two; d. height and base of triangle were equal to that of the corresponding parallelogram; area of triangle was one-half that of the corresponding parallelogram; e. area $= \frac{1}{2} \times$ base × height.

Sheet 3:

2.

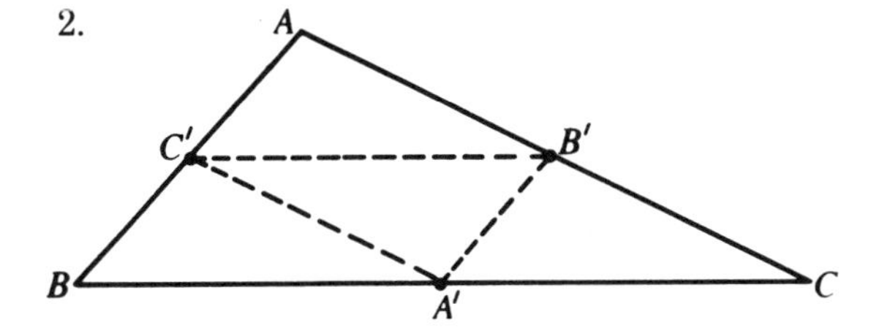

3. a. five; b. $\triangle AC'B'$ and $\triangle C'BA'$, $\triangle AC'B'$ and $\triangle B'A'C$, $\triangle AC'B'$ and $\triangle A'B'C'$, $\triangle C'BA'$ and $\triangle B'A'C$, $\triangle C'BA'$ and $\triangle A'B'C'$, $\triangle B'A'C$ and $\triangle A'B'C'$; c. $\triangle ABC$ and $\triangle AC'B'$; other possibilities include $\triangle ABC$ and $\triangle C'BA'$, $\triangle ABC$ and $\triangle B'A'C$, $\triangle ABC$ and $\triangle A'B'C'$. d. three; $BC'B'A'$, $A'C'B'C$, $B'A'C'A$. 4. a. $B'C' = \frac{1}{2}(BC)$; yes; b. $\overline{B'C'} \parallel \overline{BC}$; yes. 5. For all triangles, answers will vary from informal arguments based on the analysis of a large number of examples to a standard proof as found in a high school geometry text. ■

In the following activities, Ⓡ means to press the RETURN key.

1. Enter the following Logo procedure. Leave spaces *exactly* as shown.

```
TO SQUARE :X Ⓡ
REPEAT 4 [FORWARD :X RIGHT 90] Ⓡ
END Ⓡ
```

2. a. Execute the procedure by typing SQUARE 20 Ⓡ.
   b. Describe the resulting display. ____________________
   c. Execute the procedure again, this time typing SQUARE 50 Ⓡ.
   d. Describe the result. ____________________
   These two squares are said to be *similar*. That is, they have the same shape, but different size.

3. Type CS Ⓡ to clear the screen. Now type SQUARE 20 Ⓡ. Each side of the square is 20 units long. Therefore,
   a. the perimeter (or distance around) is ______ units, and
   b. the area is ______ square units.

4. a. Without clearing the screen, draw a second square by typing SQUARE 40 Ⓡ.
   b. In the space provided at the right, draw a sketch of both squares as they appear on the screen.
   c. Label the lengths on your sketch.
   d. The perimeter of SQUARE 40 is ______ units. Thus, the perimeter of the larger square is twice the perimeter of the smaller one.

5. What would you type to draw a square whose perimeter is five times the perimeter of SQUARE 20? ____________________
   Check the reasonableness of your answer by typing CS Ⓡ SQUARE 20 Ⓡ SQUARE ______ Ⓡ, where the blank contains your suggested input number.

6. What would you type to draw a square satisfying the given condition? Check your answers as in exercise 5.
   a. Perimeter one-half that of SQUARE 20 ____________________
   b. Area four times that of SQUARE 20 ____________________
   c. Area one-fourth that of SQUARE 20 ____________________
   d. Area one-half that of SQUARE 20 ____________________

To make sure that the Logo file GEOMETRY has been loaded into the computer, hold down the CTRL key while you press the T key. Now type POTS Ⓡ. You should see several procedure names listed, including TO PAR and TO REC.

1. a. Type CS Ⓡ and then PAR Ⓡ. Next type REC Ⓡ. Compare the height, base, and area of the parallelogram with the height, base, and area of the superimposed rectangle.

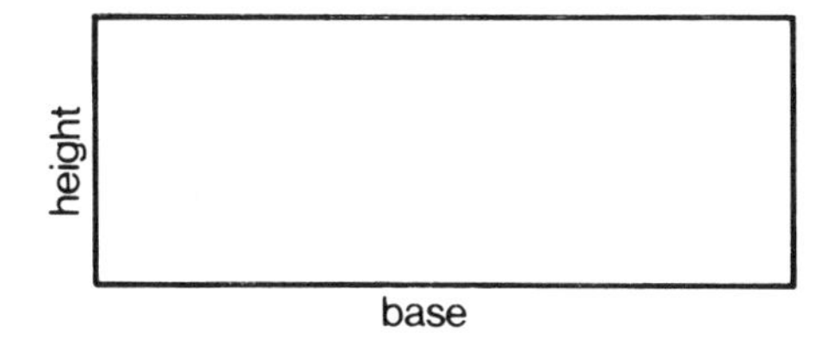

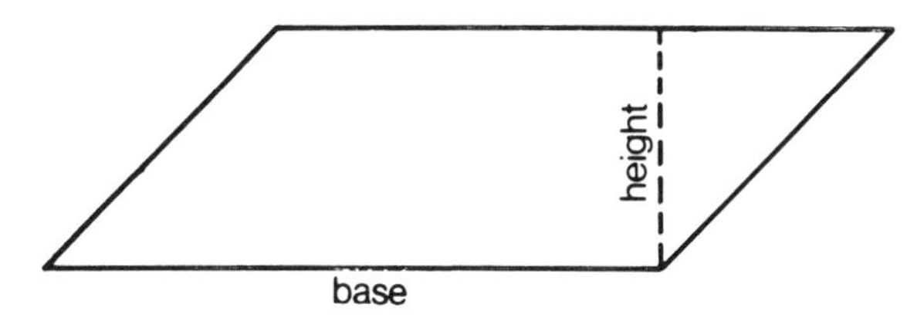

   b. Repeat step 1a several times.

   c. For each result, how did the height, base, and area of the parallelogram appear to be related to the height, base, and area of the superimposed rectangle? ______________________________

   Recall that the area of a rectangle is given by area = base × height.

   d. Write a formula for the area of a parallelogram in terms of its base and height. ______________________________

2. a. Clear the screen by typing CS Ⓡ. Now type PAR Ⓡ to draw another parallelogram.

   b. The procedure PAR2 will draw a diagonal of the parallelogram. Type PAR2 Ⓡ to see this result. The diagonal divides the parallelogram into ______ congruent triangles, that is, triangles of the same size and shape.

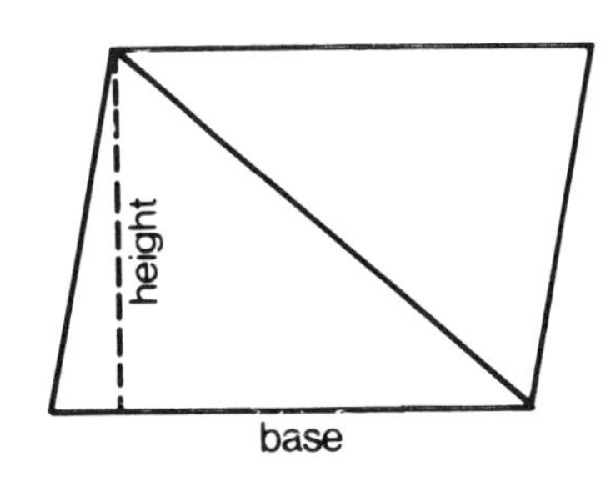

   c. Repeat steps 2a and 2b several times.

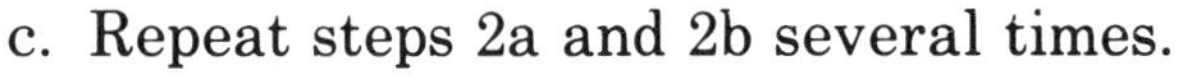

   d. For each figure, how did the height, base, and area of the lower triangle compare to the height, base, and area of the parallelogram? ______________________________

   e. Use your formula for the area of a parallelogram (exercise 1d) to write a formula for the area of a triangle in terms of its base and height.

____

Hold down the CTRL key while you press the T key. Then type POTS Ⓡ. The procedure names TO TRI and TO TRI2 should be listed. They will be used in the following activities.

1. a. Type CS Ⓡ and then TRI Ⓡ to see a random triangle. Now type TRI2 Ⓡ to see what happens when the midpoints of the sides are connected.
   b. Repeat step 1a several times. Look for properties or relationships that appear to be true for all the triangles displayed.

2. In triangle ABC at the right, A', B', and C' are the midpoints of sides $\overline{BC}$, $\overline{AC}$, and $\overline{AB}$, respectively. Complete the figure as would be done by the procedure TRI2.

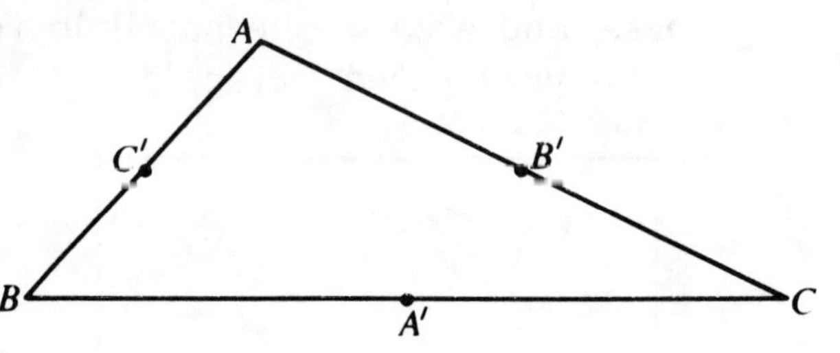

3. Refer to your sketch for exercise 2.
   a. How many triangles are in your figure? ______
   b. Name the pairs of congruent triangles. ______________________________
   c. Triangles ABC and $B'A'C$ are similar but not congruent. Name another pair of such triangles. ______________________________
   d. How many parallelograms do you see in your figure? ______________________________
   Name them: ______________________________
   e. Type CS Ⓡ TRI Ⓡ TRI2 Ⓡ. Verify your answers to 3a and 3d for this triangle.

4. Refer again to your sketch for exercise 2.
   a. What relationship appears to exist between the lengths of segments $\overline{BC}$ and $\overline{B'C'}$? ______________________________
   Does the same relationship appear to hold for the pairs of segments $\overline{AB}$ and $\overline{A'B'}$ and $\overline{AC}$ and $\overline{A'C'}$? ______________________________
   b. Identify another relationship between segments $\overline{BC}$ and $\overline{B'C'}$.
   ____
   Does this relationship also seem to hold for the pairs of segments $\overline{AB}$ and $\overline{A'B'}$ and $\overline{AC}$ and $\overline{A'C'}$? ______________________________
   c. Gather additional support for your answers to 4a and 4b by typing CS Ⓡ TRI Ⓡ TRI2 Ⓡ and studying the display.

5. Do you think the relationships discovered in exercise 4 are true for only certain triangles, or for all triangles? ______________________________
   Why do you think so? ______________________________

ACTIVITIES

USING CALCULATORS TO FILL YOUR TABLE

March 1981

By Dwayne E. Channell, Western Michigan University, Kalamazoo, MI 49008

Teacher's Guide

Grade Level: 7–10.

Materials: Copies of the three worksheets for each student, extra sheets of lined paper, and a set of calculators.

Objective: Students are to construct tables of values and use them to solve maximization problems.

Directions: Distribute copies of the three worksheets, paper, and calculators to your students.

Work with your students to be certain each understands the problem and the method being used to complete the table in problem 3. Be sure each student understands that this table lists all of the possible dimensions for gardens that can be bordered with the 120-cm sections of fence.

In problem 5, the list of possible widths for the table is not given to the students. You may want to help them complete this column before they begin work on the remainder of the table.

In problems 7 and 8, students will need to organize their own tables. Problem 8 is more difficult than the others. You may wish to work through one case (e.g., $w =$ 0.9 m where one section of fence is used for the width) to demonstrate a method for finding l and A and suggest that students use these values to begin their tables.

Additional Problems: Three different garden locations and designs are described on the activity sheets. If Tim and Kathy use a combination of the 120-cm and the 90-cm fencing sections instead of sections of just one length, could they enclose a larger garden area at any of the three locations?

A computer can be programmed to generate the tables of numbers needed to solve problems of this type. If you have access to a microcomputer or to a time-sharing terminal, you may want to try a programming approach to these problems. While at the keyboard, try this problem: Assume that an 18-m roll of fencing is available and that this fencing can be bent at 2-cm intervals. By using this single piece of fencing, what is the largest garden area that Kathy and Tim could enclose at each of the three garden locations?

BIBLIOGRAPHY

Watkins, Ann E. "The Isoperimetric Theorem." *Mathematics Teacher* 72 (February 1979):118–22.

Answers: Sheet 1: 1.a. 15; b. 20; 2.a. 13; b. 1.2 m × 15.6 m; c. 18.72;

3.a.

Width (m)	Length (m)	Area (m^2)
1.2	15.6	18.72
2.4	13.2	31.68
3.6	10.8	38.88
4.8	8.4	40.32
6.0	6.0	36.00
7.2	3.6	25.92
8.4	1.2	10.08

b. A width of 8.4 m would use 7 sections of fence on each of two sides of a garden. This leaves one section for the third side.

Sheet 2: 4.a. 40.32 m^2; b. 4.8 m × 8.4 m; c. 4; d. 7;

5.

Width (m)	Length (m)	Area (m^2)
0.9	16.2	14.58
1.8	14.4	25.92
2.7	12.6	34.02
3.6	10.8	38.88
4.5	9.0	40.50
5.4	7.2	38.88
6.3	5.4	34.02
7.2	3.6	25.92
8.1	1.8	14.58

6.a. 90 cm; b. 4.5 m × 9.0 m; c. 5; d. 10

Sheet 3: 7.c. 90-cm sections; d. ten for width, ten for length; 8.a. 2.7 m × 6.3 m; b. 26.73 m^2; c. no

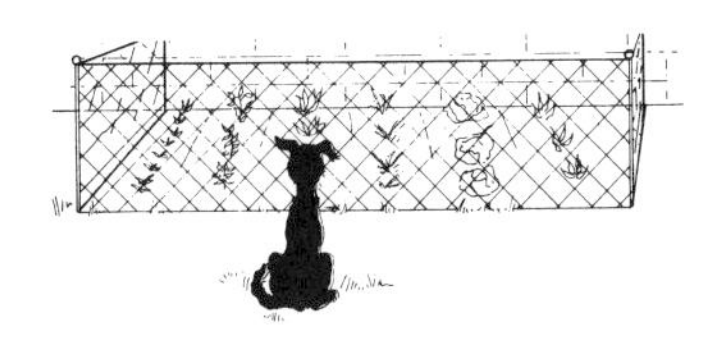

FENCING A GARDEN

Kathy and Tim are planning a small vegetable garden in their backyard. They want to enclose a rectangular garden area with a small fence so as to protect their plants from the family dog. They plan to use part of an existing wall as one of the four borders for their garden.

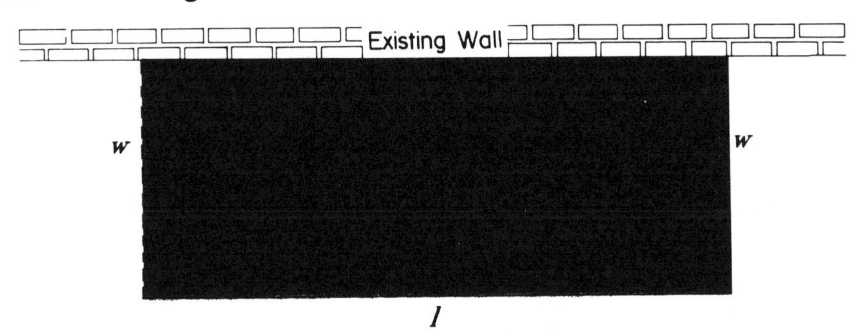

Tim and Kathy have enough money in their budget for the purchase of 18 meters (m) of fencing. Fencing sections are available in both 120-cm and 90-cm lengths. In order to grow as many vegetables as possible, they want to use the 18 m of fencing to enclose the maximum possible garden area. Which length of fencing sections should Tim and Kathy purchase?

1. a. How many 120-cm lengths of fence can they purchase? ________________

   b. How many 90-cm lengths of fence can they purchase? ________________

2. a. If they purchase the 120-cm sections and use a single section for each width of the fence, how many sections would they use for the length? (Reminder: No fencing is needed along the back of the garden.) ________________

   b. What would be the dimensions of the garden in meters? ________________

   c. How many square meters (m^2) of garden area would be enclosed by this fence? ________________

3. a. Complete the table below to find the areas enclosed by each of the different sized fences that can be made using the 120-cm sections.

Width(m)	Length(m)	Area(m^2)
1.2		
2.4		
3.6		
4.8		
6.0		
7.2		
8.4		

   b. Why is 8.4 m the longest possible width? ________________

____

____

4. a. What is the largest area that Kathy and Tim can enclose if they use the 120-cm sections of fence? ____________

   b. What are the dimensions in meters of this rectangular garden?

   ____

   c. How many sections of fence would be used for each width of this garden's border? ____________

   d. How many sections of fence would be used for the length of this garden's border? ____________

5. Could Tim and Kathy enclose even more garden area if they use the 90-cm fencing sections? Complete the table below to find the dimensions and areas of all the possible gardens that can be bordered using the 90-cm sections of fence. The list of possible widths is started for you.

Width(m)	Length(m)	Area(m^2)
0.9		
1.8		

6. a. Which length of fencing sections should Tim and Kathy purchase if they wish to enclose a garden with the largest possible area?

   ____

   b. What are the dimensions in meters of the garden they should plant?

   ____

   c. How many sections of fence should they use for each width of their garden's border? ____________

   d. How many sections should they use for the length of their garden's border?

   ____

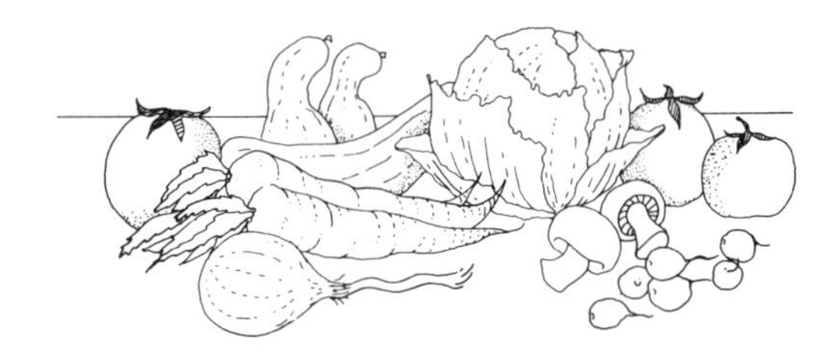

7. Kathy and Tim's parents suggested they plant their garden at a corner of the backyard so that the existing wall could serve as two of the four borders of their garden.

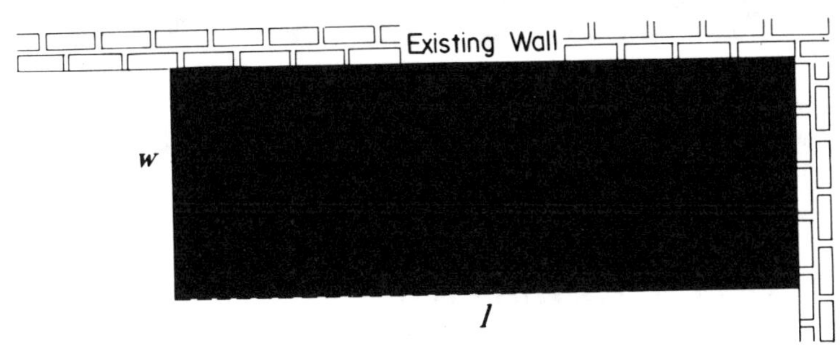

a. Construct a table of the dimensions and areas of the possible gardens bordered by the 120-cm sections of fencing.

b. Construct a table of the dimensions and areas of the possible gardens bordered by the 90-cm sections of fencing.

c. Which type of fencing should they buy to enclose the maximum garden area? ____________________

d. How many fencing sections should they use for the width and how many for the length of this garden? ____________________

8. Suppose Tim and Kathy used the 90-cm sections of fence and planted their garden at a corner of the house so that it was positioned as shown here.

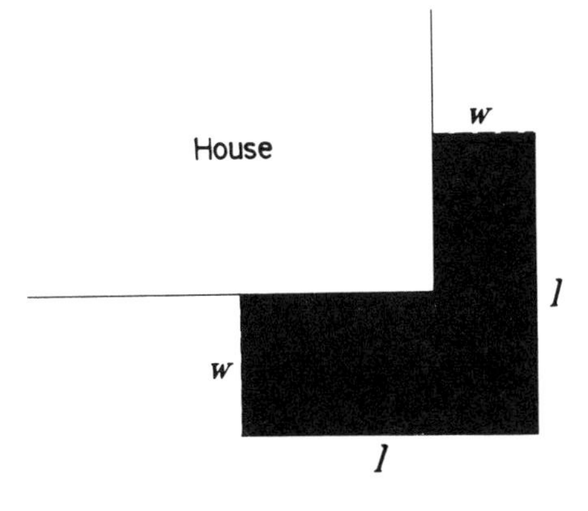

a. What are the dimensions (w and l) of the largest garden they could enclose?

____

b. What is the area of this garden? ____________________

c. Could they use the 120-cm sections to enclose a similarly shaped garden with a larger area? ____________________

ACTIVITIES

SPATIAL VISUALIZATION

November 1984

By GLENDA LAPPAN, ELIZABETH A. PHILLIPS, and MARY JEAN WINTER, Michigan State University, East Lansing, MI 48824

Teacher's Guide

Introduction: Contemporary mathematics programs must be designed to equip students with the mathematical methods that support the full range of problem solving, including the use of imagery, visualization, and spatial concepts. It seems advisable that there should be increased emphasis on using concrete representations and puzzles that aid in improving the perception of spatial relationships.

Improvement of students' ability to visualize three-dimensional shapes and to perceive spatial relationships is the focus of the four worksheets included in this activity.

Grade levels: 7–10

Materials: Ten to twelve small cubes for each student or small group of students, tape, one set of worksheets for each student, and a set of transparencies for discussion of solutions

Objectives: Students will improve their ability to visualize by building, drawing, and evaluating three-dimensional figures. These three activities are used in various combinations throughout the worksheets. A student will see how a solid and a drawing of it are related to each other and will be able to build a solid and draw a two-dimensional representation of it.

Background: Most of a student's mathematical experience with the three-dimensional world is obtained from two-dimensional pictures. Yet many students cannot "read" these two-dimensional pictures well enough to determine needed information about the solid objects. For example, how many cubes are needed to build the solid shown here?

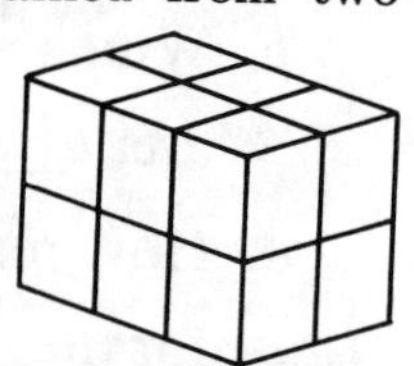

Students usually make two types of errors. They count the faces of every cube showing and answer "sixteen," or they count only the cubes showing and answer "ten." To answer correctly, the students must be able to visualize the hidden corner of the solid. It is important for students to explore this situation from both directions—not only reading information in two-dimensional pictures of the real world but also representing information about the real world with two-dimensional pictures. To accomplish this goal it is necessary to move back and forth from concrete experiences (solids) to abstractions (drawings).

Because of their flexibility and overall usefulness, cubes are used as the basic building units for the three-dimensional objects. As the students build solids from the cubes and look at them from different perspectives, they are developing more discriminating spatial skills. Isometric dot paper (paper with dots arranged in diagonals rather than in rows) is used to make the drawing of these views easier for the students. Using the dot paper requires that a solid be turned so that the student is looking at a corner. In the following activities, the students learn to relate a solid to a drawing of it, to make an isometric drawing of a solid, and to build a solid from an isometric drawing.

Comments: Distribute the worksheets one at a time to each student. Discuss solutions for each sheet before going on to the next one.

Sheet 1: This sheet is designed to acquaint students with representations made on isometric dot paper. The students build simple solids from cubes, copy isometric drawings of the solids, and answer questions about the numbers of cubes needed to build the solids. A discerning student will notice that cubes can sometimes be hidden by other cubes and thus not show in the isometric drawing. For example, problem 3 can be built with twelve or thirteen cubes. Notice that problems 4 and 5 are different corner views of the same solid. On sheet 3 students will be asked to draw four different views of a given solid.

Sheet 2: In this activity students look at a drawing, build a solid from the drawing, add or remove cubes, and then draw the modified solid. The emphasis is on going back and forth visually from a solid to a picture. Some students notice that solid 4 can be built in two ways. If the situation arises, tell the student to build the solid using the fewer cubes or challenge the student to draw both possibilities.

Sheet 3: In this worksheet students look at a solid from a corner to see the solid as it can be drawn on isometric dot paper. As an introduction, students look at the corners of a solid and match these to the appropriate drawings. This matching requires a great deal of eye movement back and forth from solids to drawings. Suggest to students who are having difficulty that they view the solid with one eye shut. Once these views have been matched, the students are asked to draw isometric views of a simple solid.

Sheet 4: In this activity two simple solids, called puzzle pieces, are put together to match a drawing of a solid. Students are then asked to shade the drawing to show each puzzle piece. Masking or transparent tape can be used to hold cubes together temporarily to form the puzzle pieces. To be successful, a student must put the pieces together not only in the right combination but also with the correct orientation. Finally, students are asked to create a new solid from the pieces and to draw a two-dimensional representation of this new solid. This last, open-ended problem allows the student to create, to represent, and then to evaluate the picture against the solid object.

Answers: Answers can be found on the next two pages.

Answers

Sheet 1

1. 5; 2. 8; 3. 12 or 13; 4. 10; 5. 10

Sheet 2

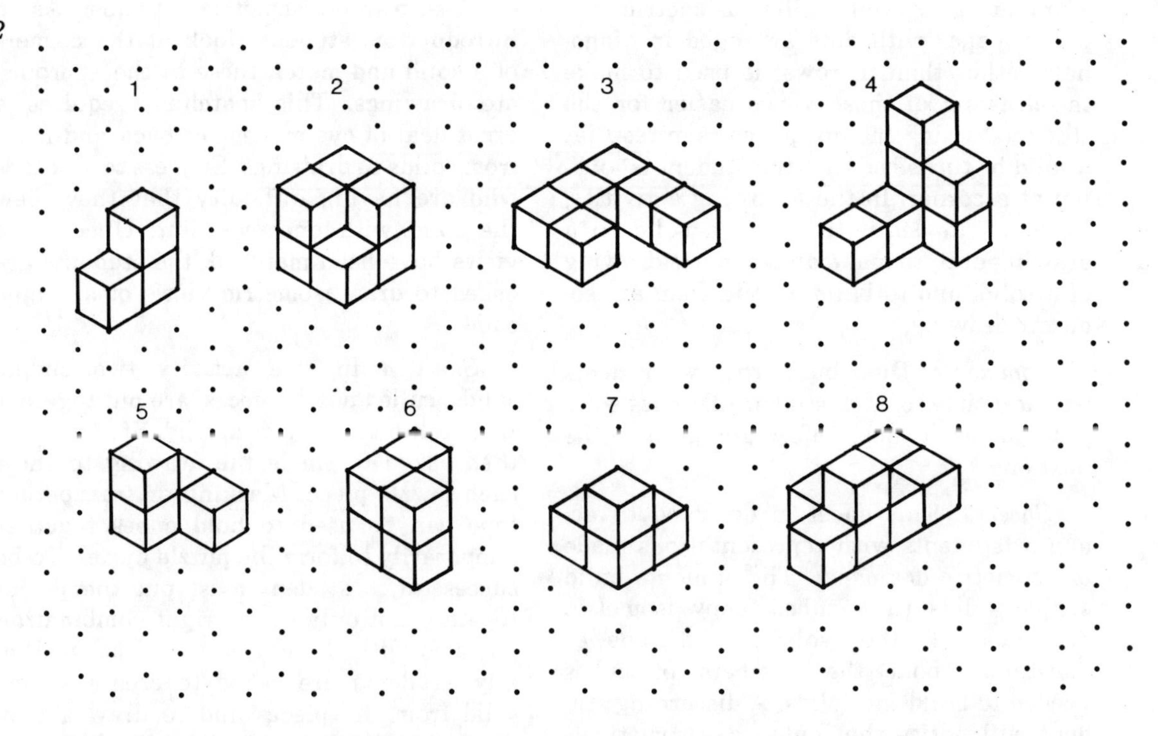

Sheet 3

Corner C; Corner B; Corner A; Corner D

Sheet 4

A.

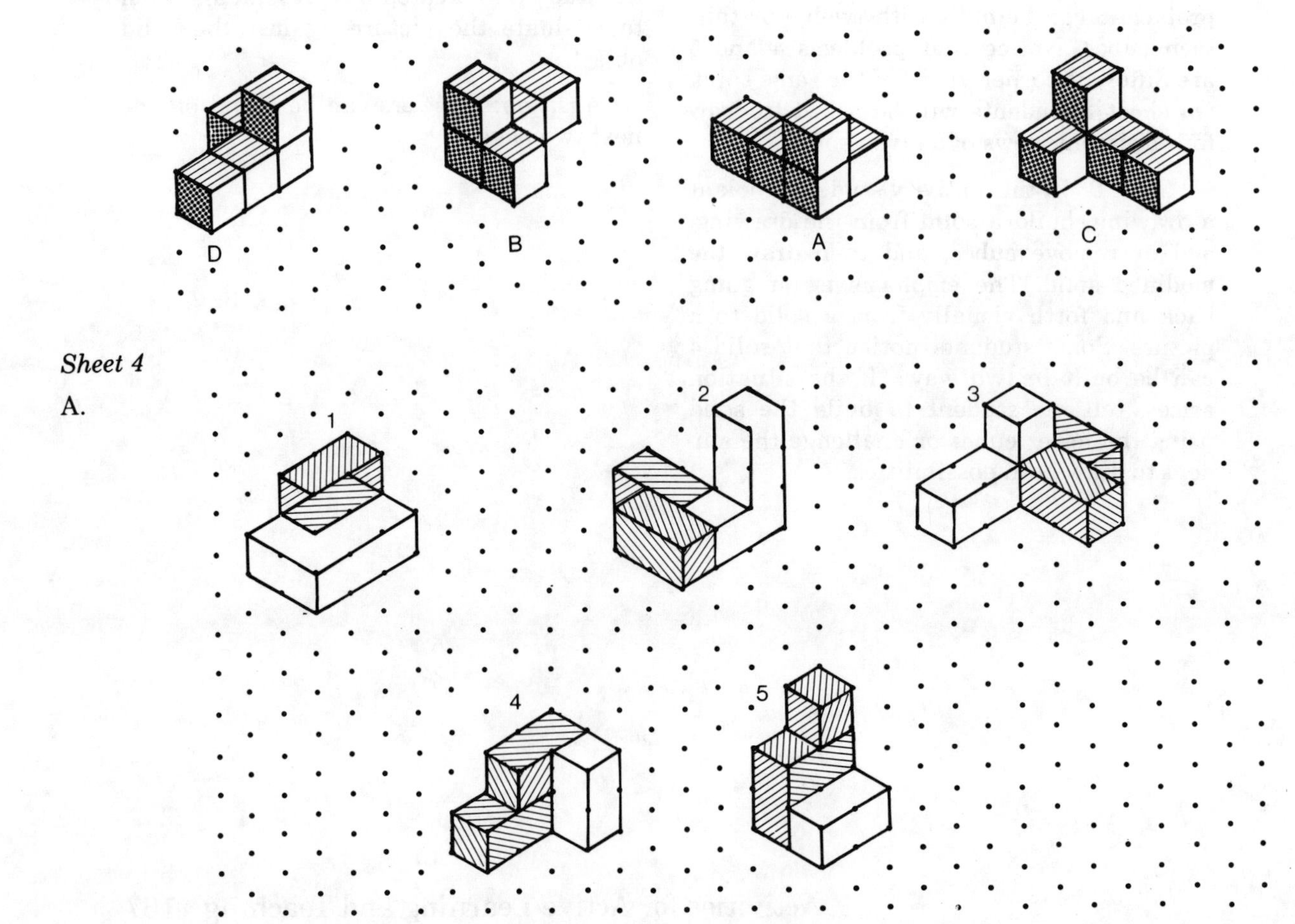

B.

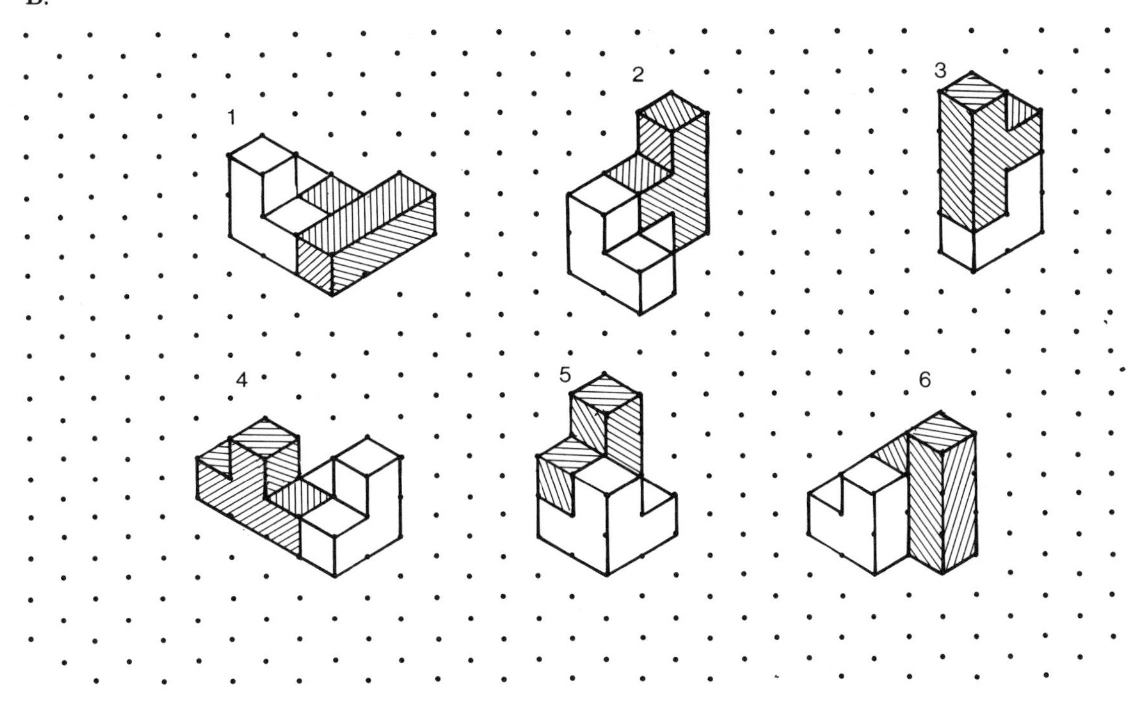

These activities are based on a unit entitled *Spatial Visualization*, developed as part of the Middle Grades Mathematics Project, a curriculum development project supported by funds from the National Science Foundation. For further information contact Glenda Lappan.

For each solid shown, do the following:

- Build the solid from cubes.
- Copy the drawing.
- Count the number of cubes used in the drawing.
- Check your count from the solid.

1

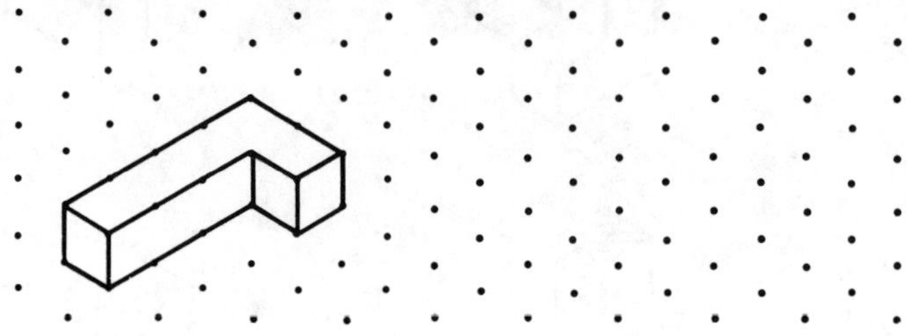

Number of cubes ______

2

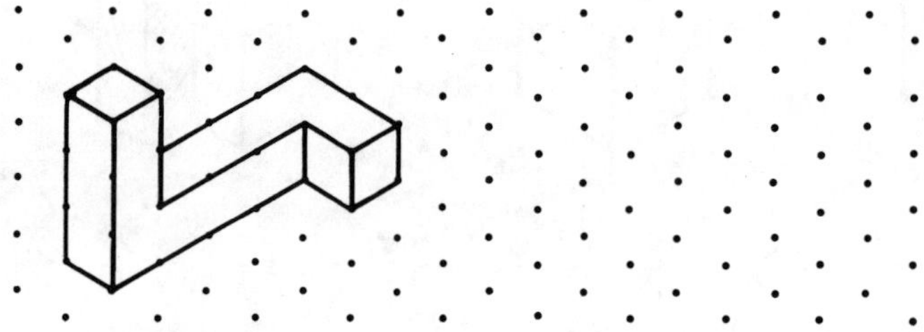

Number of cubes ______

3

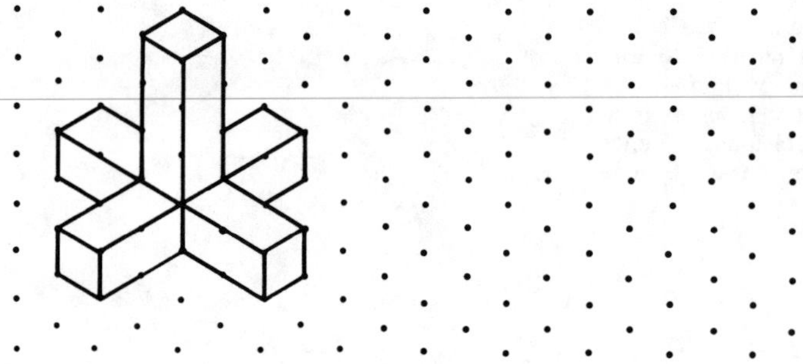

Number of cubes ______

4

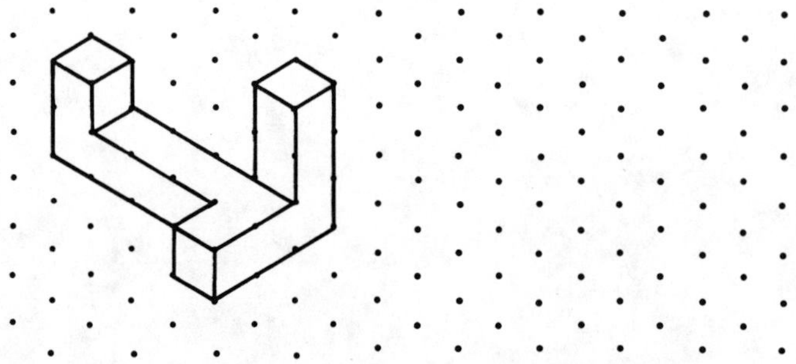

Number of cubes ______

5

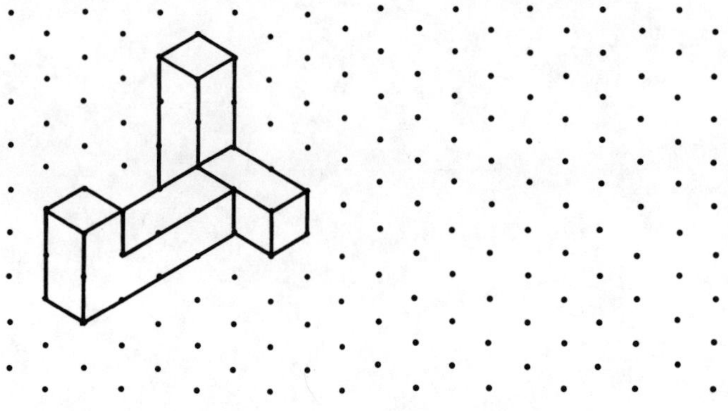

Number of cubes ______

For each of the four solids 1–4, do the following:

- Build the solid.
- Take away the shaded cube or cubes and then draw the remaining solid.

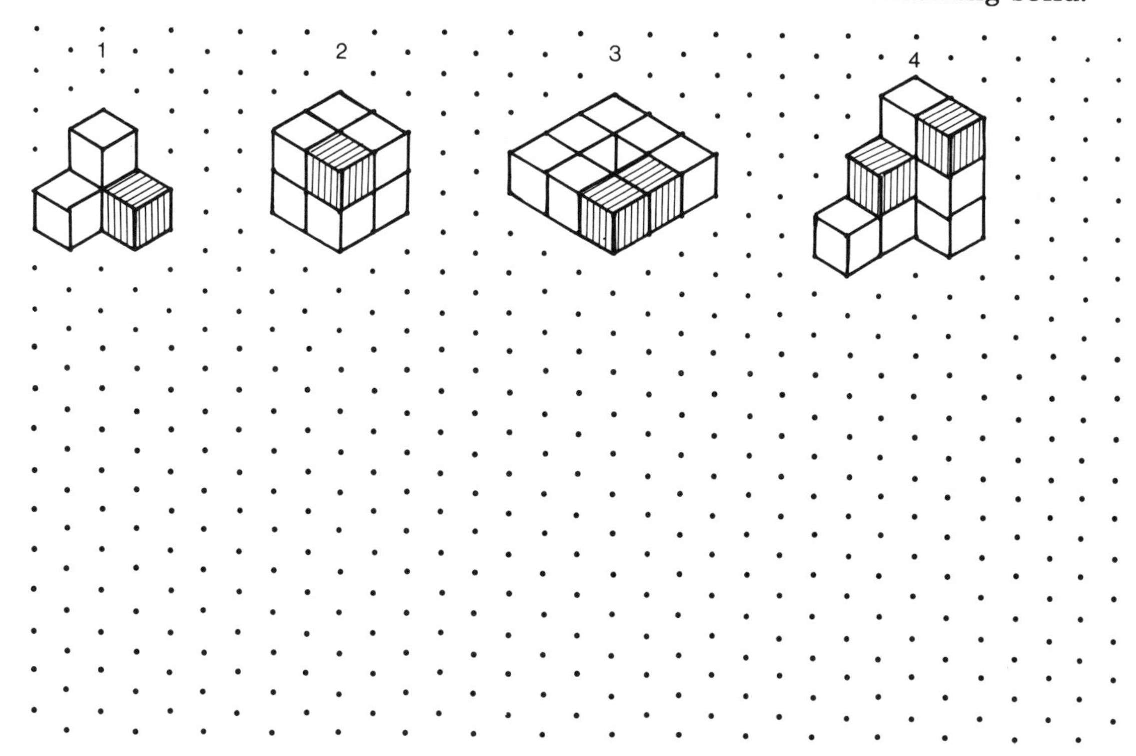

For each of the solids 5–8, do these steps:

- Build the solid.
- Add a cube to each shaded face and then draw the new solid.

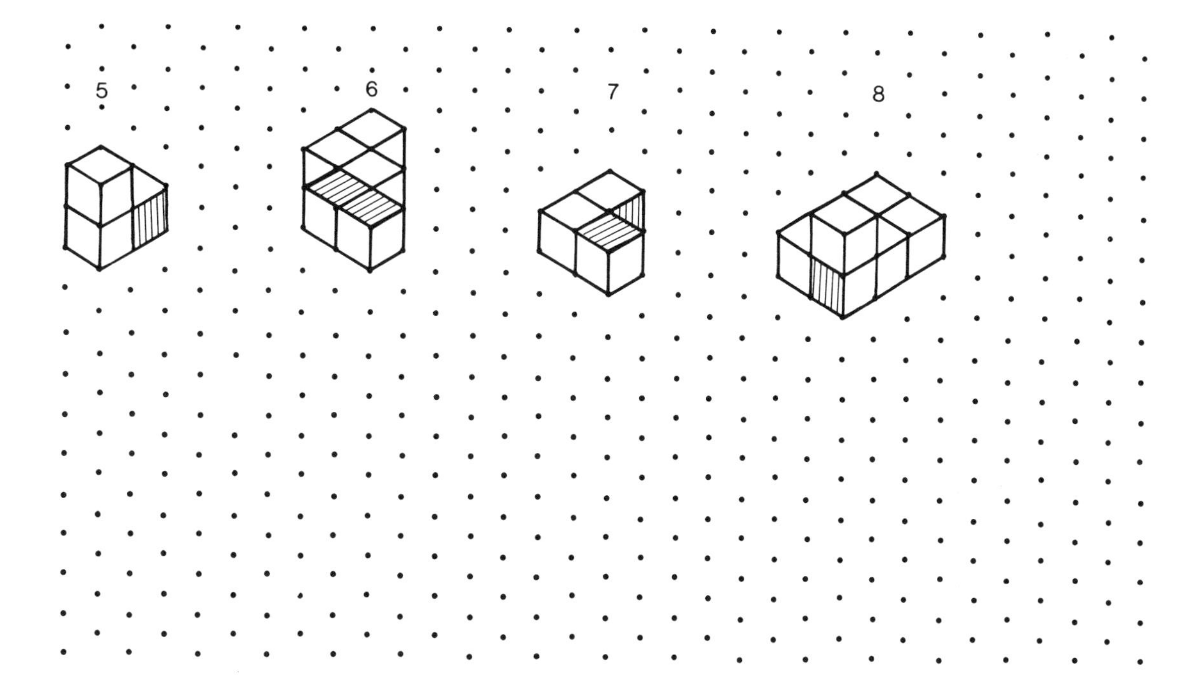

Build a solid on a piece of paper using the following plan:

- Label the corners of the paper *A*, *B*, *C*, and *D*.
- Position the paper as shown, with corners *A* and *B* at the bottom.
- Build the solid using cubes. The numbers tell you how high each stack of cubes should be.

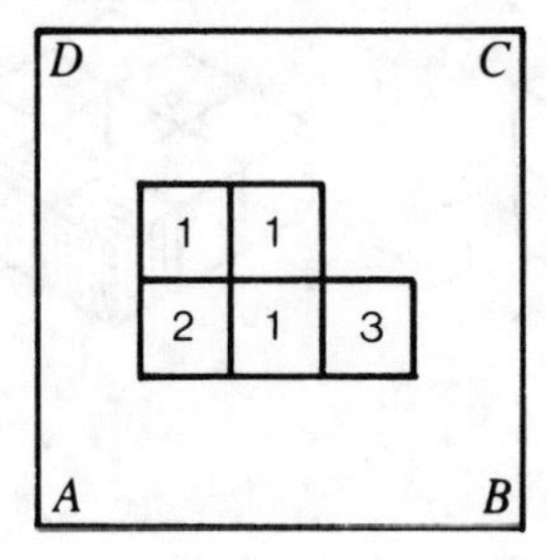

The following drawings represent the four corner views of the solid you built. Turn the paper on which your solid is built and look at the solid from each corner. Match the letter of each corner to the appropriate drawing.

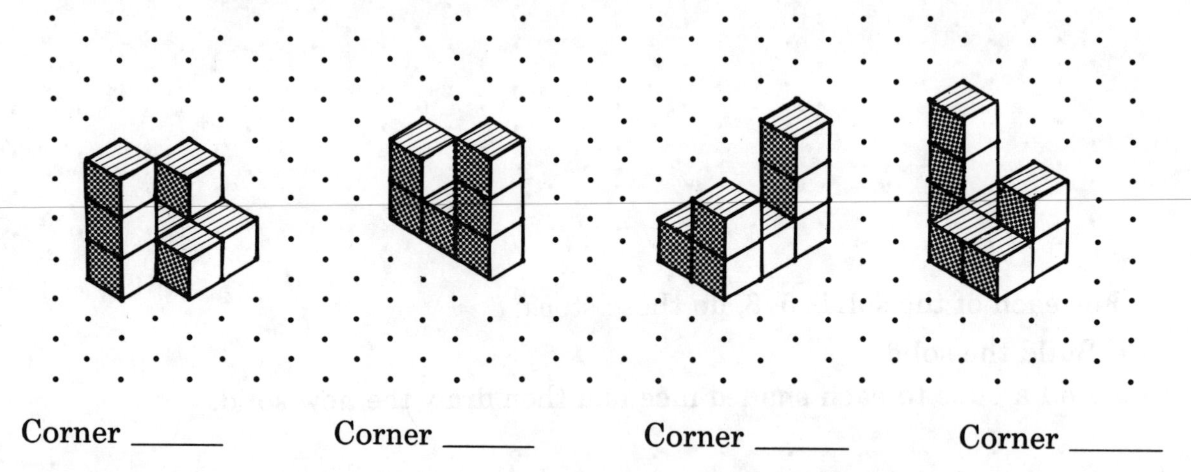

Corner ______ Corner ______ Corner ______ Corner ______

Build the solid shown below and draw views of it from two different corners. For each drawing, indicate the letter of the corner that it represents.

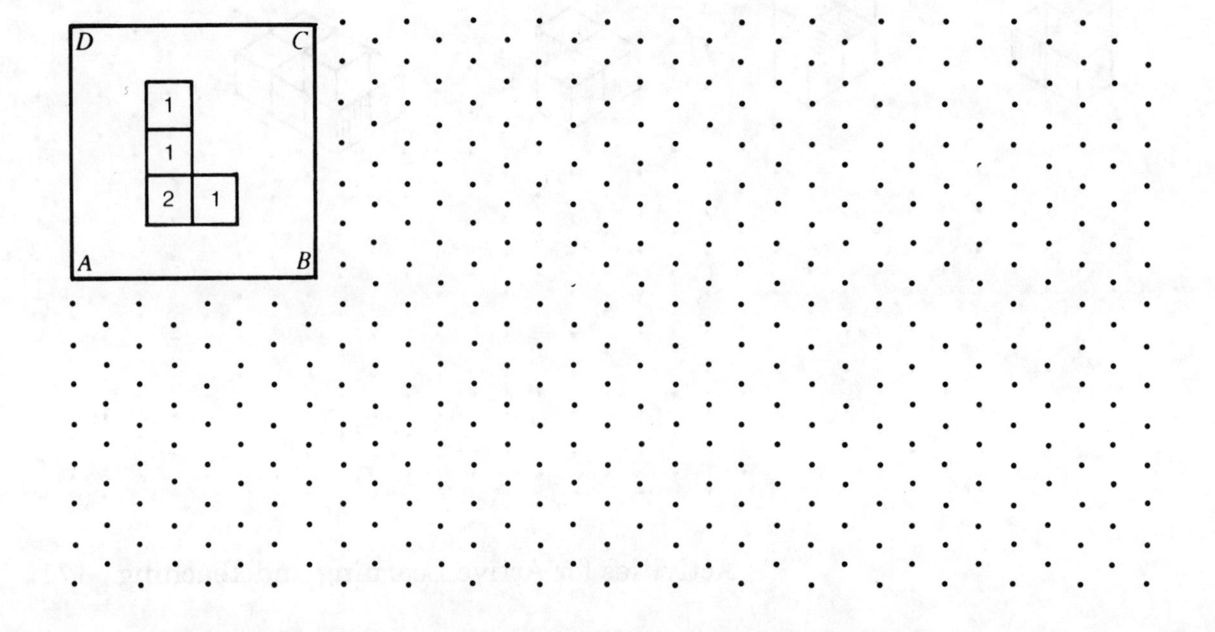

For parts A and B, construct the puzzle pieces indicated from cubes attached with tape.

- Use the two puzzle pieces to build each solid.
- Show how you built them by shading one puzzle piece in each drawing.

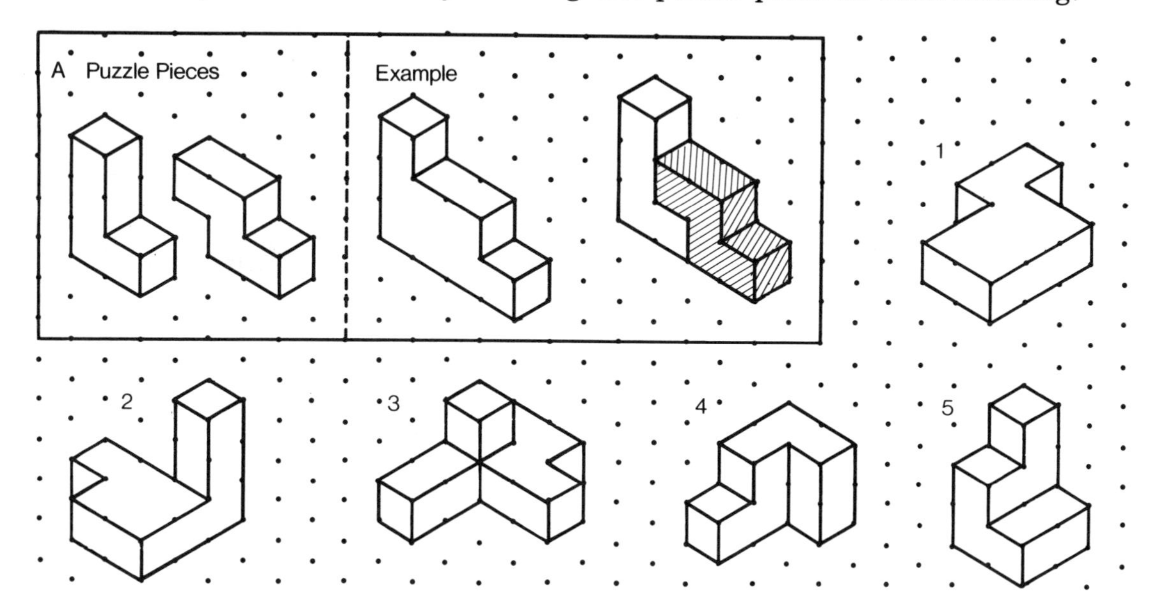

Find a different way to make a solid from the two puzzle pieces. Draw your solid here. Ask a friend to solve your puzzle. Build and draw a different solid.

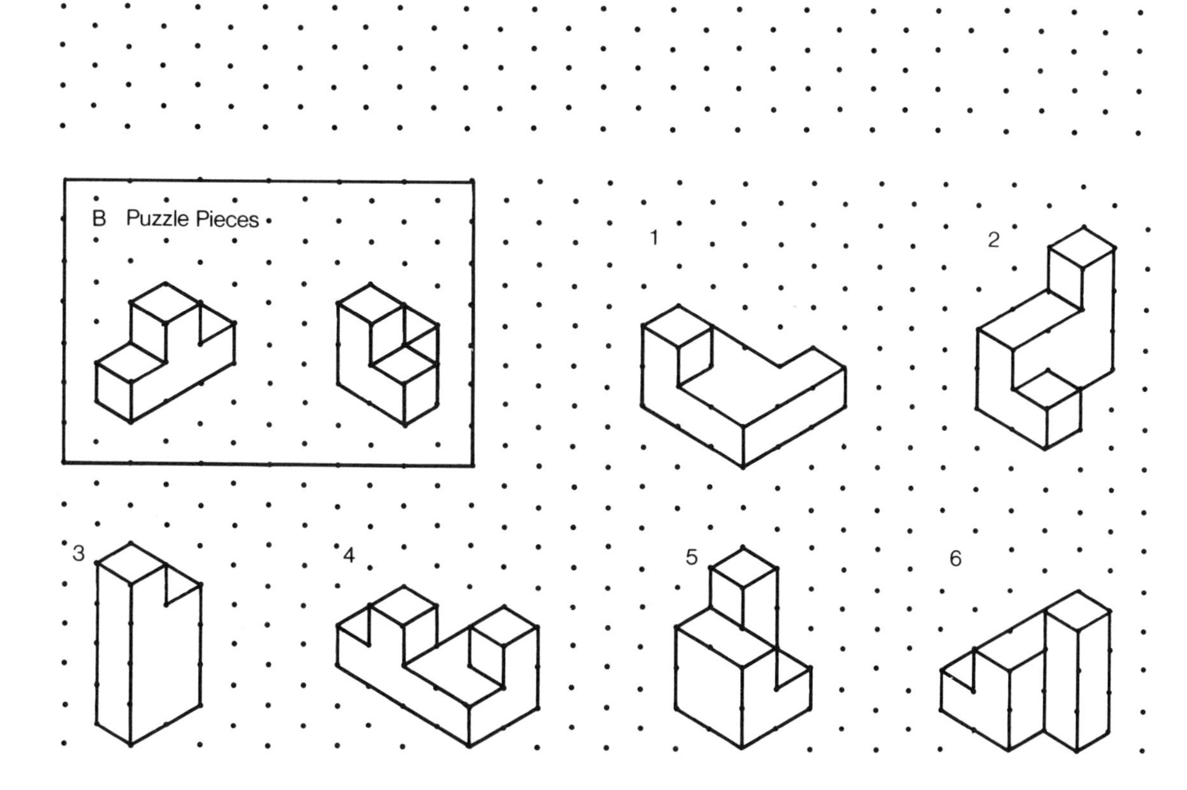

ACTIVITIES

SEMIREGULAR POLYHEDRA

October 1982

By Rick N. Blake and Charles Verhille, University of New Brunswick, Fredericton, NB E3B 6E3

Teacher's Guide

Grade level: 7–12

Objectives:

1. To provide students with the opportunity to count deductively, to collect data, and to look for patterns using polyhedra
2. To check and modify, if necessary, a given pattern after obtaining further data
3. To establish patterns and test their validity

Materials: One set of worksheets for each student; models of the five regular polyhedra for each student or for small groups of students. Models of the truncated regular polyhedra should be available for classroom demonstrations and for some students who have difficulty visualizing the faces of truncated polyhedra. Templates for those models can be found in Wenninger (1975).

Prerequisites: Students should have had some experience with three-dimensional objects. These include the following:

1. Using models of the regular polyhedra to—
   a) count faces, vertices, and edges
   b) discover Euler's formula, $F + V = E + 2$
2. Being aware that the faces of regular polyhedra are congruent
3. Finding cross sections of three-dimensional figures

Procedure: Prior to distributing the worksheets, use a model of a hexahedron to identify with the class the faces, vertices, edges, and shapes of faces. Discuss truncating the hexahedron and then use a truncated model with the class to identify the vertices, edges, and the shapes of the faces.

Distribute all three worksheets to each student. Models of the regular polyhedra should be available to individual students or to small groups of students. Students should be encouraged to use the models to help them complete the worksheets.

Copies of the completed worksheets (except questions 10, 12, 14, and 15) could be posted about the room to aid students in checking question 1 as well as data they collect when completing the rest of the sheets. It is important that the information for problem 1 be correct before proceeding.

Allow students sufficient time to find the pattern in problem 10. If students cannot find the patterns in problems 10, 12, and 14, the teacher could work through the reasoning in problem 10 and have students do problems 12 and 14.

Extension:

1. Star polyhedra can be formed by placing pyramids on the faces of the regular polyhedra. Thinking strategies similar to the ones used in this activity can be used to find their faces, vertices, and edges. See Wenninger (1975) for templates and models.

2. Some patterns involving faces, vertices, and edges can be found by truncating regular prisms and pyramids. These can be generalized to prisms and pyramids with n-gonal bases.

Solutions:

1. See table. $F + V = E + 2$ (Euler's formula).
3. *a*) 6, square; *b*) becomes an octagon; *c*) 8, truncating the vertices, triangle; *d*) $6 + 8 = 14$.
4. *a*) 8; *b*) zero; *c*) 3; *d*) $8(3) = 24$.
5. *a*) 12; *b*) 12; *c*) 3; *d*) $12 + 8(3) = 36$.
6. *a*) 4, triangle; *b*) becomes a hexagon; *c*) 4, truncating the vertices, triangle; *d*) $4 + 4 = 8$.
7. *a*) 4; *b*) zero; *c*) 3; *d*) $4(3) = 12$.
8. *a*) 6; *b*) 6; *c*) 3; *d*) $6 + 4(3) = 18$.
9. *a*) decagons and triangles; *b*) $12 + 20 = 32$.
10. Number of faces plus number of vertices on a regular polyhedron equals the number of faces on the truncated polyhedron.
11. $20(3) = 60$.
12. (Number of vertices on a regular polyhedron) × (number of vertices per truncated vertex)
13. $30 + 20(3) = 90$.
14. (Number of edges of a regular polyhedron) + (number of vertices of the regular polyhedron) × (number of vertices per truncated vertex)
15. *a*) $F + V = E + 2$; *b*) same.
16.

Polyhedron	*F*	*V*	*E*	*F*	*V*	*E*
Tetrahedron	4	4	6	8	12	18
Hexahedron	6	8	12	14	24	36
Dodecahedron	12	20	30	32	60	90
Octahedron	8	6	12	14	24	36
Icosahedron	20	12	30	32	60	90

17. Yes

BIBLIOGRAPHY

Butler, Ruth, and Robert W. Clark. "Faces of a Cube." *Mathematics Teacher* 72 (March 1979): 199–202.

Davis, Edward J., and Don Thompson. "Sectioning a Regular Tetrahedron." *Mathematics Teacher* 73 (February 1980):121–25.

O'Daffer, Phares, and S. Clements. *Geometry: An Investigative Approach.* Menlo Park, Calif.: Addison-Wesley, 1976.

Wenninger, Magnus. *Polyhedron Models for the Classroom.* 2d ed. Reston, Va.: National Council of Teachers of Mathematics, 1975.

1. *a*. Use your models of regular polyhedra to complete the first half of the table at the bottom of sheet 3.

   b. Write a formula relating the numbers F, V, and E. ______________ (Check your results before proceeding.)

2. If you cut off the vertices of a polyhedron, you obtain a *truncated polyhedron*. Let the cuts be made approximately one-third the distance from each vertex.

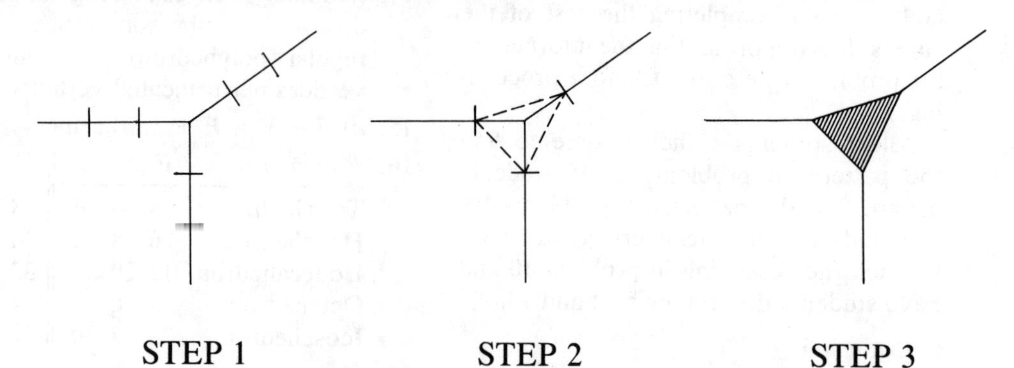

STEP 1 STEP 2 STEP 3

3. *a*. How many faces are on a hexahedron? ____

   What is the shape of each face? __________

   b. What happens to the shape of the face of the hexahedron when it is truncated?

   ____

   c. When you truncate a hexahedron, how many additional faces do you get? ________

   Where do these faces come from?

   ____

   What is their shape? ______________

   d. What is the total number of faces on a truncated hexahedron? ________

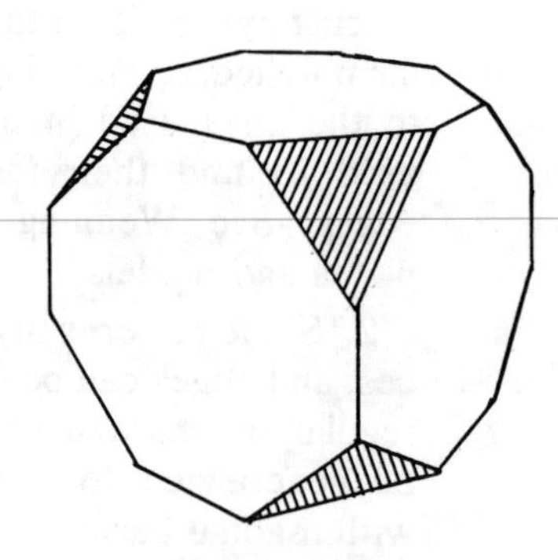

TRUNCATED HEXAHEDRON

   e. Record your information in the table on sheet 3.

4. *a*. How many vertices are on a hexahedron? ______________

   b. How many of these are on a truncated hexahedron? ______________

   c. How many vertices are formed at each additional face? ______________

   d. What is the total number of vertices on a truncated hexahedron? __________

   e. Record your information in the table on sheet 3.

5. *a*. How many edges are on a hexahedron? ______________

   b. How many of these are on a truncated hexahedron? ______________

   c. How many edges are formed with each additional face? ______________

   d. What is the total number of edges on a truncated hexahedron? __________

   e. Enter your data in the table on sheet 3.

6. *a*. How many faces are on a tetrahedron? ____
What is the shape of each face? ____
b. What happens to the shape of the face of a tetrahedron when it is truncated?
____
c. When you truncate a tetrahedron, how many additional faces do you get? ____
Where do these faces come from? ____
What is their shape? ____
d. What is the total number of faces on a truncated tetrahedron? ____
e. Record your information in the table on sheet 3.

TRUNCATED TETRAHEDRON

7. *a*. How many vertices are on a tetrahedron? ____
b. How many of these are on a truncated tetrahedron? ____
c. How many vertices are formed with each additional face? ____
d. What is the total number of vertices on a truncated tetrahedron? ____
e. Record your information in the table on sheet 3.
8. *a*. How many edges are on a tetrahedron? ____
b. How many of these are on a truncated tetrahedron? ____
c. How many edges are formed with each additional face? ____
d. What is the total number of edges on a truncated tetrahedron? ____
e. Record your information in the table on sheet 3.
9. Use the reasoning in problems 3 and 6 to determine the following:
a. What is the shape of each face of a truncated dodecahedron? ____
b. How many faces does it have? ____
c. Enter this information in the table on sheet 3.

TRUNCATED DODECAHEDRON

10. What is a rule for finding the number of faces when you truncate regular polyhedra?
____
____
11. Use the reasoning in problems 4 and 7 to determine the number of vertices of a truncated dodecahedron and enter this information in the table on sheet 3.
12. What is a rule for finding the number of vertices of truncated regular polyhedra?
____

13. Use the reasoning in problems 5 and 8 to determine the number of edges of a truncated dodecahedron and enter this information in the table on sheet 3.

14. What is a rule for finding the number of edges of truncated regular polyhedra? ____
____

15. *a*. What is the relationship among the number of faces, vertices, and edges for a truncated polyhedron? ____________________

    b. How does it compare to the relationship you found in problem 1? ______
____

16. Determine and record the number of faces, vertices, and edges for a truncated octahedron and icosahedron.

17. Does the relationship from problem 15 still hold? ____________________

18. Test your rules from problems 10, 12, and 14 for both the truncated octahedron and icosahedron.

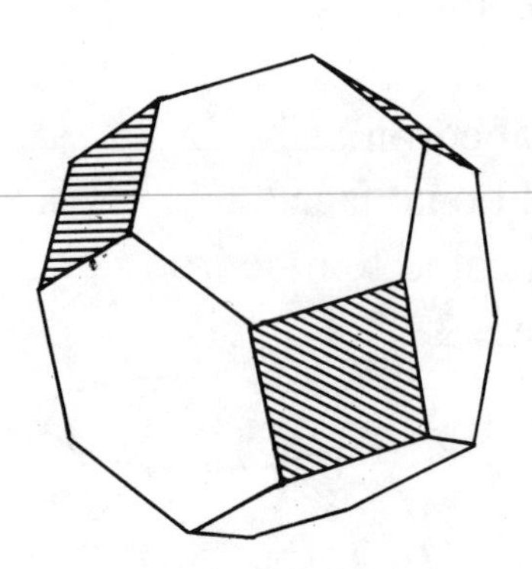

TRUNCATED OCTAHEDRON

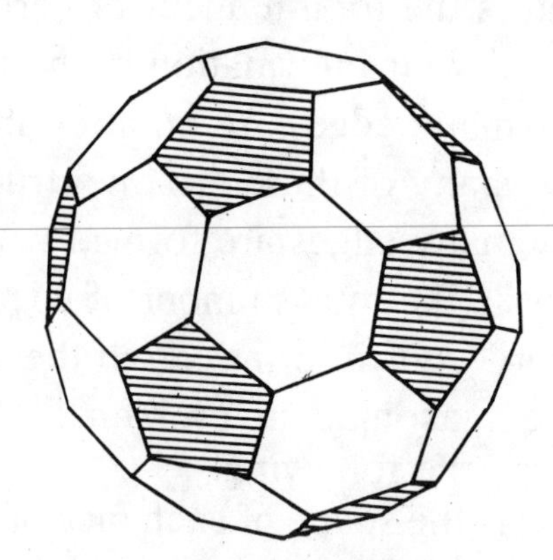

TRUNCATED ICOSAHEDRON

Polyhedron	Regular				Truncated			
	Faces F	Vertices V	Edges E	Shapes of Faces	Faces F	Vertices V	Edges E	Shapes of Faces
Tetrahedron								
Hexahedron								
Dodecahedron								
Octahedron								
Icosahedron								

ACTIVITIES

VISUALIZATION, ESTIMATION, COMPUTATION

December 1982

By Evan M. Maletsky, Montclair State College, Upper Montclair, NJ 07043

Teacher's Guide

Grade Level: 7–12

Materials: Scissors, paper clips, and copies of the worksheets. Access to a microcomputer would be helpful for the last worksheet, but it is not necessary.

Objectives: To construct a model of a cone; to visualize how the radius, height, area, and volume change as the cone changes shape; to estimate the position for maximum volume; and to develop problem-solving skills by measurement, computation, graphing, and analysis of data.

Directions: Distribute sheet 1 and have every student make the model. Follow this with sheet 2, sheets 2 and 3, or sheets 2, 3, and 4, depending on the level and ability of the class.

Sheet 1. Once the circular piece of paper is cut out and the radius cut, watch to see that each student forms the cone by placing the edge marked "top" over the other side lettered A, B, C, ...

Sheet 2. This activity focuses on the changing dimensions of the model as the paper cone is curled up and opened. Encourage each student to visualize how the radius and height change. Note that only the lateral area of the cone is formed by the paper. The circular base is located by, but not actually part of, the paper model. Be sure that each student moves the model through all possible positions in studying the changing volume, as it first will increase but then begin to decrease. This is most obvious when one notes the extreme positions of the cone.

Sheet 3. Have students compare the positions they chose for maximum volume with the choices of others in the class. The reader may want to make a frequency distribution of the numbers choosing the various lettered positions.

The model has been marked so that the diameters for the lettered positions are in centimeters and the radii in 0.5 centimeters. Be sure the heights are also recorded in 0.5 centimeters so that the volume will be in cubic centimeters.

Plot enough points on the graph that some trend in the changing volume can be observed. This will help give a better collective estimate of the maximum position. If you do not use sheet 4, be sure to note here that the best choice of letters for the maximum volume occurs at *C*, radius 6.5 cm. This is certain to surprise most students.

Sheet 4. The major goal here is to have students interpret the program in BASIC and analyze the printed output but not to write a program. Access to a microcomputer is not necessary. However, the suggestion at the end would be interesting to follow with those who have access to, and are familiar with, a microcomputer. Having the radius move from 6.0 to 7.0 cm by increments of 0.1 cm will give these results. The program is written for the Apple but can be easily modified for other computers.

RADIUS	HEIGHT	VOLUME
6	5.29	199.49
6.1	5.18	201.69
6.2	5.06	203.51
6.3	4.93	204.93
6.4	4.8	205.89
6.5	4.66	206.34
6.6	4.52	206.23
6.7	4.37	205.5
6.8	4.21	204.07
6.9	4.05	201.84
7	3.87	198.73

Answers:

Sheet 2

1. Decrease
2. Remain fixed at 8 cm
3. 8 cm
4. 8 cm
5. 0 cm
6. 8 cm

(4–6: These are limiting, degenerate cases. Accept answers near these values.)

7. increases; increases
8. Yes; no. The total area approaches twice the area of the original 8-cm circle.
9. The radius first increases and then decreases.

Sheet 3

Choices will vary. The correct values for each letter, however, can be found in the printout on sheet 4.

Sheet 4

1. 30 (and 100)
2. 50
3. 60
4. 70 and 80 (using the greatest integer function)
5. 206.34 cm^3 at letter *C*

One possible modification of the program as suggested would be to change line 30 to

```
30 FOR R = 6 TO 7 STEP 0.1
```

The letter identification would then have to be dropped by deleting lines 40 and 110 and dropping the string variable X$ from line 90.

Modifications and Extensions: In certain classes you might want to fill the variously shaped cones with sand or rice and compare their volumes by that method.

For classes familiar with programming microcomputers, you might want to have them write and run their own programs rather than use sheet 4.

The exact maximum volume possible for a cone with a slant height of 8 cm can be found using calculus. The greatest volume is 206.37 cm^3, and it occurs at a radius of 6.53 cm.

For this activity you'll need a pair of scissors and a paper clip.

Step 1: Cut out the circle below. Then cut through to the center on the 8-cm radius line.

Step 2: Place the side of the cut line on top of the lettered surface of the circle.

Step 3: Move the edge of the cut line to one of the lettered positions and use a paper clip to hold it in place.

The circle now forms the lateral surface of a cone.

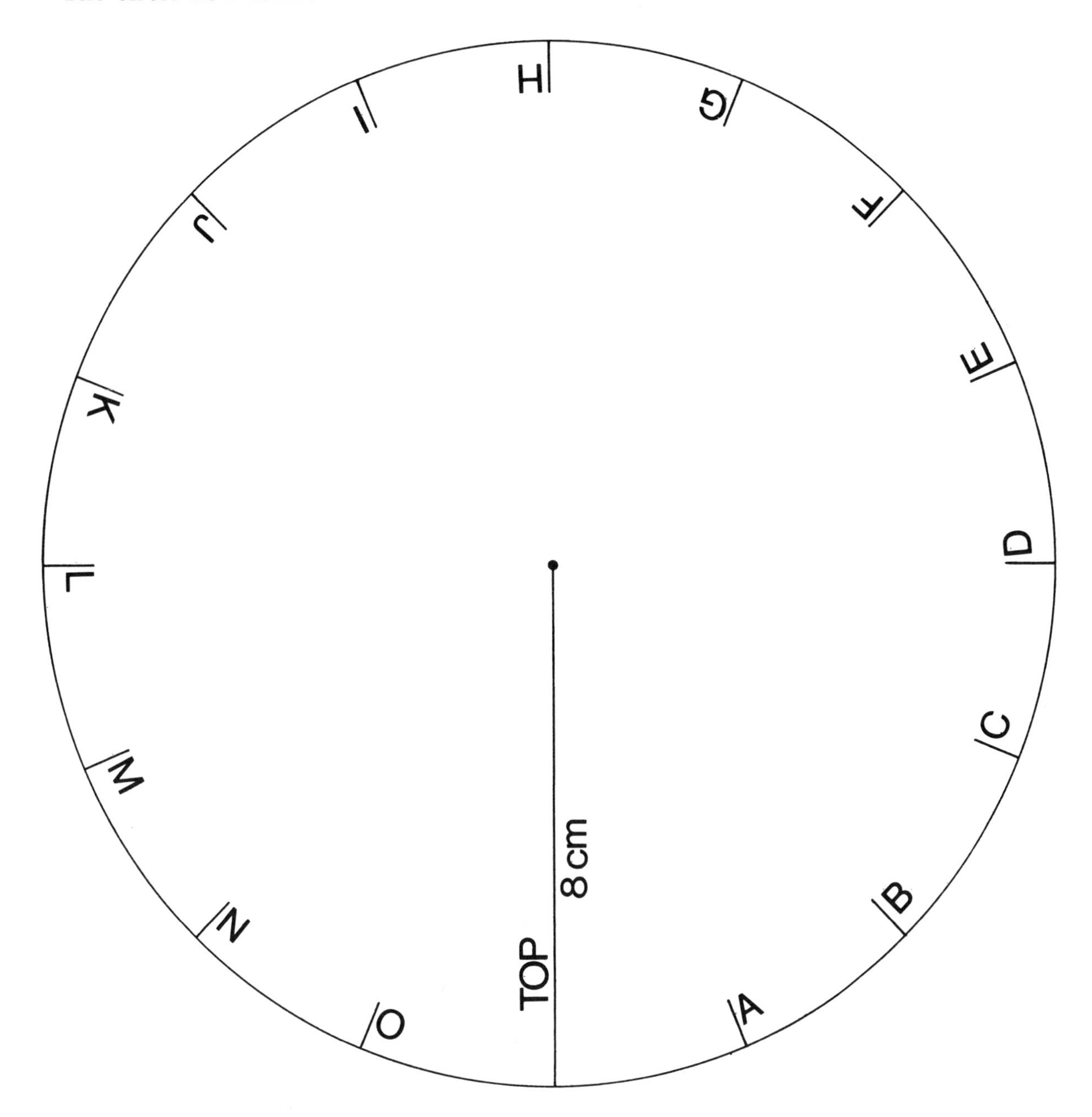

A cone has three important linear dimensions. They are the *radius*, *height*, and *slant height*.

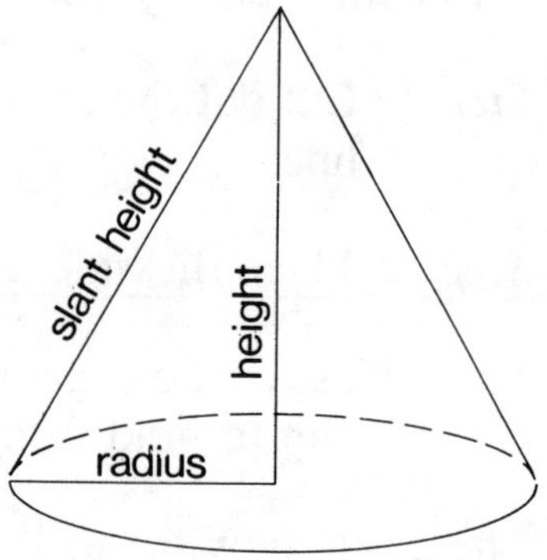

Slowly move the cut edge from letter *A* to *B* to *C*, and so on. The height of the cone will increase.

1. Does the radius increase, decrease, or remain the same? ____________
2. What does the slant height do? ____________
3. What is the longest radius possible? ____________
4. What is the greatest height possible? ____________

This time open the model of the cone from a tightly overlapped position by moving the edge back toward the letters *C*, *B*, and *A*. The height of the cone will decrease.

5. At what height is the lateral area the greatest? ____________
6. At what height is the area of the base the least? ____________
7. As the height of the cone decreases, the lateral area increases.

   What happens to the area of the base? ____________

   What happens to the total area? ____________
8. As the cone is opened, the area of the base approaches the area of the original 8-cm circle.

   Does the lateral area approach the same value? ____________

   Does the total area approach the same value? ____________

Make the paper cone tighter again, moving the edge from letter *A* to *B* to *C*, and so on. The height will increase and the total area will decrease.

9. Describe in your own words what appears to happen to the volume.

____

____

As the paper model moves to narrower and wider positions, the volume of the cone formed changes.

1. Take your model and set it at the lettered position that you think forms the cone with the greatest volume. Paper clip it in place. LETTER __________

2. Measure, to the nearest 0.5 centimeters, the diameter of the cone in that position. Divide by 2 to find the radius, r. RADIUS __________

3. Carefully measure to the nearest 0.5 centimeters the height, h, of the cone in that same position. HEIGHT __________

4. Substitute your values in the formula to find the volume for that position. Use 3.14 for π. VOLUME __________

$$V = \frac{1}{3}\pi r^2 h$$

5. Plot your radius and volume on the same axes below. Then plot the values for some other radii. Use results others in the class have recorded or compute more volumes yourself. Do the different results help you better estimate the position for the maximum volume?

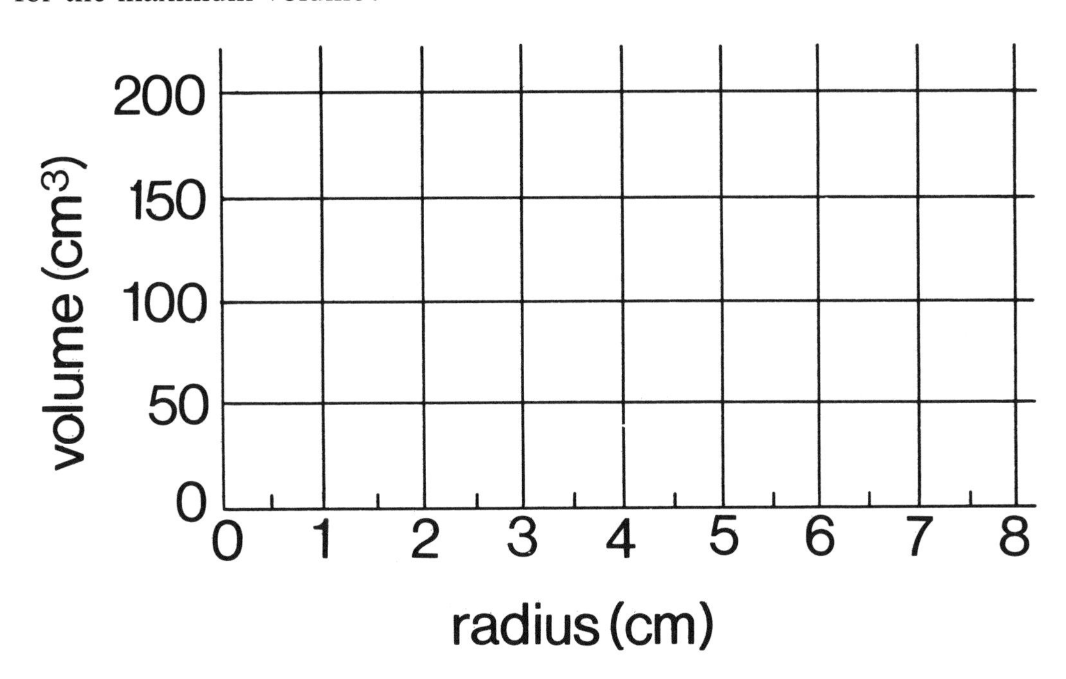

A microcomputer can be especially helpful in solving problems of this type. With just one program, it can compute many different volumes very rapidly. Here is one such program written in BASIC.

```
10  PRINT "      RADIUS","HEIGHT","VOLUME"
20  PRINT
30  FOR R = 0 TO 8 STEP .5
40  READ X$
50  LET H =  SQR (64 - R * R)
60  LET V = 3.1416 * R * R * H / 3
70  LET H =  INT (H * 100 + .5) / 100
80  LET V =  INT (V * 100 + .5) / 100
90  PRINT X$;"    ";R,H,V
100  NEXT R
110  DATA  -,O,N,M,L,K,J,I,H,G,F,E,D,C,B,A,-
120  END
```

Give the number of the line(s) that—

1. selects values of the radius R from 0 to 8 by increments of 0.5 cm, ________
2. computes the height H using the Pythagorean theorem, ________
3. computes the volume V using the formula $V = 1/3\ \pi R^2 H$, ________
4. rounds the values of H and V to two decimal places. ________

Here is the printout when the program is run.

	RADIUS	HEIGHT	VOLUME
-	0	8	0
O	.5	7.98	2.09
N	1	7.94	8.31
M	1.5	7.86	18.52
L	2	7.75	32.45
K	2.5	7.6	49.74
J	3	7.42	69.9
I	3.5	7.19	92.28
H	4	6.93	116.08
G	4.5	6.61	140.26
F	5	6.24	163.49
E	5.5	5.81	184.03
D	6	5.29	199.49
C	6.5	4.66	206.34
B	7	3.87	198.73
A	7.5	2.78	163.98
-	8	0	0

5. What is the greatest volume listed? ________ cm^3. At what letter does it occur? ________

6. Compare these results with yours from sheet 3. Then plot all the radii and volumes listed on the graph on sheet 3.

The maximum volume shown occurs between a radius of 6 and 7 cm. Try modifying the program to print the volumes for radii from 6 to 7 cm by increments of 0.1 cm, and thereby obtain a more accurate estimate of the critical radius for maximum volume.

ACTIVITIES

GENERATING SOLIDS

October 1983

By EVAN M. MALETSKY, Montclair State College, Upper Montclair, NJ 07043

Teacher's Guide

Grade level: 8–11

Materials: Worksheets for each student and several polygons cut from cardboard to appropriate dimensions for purposes of demonstration

Objectives: To visualize, identify, and describe the solids generated by rotating polygons about axes located in various positions; to relate these figures to cylinders and cones; and to compute their surface areas and volumes by first finding the radii and heights needed

Directions: Use cardboard models to illustrate how a rectangle rotated about a side and a right triangle rotated about a leg will generate a cylinder and a cone, respectively. Review key terms such as *radius, height,* and *slant height* as well as the related formulas for surface area and volume. Use the worksheets one at a time.

Sheet 1. Be sure each student properly identifies the radius and height for each figure generated. Discuss how changing the location of the axis of rotation changes the dimension of the solid formed. Note that when the axis cuts the rectangle and is parallel to a side, a cylinder is formed. The last illustration produces a figure that might best be described as a large cylinder with a smaller one having the same axis cut from its center.

Sheet 2. When the first right triangle is rotated about its legs, two different cones are formed. The radius (4) and height (3) of the first cone are interchanged to become the height (4) and radius (3) of the second cone. However, the slant height (5), found by the Pythagorean theorem, remains the same for both cases.

Encourage students to describe the results in the remaining two illustrations. If necessary, show the solids generated by rotating a cardboard cutout of the triangle. In the second illustration, two cones are formed extending in opposite directions from a common circular base. The third illustration might be described as a cylinder with a cone of the same radius and height cut away.

Sheet 3. [illegible] is needed here to see the figures generated and how they are rel[illegible]ones. Encourage students to do their own thinking on these proble[illegible] want to show each generated solid to slower classes by spinning a c[illegible] a wire taped in the appropriate place for the axis.

If visual[illegible]nphasis, you may want students to give answers simply in terms of [illegible]eet 3. For other classes, these problems can serve as valuable co[illegible]s using a calculator.

Answer[illegible] given to two decimal places using 3.14 for π. Volumes are given in cu[illegible]eas in square units.

Sheet 1:

1. a) $r = 3$, $h = 4$ b) $r = 4$, $h = 3$
2. about the x-axis $V = 113.04$
   about the y-axis $V = 150.72$
3. a) cylinder   b) cylinder
4. about the x-axis $S = 75.36$
   about the y-axis $S = 113.04$
5. A cylinder ($r = 5$, $h = 3$) with another cylinder ($r = 1$, $h = 3$) cut from its center
6. $V = 3.14 \cdot 5^2 \cdot 3 - 3.14 \cdot 1^2 \cdot 3 = 226.08$
   $S = 2 \cdot 3.14 \cdot 5 \cdot 3 + 2 \cdot 3.14 \cdot 1 \cdot 3 + 2(3.14 \cdot 5^2 - 3.14 \cdot 1^2) = 263.76$

Sheet 2:

1. a) $r = 4$, $h = 3$ b) $r = 3$, $h = 4$
2. about the x-axis $S = 87.92$
   about the y-axis $S = 65.94$
3. A cone ($r = 3$, $h = 4$) joined base to base with another cone ($r = 3$, $h = 2$)
4. $V = \frac{1}{3} \cdot 3.14 \cdot 3^2 \cdot 4 + \frac{1}{3} \cdot 3.14 \cdot 3^2 \cdot 2 = 56.52$
5. A cylinder ($r = 4$, $h = 4$) with a cone ($r = 4$, $h = 4$) cut from it
6. $V = 3.14 \cdot 4^2 \cdot 4 - \frac{1}{3} \cdot 3.14 \cdot 4^2 \cdot 4 = 133.97$

Sheet 3:

1. A cylinder ($r = 3$, $h = 2$) with two cones ($r = 3$, $h = 1$ and $r = 3$, $h = 2$) joined on its bases
   $V = \pi \cdot 3^2 \cdot 2 + \frac{1}{3} \cdot \pi \cdot 3^2 \cdot 1 + \frac{1}{3} \cdot \pi \cdot 3^2 \cdot 2 = 27\pi$
2. A cylinder ($r = 3$, $h = 3$) with a cone ($r = 1$, $h = 3$) cut down the center
   $V = \pi \cdot 3^2 \cdot 3 - \frac{1}{3} \cdot \pi \cdot 1^2 \cdot 3 = 26\pi$
3. A cylinder ($r = 2$, $h = 4$) with a cone ($r = 2$, $h = 2$) cut from it
   $V = \pi \cdot 2^2 \cdot 4 - \frac{1}{3} \cdot \pi \cdot 2^2 \cdot 2 = \frac{40}{3}\pi$
4. A cone ($r = 3$, $h = 5$) with another cone ($r = 3$, $h = 2$) cut from it
   $V = \frac{1}{3} \cdot \pi \cdot 3^2 \cdot 5 - \frac{1}{3} \cdot \pi \cdot 3^2 \cdot 2 = 9\pi$

Extensions: Better classes may want to use the Pythagorean theorem with the solids on sheet 3 to find their surface areas. Other interesting solids to explore include—

1. rotating a trapezoid about the shorter of its two parallel sides;
2. rotating a rectangle about a diagonal;
3. rotating a rhombus about a diagonal. ■

When this rectangle is rotated about the line shown, it generates a cylinder. The volume and surface area of the cylinder depend on the dimensions of the rectangle.

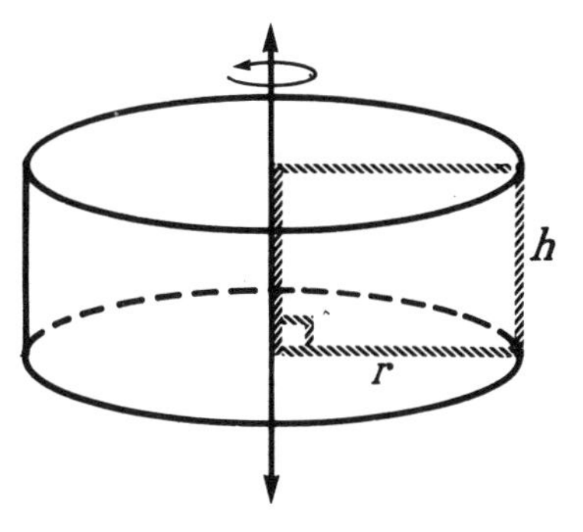

1. Find the radius and height of the cylinder formed when this rectangle is rotated about

   a) the x-axis. $r =$ ____ $h =$ ____

   b) the y-axis. $r =$ ____ $h =$ ____

2. Which cylinder will have the greater volume? Guess first and then compute and compare.

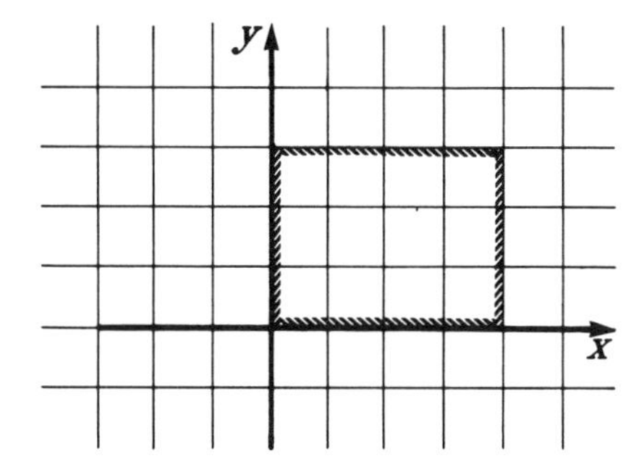

3. Name the solid formed when this rectangle is rotated about

   a) the x-axis. ____________

   b) the y-axis. ____________

4. Which solid will have the greater surface area? Guess first and then compute and compare.

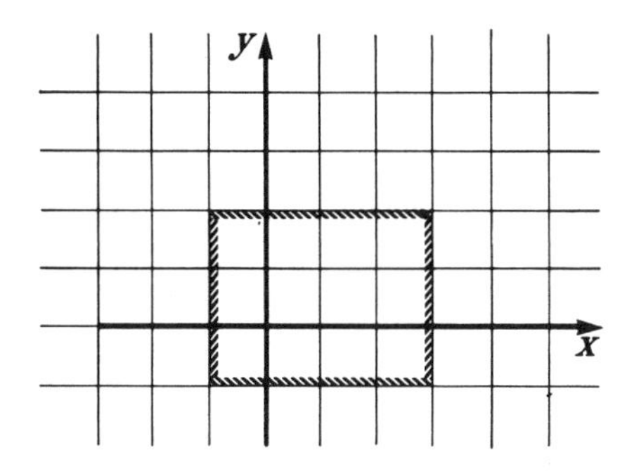

5. Describe the solid formed when this rectangle is rotated about the y-axis.

6. Find its volume and surface area.

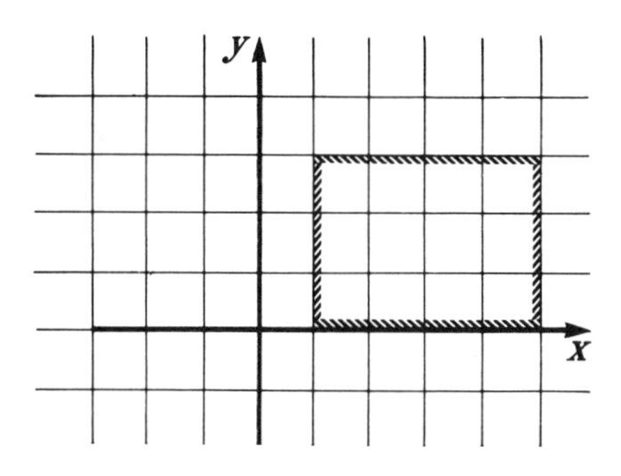

When this right triangle is rotated about the line shown, a cone is formed. The radius and height of the cone depend on the dimensions of the right triangle and the location of the axis of rotation.

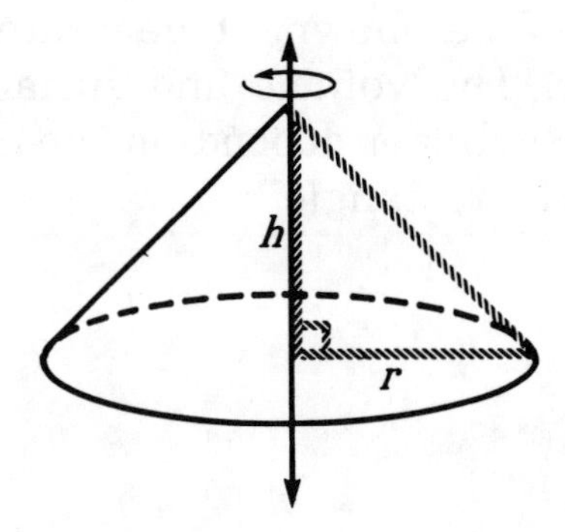

1. Find the radius and height of the cone formed when this triangle is rotated about

   a) the x-axis. $r =$ _____ $h =$ _____

   b) the y-axis. $r =$ _____ $h =$ _____

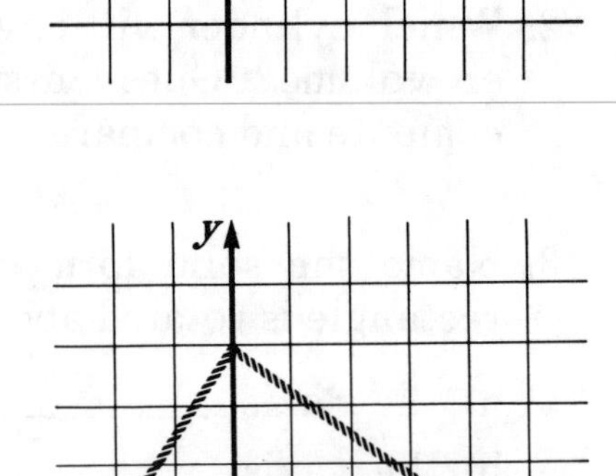

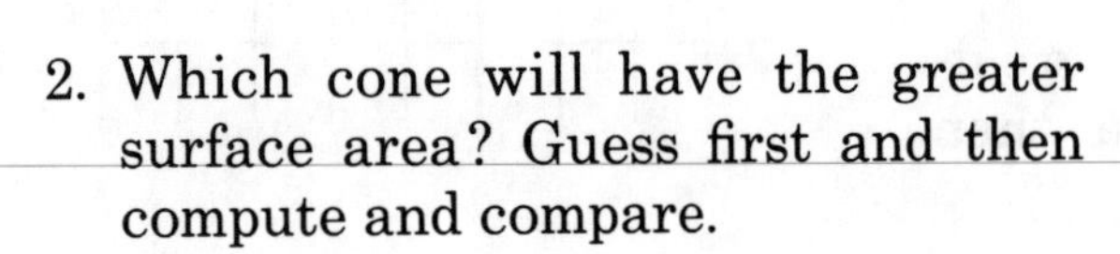

2. Which cone will have the greater surface area? Guess first and then compute and compare.

3. Describe the solid formed when this triangle is rotated about the x-axis.

4. Find the volume of the solid formed.

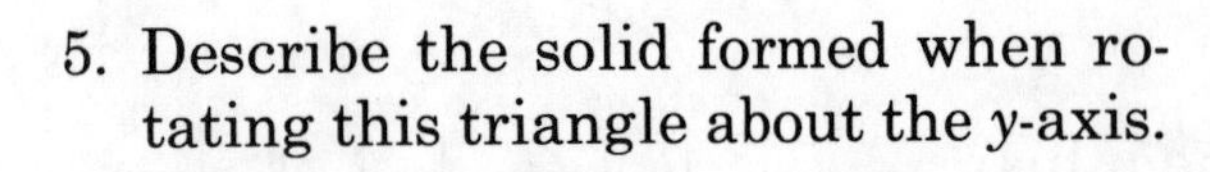

5. Describe the solid formed when rotating this triangle about the y-axis.

6. Use your knowledge of both cylinders and cones to find the volume of the solid formed.

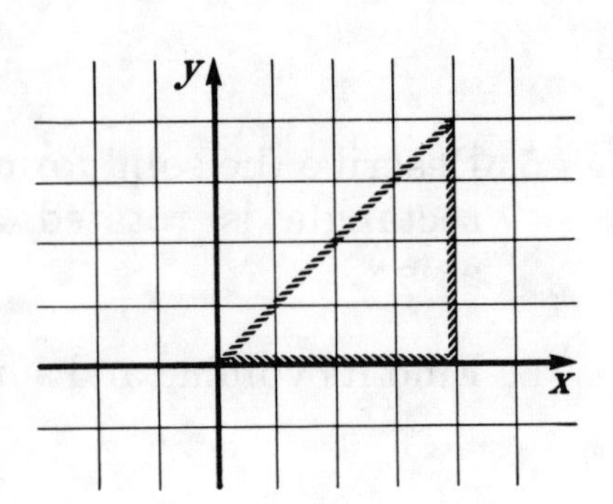

Some solids of revolution are not as easy to visualize as others. These involve parts of both cylinders and cones.

Describe the shape of the solid formed by rotating the polygon about the axis shown.

Then find the necessary dimensions and use them to compute the volume. Leave your answers in terms of π.

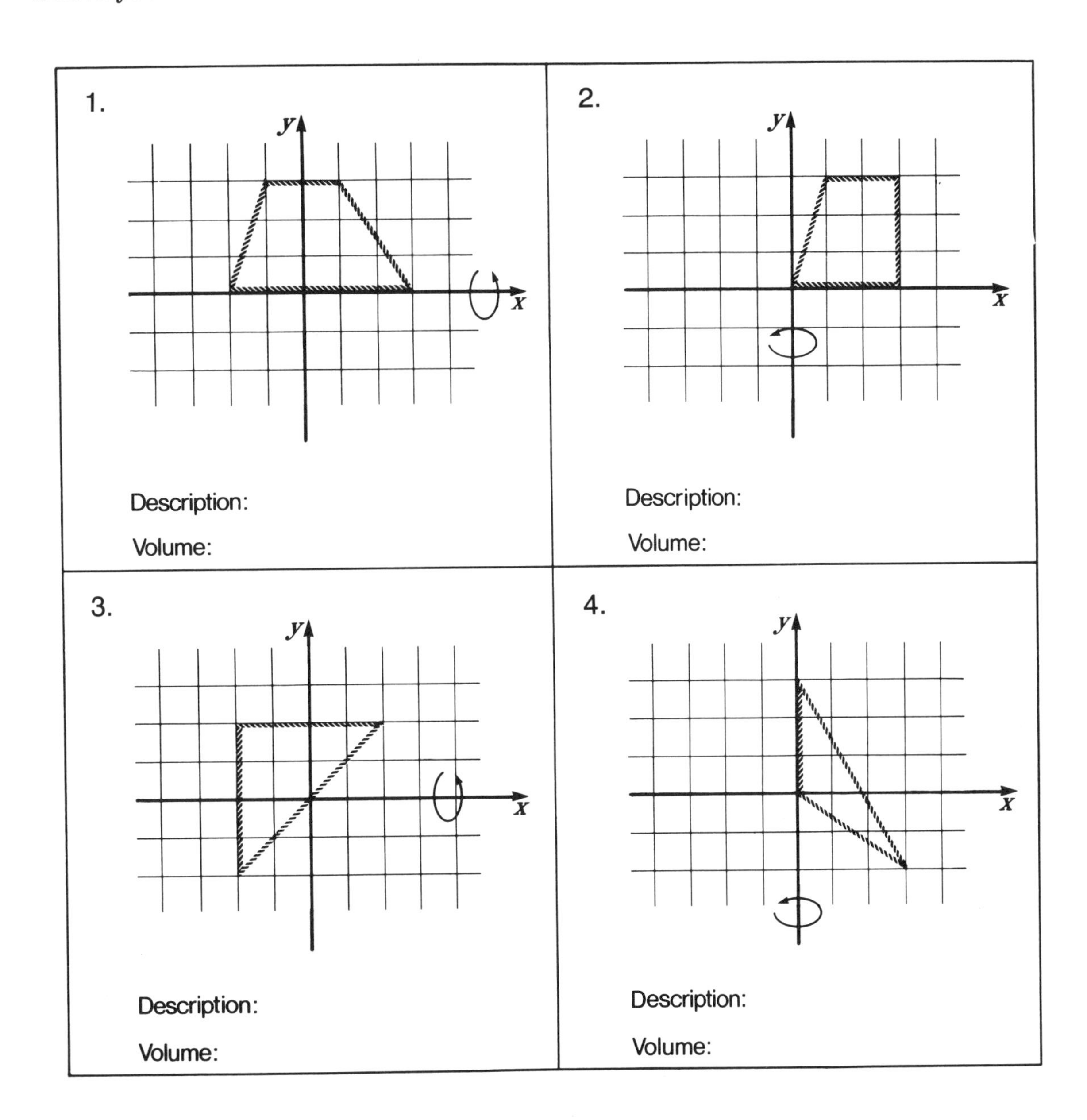

Activities for

Data Analysis and Probability

The role of data analysis and probability in school mathematics is rapidly changing both in emphasis and in methods of instruction. The elevation of these two strands to a more prominent position in the curriculum is accompanied by a greater emphasis on developing statistical thinking as opposed to simply learning formulas and procedures. The activities in this section can help shift instruction from direct teaching of isolated techniques to more student-centered experiences featuring data collection and representation, interpretation, prediction, and decision making.

The first four activities focus on the analysis of published information and data collected by students themselves. It is advisable that these activities be completed in sequence. "Stem-and-Leaf Plots" uses interesting real-world data (e.g., best pop albums) to introduce students to an effective, relatively new method for organizing and displaying information. In addition to making stem-and-leaf plots, pupils interpret their displays and communicate their findings in writing. Box-and-whisker graphs are generally more appropriate than stem-and-leaf plots when comparing several sets of data. "Exploring Data with Box Plots" introduces students to this visual display method in the context of analyzing data about automobiles. The next activity, "Plotting and Predicting from Pairs," revolves around the use of scatterplots to analyze possible relationships shown by bivariate data such as height and shoe size or the number of rebounds and assists by a basketball team. In the process of graphing bivariate data and analyzing their graphs, students are exposed to the concepts of positive relationship, negative relationship, and no relationship. These concepts are in turn applied to the analysis of time-related data. "Data Fitting without Formulas" introduces the median-fit technique as a simple method to fit a straight line to a scatterplot for purposes of aiding prediction from data. Again, the emphasis is on the application of the technique to data collected from real-world situations.

The remaining three activities in this section focus on two powerful, broadly useful methods for dealing with uncertainty—geometric probability models and simulations. In "Area Models for Probability," students use square regions to analyze probabilities associated with games of chance. Use of the area model also enhances students' understanding of fraction-related concepts and their facility with mental computations with fractions. The next activity, "Area Models and Expected Value," furnishes additional opportunities for students to practice using area models to analyze compound situations. Questions are extended to include expected value. The final activity, "Problem Solving with Simulation," engages students in using a physical model and a computer program to simulate and help them to solve a quality assurance problem. Pupils also gain additional experience in the use of stem-and-leaf and box-and-whisker plots to summarize and interpret data.

ACTIVITIES

STEM-AND-LEAF PLOTS

October 1985

By JAMES M. LANDWEHR, AT&T Bell Laboratories, Murray Hill, NJ 07974
ANN E. WATKINS, Los Angeles Pierce College, Woodland Hills, CA 91371

Teacher's Guide

Introduction: Methods of gathering, organizing, and interpreting information, drawing and testing inferences from data, and communicating results should be central to the secondary school mathematics curriculum.

The American Statistical Association–National Council of Teachers of Mathematics Joint Committee on the Curriculum in Statistics and Probability with funding from the National Science Foundation has developed instructional materials for implementing the spirit of the recommendations above. The Quantitative Literacy Project has produced four booklets for teaching aspects of statistics and probability: *Exploring Data, Introduction to Probability, The Art and Techniques of Simulation,* and *Information from Samples.*

The activity sheets that follow are similar to those in *Exploring Data.* That booklet is intended to teach students effective new techniques for organizing and presenting data, interpreting these displays, and, especially, communicating the results.

One particularly effective method for displaying data is the stem-and-leaf plot, invented less than twenty years ago by John Tukey of Princeton University and AT&T Bell Laboratories. It is fast and easy to construct. If turned on its side, the stem-and-leaf plot resembles a histogram, but with an important difference—none of the original data are lost. Making a plot like this enables students to see such important features of the data as outliers, gaps, the location of the center, the range, and whether or not the distribution is skewed to smaller or larger values.

We have tried to present data that are interesting to secondary school students. For example, most students are likely to have an opinion about the best pop album. Interest in the data inspires interest in the mathematics.

Student communication of ideas and obtained results is an expectation of this activity. Emphasize to your students that a numerical result is of little use in the real world unless its meaning can be communicated to others. The activity sheets contain questions that ask students to summarize the information seen in a plot. They will resist writing these summaries and complain, "But this isn't an English class!" Insist that they do the writing and that they use good grammar, correct spelling, and complete sentences. Clarity in writing will encourage clarity in thought. These exercises will also give those students who have never had any success in mathematics a chance to shine. Often the students weakest in mathematics make the most perceptive comments.

Grade levels: 7–12

Materials: Calculators, a map of the United States, and a set of activity sheets for each student

Objectives: Students will (1) make stem-and-leaf plots of real-world data and (2) write accurate, perceptive analyses of the information displayed in the plots.

Directions: Distribute the activity sheets one at a time. Students can complete the activities either working independently or in small groups. Class discussion and a summary of the results should follow the completion of each sheet. The completion of the three worksheets will require two or three instructional periods, depending on the level of your class.

Sheet 1: No unique way has been found to construct stem-and-leaf plots. We have placed the smaller numbers at the top so that when the plot is turned counterclockwise ninety degrees, it resembles a histogram. Students will initially require some direction in analyzing the salient features of the displayed data. You might suggest that they consider, among other things, the range of percentages of pedestrian deaths, a percentage interval that typifies most of the cities, and any unusually high or low percentages.

Students may want to know the total number of pedestrian deaths in each city and also how vigilant the police are in citing jaywalkers and then discuss whether or not they think variables like these might be related to the percentage of pedestrian traffic deaths. This information is given in table 1 for 1983.

TABLE 1

City	Number of Pedestrian Deaths	Number of Pedestrian Citations per 100 000 Residents
New York	299	7.3
Los Angeles	131	1349
Chicago	97	0
Houston	100	21
Philadelphia	39	3
Detroit	42	389
Dallas	47	164
San Diego	33	295
Phoenix	33	100
San Antonio	24	0
Honolulu	18	224
Baltimore	19	0
Indianapolis	9	49.5
San Francisco	31	255
San Jose	20	95

Source: *Los Angeles Times*, 8 September 1984

Some questions that students can be encouraged to research are these:

1. What proportion of pedestrian deaths are children?
2. Are the pedestrian deaths usually the fault of the driver or of the pedestrian?
3. What percentage of traffic deaths are pedestrian deaths in the area of your school?

Sheet 2: After discussing solutions to the exercises on this worksheet, you may want to point out some of the problems associated with the method of balloting used by the music critics. For example, suppose two albums, *A* and *B*, are very good and only two people are voting. The first voter wants album *A* to win, so she ranks *A* first and *B* second. The second voter wants *B* to win, so he ranks *B* first and some mediocre album second. The second voter has made his choice, *B*, the winner by a rather unethical method.

Sheet 3: Some students may need assistance in completing the third stem-and-leaf plot because of the large number of stems involved. As a follow-up to this final activity, you might assign the class the project of making a stem-and-leaf plot of the populations of the fifty states. This will lead to questions of how to handle numbers of three or more digits. Possible solutions include truncating or rounding. Another useful project would be to have the class make a stem-and-leaf plot of their heights in meters.

Answers:

Sheet 1: 1. 25%; 2. 35%; 3. Answers will vary. 4. It is often helpful, but not necessary, to put the leaves in numerical order.

1	8
2	0 5 7 9
3	2 2 4 4 5 8 9
4	0
5	7
6	0

5. a. San Francisco and New York; b. Possible explanations include the fact that both cities are densely populated, so traffic moves slowly. Consequently, a collision between two cars is unlikely to result in a death. Both cities have a large number of pedestrians, who are more likely to die than the occupant of a car if involved in an accident. San Francisco has a high alcoholism rate, which may further account for pedestrians being hit. 6. The percentage of traffic deaths that are pedestrian deaths in the fifteen largest U.S. cities ranges from 18 percent in San Antonio to 60 percent in San Francisco. Most cities have between 20 percent and 40 percent. Two cities, San Francisco and New York, have unusually high percentages. 7. The stem-and-leaf plot is better than the table for displaying the data because we can easily find the largest and smallest values; we can see whether any values are unusually large or small; we can see how spread out the values are; we can see where the majority of values are located; and we can see if any gaps occur in the data.

A weakness of the stem-and-leaf plot is that the names of the cities are missing. To find out which city corresponds to 60 percent, we have to go back to the table.

Sheet 2: 1. 17 points; 2. 40 points; 3. Answers will vary. (Actually, it was third on one list and fourth on another.) 4. Answers will vary. (This album was, in fact, first on five lists, second on two lists, and third, fourth, fifth, and sixth on one list each.)

5.

1	2 3 4 5 7 9
2	4 4 5 6
3	
4	0 6
5	5
6	
7	
8	3
9	4

6. The top fifteen albums have between 12 and 94 points each. They cluster into three groups. Two albums, "Born in the U.S.A." and "Purple Rain," have many more points than the middle group of three albums with 40, 46, and 55 points. The nine lowest-rated albums are all clustered between 12 and 25 points. One suspects that the albums in the lowest group beat out other candidates by a very small margin. 7. It was possible to be a "top album" with only 12 points. An album could do this by being fifth on but two of the fifteen lists. "Lush Life" came in eleventh and was listed only twice. (In fact, seventy-six different albums were listed by the fifteen pop music critics. One of the critics said that such a variety were named because no other album "exhibited the unbending command or vision of the two premier works.")

Sheet 3:

1.

5	7 8
6	4 5 6 6 6 7 7 8 8 9 9
7	1 3 3 5 5 5 6 6 6 6 7 7 7 7 8 8 8 8 8 9
8	0 0 1 2 2 2 3 3 3 4 4 5 5 5 8 9
9	1 5

2.

5	
·	7 8
6	4
·	5 6 6 6 7 7 8 8 9 9
7	1 3 3
·	5 5 5 6 6 6 6 7 7 7 7 8 8 8 8 8 9
8	0 0 1 2 2 2 3 3 3 4 4
·	5 5 5 8 9
9	1
·	5

3.

Stem	Leaves
5	
·	
·	
·	7
·	8
6	
·	
·	4 5
·	6 6 6 7 7
·	8 8 9 9
7	1
·	3 3
·	5 5 5
·	6 6 6 6 7 7 7 7
·	8 8 8 8 8 9
8	0 0 1
·	2 2 2 3 3 3
·	4 4 5 5 5
·	
·	8 9
9	1
·	
·	5
·	
·	

4. We prefer the last plot because the shape of the distribution is easier to see. The first and second plots have too many leaves on some lines. 5. Answers will vary. 6. The percentage of students who graduate from high school ranges from 57 percent in Louisiana to 95 percent in North Dakota. Louisiana and Washington, D. C., have unusually low graduation rates. In general, southern states have the lowest rates of graduation. The four highest rates are 95, 91, 89, and 88 percent, respectively. Three of these are from midwestern states. The rates are centered at 77 percent and are spread out rather uniformly between 64 percent and 85 percent.

BIBLIOGRAPHY

Landwehr, James M., and Ann E. Watkins, *Exploring Data*. Washington, D. C.: American Statistical Association, 1984.

MacDonald, A. D., "A Stem-Leaf Plot: An Approach to Statistics." *Mathematics Teacher* 75 (January 1982): 27–28, 25.

Velleman, Paul F., and David C. Hoaglin. *Applications, Basics, and Computing of Exploratory Data Analysis*. Boston: Duxbury Press, 1981.

The chart below gives the percentage of traffic deaths that were pedestrian deaths for the fifteen largest U.S. cities.

New York	57%	Phoenix	29%
Los Angeles	34%	San Antonio	18%
Chicago	39%	Honolulu	20%
Houston	32%	Baltimore	32%
Philadelphia	__%	Indianapolis	__%
Detroit	40%	San Francisco	60%
Dallas	27%	San Jose	38%
San Diego	34%		

Source: *Los Angeles Times*, 8 September 1984

1. Indianapolis had 36 traffic deaths. Of these, 9 were pedestrians. Fill in the percentage for Indianapolis.

2. Philadelphia had 111 traffic deaths. Of these, 39 were pedestrians. Fill in the percentage for Philadelphia.

3. a. What is the percentage for the city nearest you? ______
   b. If there were 200 traffic deaths in this city, how many were pedestrian deaths? ______

4. Below is the beginning of a stem-and-leaf plot of these data:

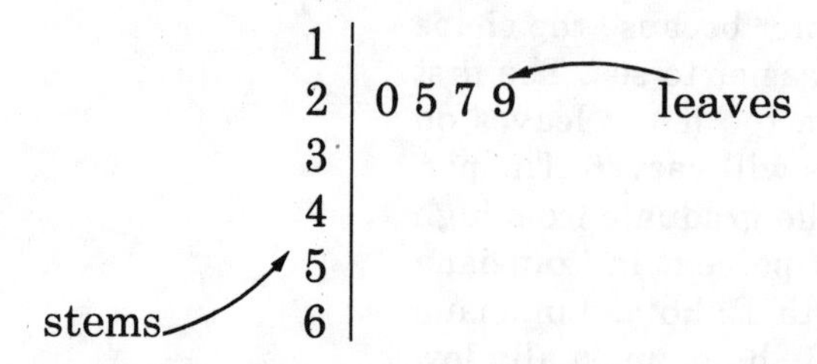

The line 2 | 0 5 7 9 represents the data for four cities: Honolulu with 2<u>0</u> percent, Indianapolis with 2<u>5</u> percent, Dallas with 2<u>7</u> percent, and Phoenix with 2<u>9</u> percent. Finish this stem-and-leaf plot for the remaining eleven cities.

5. a. Which cities have an unusually high percentage?______
   b. What is a possible explanation for these high percentages?______

Complete exercises 6 and 7 on a separate sheet of paper.

6. Write a one-paragraph summary of the information you can see from looking at this plot.

7. In what ways is the stem-and-leaf plot better than the chart for displaying the data? In what ways is it worse?

Fifteen of the pop music critics for the *Los Angeles Times* made a list of what they felt were the ten best albums of 1984. The albums were ranked by giving 10 points for each first place vote, 9 points for a second place vote, and so on. Below is the list of the top fifteen albums.

	Album	Artist	Points
1.	*Born in the U.S.A.*	Bruce Springsteen	94
2.	*Purple Rain*	Prince	83
3.	*How Will the Wolf Survive?*	Los Lobos	55
4.	*Reckoning*	R.E.M.	46
5.	*Private Dancer*	Tina Turner	—
6.	*Let It Be*	Replacements	26
7.	*Learning to Crawl*	Pretenders	25
8.	*Double Nickels on the Dime*	Minutemen	24
9.	*The Magazine*	Rickie Lee Jones	24
10.	*The Unforgettable Fire*	U2	19
11.	*Lush Life*	Linda Ronstadt	—
12.	*Zen Arcade*	Hüsker Dü	15
13.	*Soul Mining*	The The	14
14.	*Meat Puppets II*	Meat Puppets	13
15.	*Sparkle in the Rain*	Simple Minds	12

1. *Lush Life* was third on Paul Green's list, second on Dennis Hunt's list, and did not appear on any other list. Fill in the total points for this album.

2. *Private Dancer* was listed tenth, first, third, fourth, third, and fifth. Fill in the total points for this album.

3. *Zen Arcade* was on only two critics' lists. Describe a way it could have earned its 15 points.____________________

4. *Born in the U.S.A.* was listed by eleven critics—the most of any album. Describe a way this album could have earned its 94 points.____________________

____

Complete exercises 5 and 6 on a separate sheet of paper.

5. Make a stem-and-leaf plot of the points for the fifteen albums. The first stem should be a 1 and the last should be a 9.

6. Study your stem-and-leaf plot and then write a summary of what you learned.

7. The lists were extremely varied this year. How can you tell?____________________

____

____

The percentage of students that graduate from high school is given below for each state and the District of Columbia.

Ala.	67	Ill.	77	Mont.	83	R.I.	75
Alaska	78	Ind.	78	Nebr.	84	S.C.	66
Ariz.	68	Iowa	88	Nev.	75	S.Dak.	85
Ark.	76	Kans.	83	N.H.	77	Tenn.	65
Calif.	75	Ky.	68	N.J.	83	Tex.	69
Colo.	79	La.	57	N.M.	71	Utah	85
Conn.	78	Maine	77	N.Y.	67	Vt.	85
Del.	89	Md.	81	N.C.	69	Va.	76
D.C.	58	Mass.	78	N.Dak.	95	Wash.	76
Fla.	66	Mich.	73	Ohio	82	W.Va.	77
Ga	66	Minn.	91	Okla.	80	Wis.	84
Hawaii	82	Miss.	64	Oreg.	73	Wyo.	82
Idaho	78	Mo.	76	Pa.	80		

Source: *USA Today*, 19 December 1984

1. On a separate sheet of paper, make a stem-and-leaf plot of the percentages using 5 as the first stem and 9 as the last stem.

2. Make a new stem-and-leaf plot using these stems. Put the percentages 50–54 on the 5 line and the percentages 55–59 on the line below. North Dakota with 95 percent and Minnesota with 91 percent are done for you.

```
5 |
· |
6 |
· |
7 |
· |
8 |
· |
9 | 1
· | 5
```

3. The stem-and-leaf plot can be spread out even further. On a separate sheet of paper, make a third plot using the stems that are begun at the right. The percentages 50–51 go on the first line, 52–53 on the second, 54–55 on the third, 56–57 on the fourth, 58–59 on the fifth, and so on.

```
5 |
· |
· |
· |
· |
6 |
· |
· |
· |
· |
```

4. Which of the three plots do you think best displays the data?__________
   Why?____________________

5. a. What is the percentage of high school graduates in your state?________
   b. Is this percentage relatively high or relatively low?________
   c. Can you think of any reasons for this ranking?________

6. Study your last stem-and-leaf plot. On the same sheet of paper, write a summary of conclusions you can draw from this plot.

ACTIVITIES

EXPLORING DATA WITH BOX PLOTS

November 1988

By ELIZABETH H. BRYAN, Augusta College, Augusta, GA 30910

Teacher's Guide

Grade levels: 7–12

Materials: One set of activity sheets for each student; a set of transparencies for class discussion

Objectives: Students will organize and display data with a box-and-whiskers graph. They will use the plot as a summary display to detect patterns and to highlight the important features of the data for purposes of comparison.

Procedures: These activities assume that the students understand how to determine the median of a data set and how to construct stem-and-leaf plots. This background is useful for computing the statistics needed to construct the box plot and can be found in the activity "Stem-and-Leaf Plots" (Landwehr and Watkins 1985).

Distribute sheet 1 first. You may find it helpful to do sheet 1 together with your class and discuss the answers thoroughly before assigning sheets 2 and 3. The procedures for drawing a box-and-whiskers plot are described on sheet 1 and include the following:

1. Draw and label a number line.
2. Find the median, lower quartile (LQ), upper quartile (UQ), lowest value (LV), and highest value (HV).
3. Draw the box.
   a) The length of the box is the interquartile range (from LQ to UQ).
   b) The width of the box can be anything.
   c) The median is marked with a line widthwise across the box.
4. Draw the whiskers—
   a) from LV to LQ;
   b) from HV to UQ.

You should stress that box plots are appropriate when dealing with large sets of data and are helpful in comparing several sets of data because they focus attention on only a few characteristics of the data. However, students often have a difficult time interpreting them, and they are not as useful as a stem-and-leaf graph for showing details.

The major objective is to help students learn how to interpret data. Constructing the graphs and plots is only a secondary goal. You should encourage the students to make conjectures about the data on the basis of examining the plots. Be sure to emphasize the fact that "one" correct answer does not always occur. The idea is to have students explore the data, ask questions, and make observations. They should discuss their individual interpretations and attempt to support them when differing conclusions are reached.

Finally, the students should organize their results into a paragraph describing what they have learned. Writing interpretations of statistical data is often difficult for students. It is important for you to require them to complete the activity with a written summary, since this approach teaches the students to follow a model for data analysis in the same way a good statistician would.

Answers:

Sheet 1

1.

4	9
5	2 3 3 3 4 6 6 7 7 8 8
6	0 0 1 1 2 4 9
7	1 1 3 5 8
8	2

2. *a*) 60 mph
   b) 55 mph
   c) 70 mph
   d) 49 mph and 82 mph

3., 4.

30 40 50 60 70 80 90

5. *a*) 50 percent (48% actual)
   b) 25 percent (24%)
   c) 75 percent (76%)
   d) 50 percent (52%)
   e) 25 percent (24%)

6. Yes. It means that the faster speeds (those in the top 25%) are more spread out than those in the bottom 25 percent.

7. It is not in the center because the values in the second and third quarters are not equally spread out.

8. 25 percent (24%)

9. Answers will vary.
   Example: The average car in the sample was traveling at 60 mph with the middle 50 percent of them traveling at a rate between 55 mph and 70 mph. The six fastest cars (those in the top 25%) were exceeding 70 mph, whereas the six slowest (those in the bottom 25%) were traveling less than 55 mph. The rates of the cars traveling in excess of 60 mph displayed much more variability than those traveling below 60 mph.

Sheets 2, 3

1. New cars:
   Median = 2; LQ = 1; UQ = 2.5; LV = 1; HV = 6.
   Three- to four-year-old cars:
   Median = 6; LQ = 5; UQ = 9; LV = 3; HV = 14.
   Six-year-old cars:
   Median = 11; LQ = 10.5; UQ = 15.5; LV = 5; HV = 18.

2.

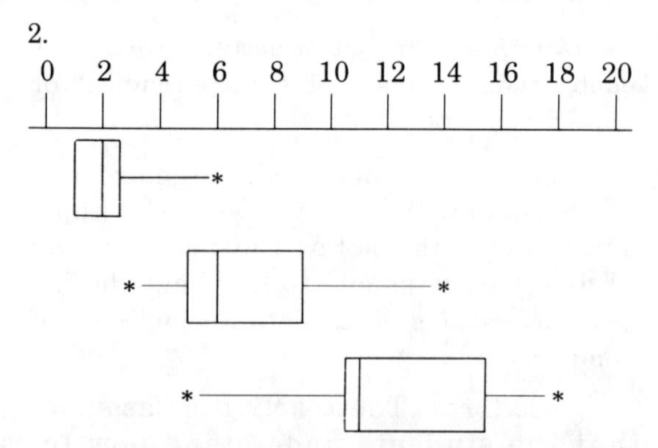

3. It has only one whisker because more than 25 percent of the categories in the new car data displayed a required repair rate of 1 percent (which was the lowest percentage).

4. *a*) New: IQR = 1.5%; three to four years old: IQR = 4%; six years old: IQR = 5%.
   b) Older cars evidence greater variability of reliability. This fact may be attributed to the quality of the product, the treatment the vehicle has received since its purchase, or a combination of both.

5. The new car age group has the smallest percentage of required repairs and would, based on that criterion, be the most reliable.
6. The six-year-old cars have the largest median and would, based on that criterion, be the least reliable.
7. Most of the detail of individual values is omitted in box plots, and we can therefore compare relative reliability ratings better. In addition, it is not possible to compare three sets of data on a back-to-back stem-and-leaf plot.
8. No. When you use a box plot to display information, only the five summary measures remain. Box plots do offer a way to focus on the relative positions of different sets of data for the purpose of making comparisons, and they are especially useful when the data sets are very large.

BIBLIOGRAPHY

Burrill, Gail. "Statistical Decision Making." NCTM *Student Math Notes*, May 1988.

Landau, Irwin, ed. "Frequency-of-Repair Records, 1982–87." *Consumer Reports* 53 (April 1988):248–49.

Landwehr, James M., and Ann E. Watkins. *Exploring Data*. Palo Alto, Calif.: Dale Seymour Publications, 1987.

———. "Stem-and-Leaf Plots." *Mathematics Teacher* 78 (October 1985):528–38.

Moser, James M., ed. *Teaching Quantitative Literacy—a manual for Workshop Leaders*. Madison: Wisconsin Department of Public Instruction, 1987.

The speed (in mph) of a sample of twenty-five cars checked by radar on an expressway is listed at the right:

Speed of Twenty-five Cars				
57	53	53	71	73
54	69	56	58	49
56	53	52	82	62
61	60	71	75	60
57	61	58	78	64

1. Make a stem-and-leaf plot for these data.

2. Use your stem-and-leaf plot to find each of the following:
   a) The median speed
   b) The median of the lower half (the lower quartile)
   c) The median of the upper half (the upper quartile)
   d) The extremes (the lowest and highest values)

3. Mark dots for the median, quartiles, and extremes beneath the number line at the right.

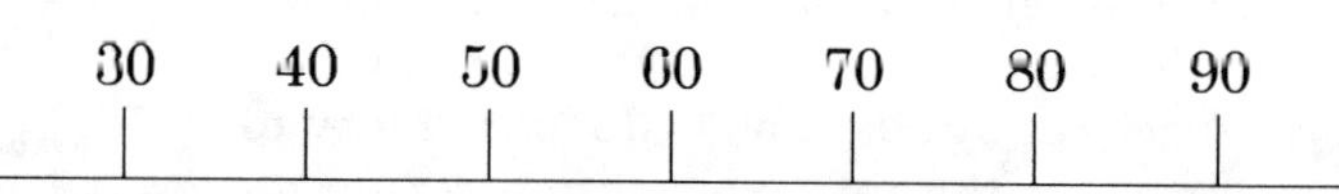

4. Draw a box between the two quartiles. Mark the median with a line vertically across the box. Draw two "whiskers" from the quartiles to the extremes.

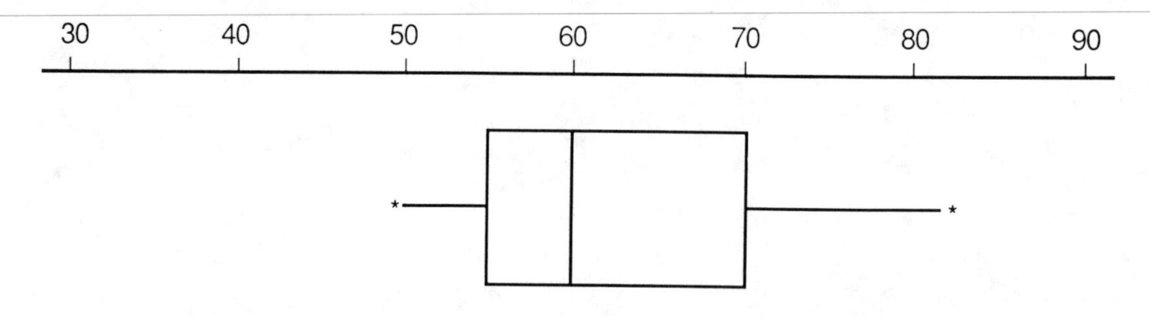

5. About what percent of the speeds are—
   a) below the median?
   b) below the lower quartile?
   c) above the lower quartile?
   d) in the box?
   e) on each whisker?

6. Is one whisker longer than the other? What do you think this result means?

7. Why isn't the median in the center of the box?

8. If an officer is writing tickets to each driver in the sample whose speed is more than 70 mph, about what percentage of the drivers will be ticketed?

9. Write a description summarizing the information in the box plot.

Consumer Reports (Landau 1988) generated a reliability report designed to demonstrate how the "average" car ages. The data below give the percentage of problems requiring repairs in seventeen different categories for new cars, for cars three to four years old, and for cars six years old. The figures in the table are based on data obtained from over 544 000 responses to Consumer Union's annual questionnaire. The study included 252 cars for the model years 1982–1987.

How the "Average" Car Ages

	Percentage of Cars Needing Repairs		
	New	3–4 Years Old	6 Years Old
Electrical system, chassis	6	14	18
Body integrity	6	9	11
Body hardware	5	9	12
Fuel system	3	11	17
Transmission, automatic	2	6	10
Transmission, manual	2	4	5
Body exterior, paint	2	6	11
Brakes	2	9	16
Air-conditioning	2	7	17
Suspension	1	8	11
Ignition system	1	5	8
Exhaust system	1	5	15
Engine mechanical	1	6	11
Engine cooling	1	5	12
Drive line	1	3	5
Clutch	1	6	13
Body exterior, rust	1	3	11

1. Find the median, the quartiles, and the extremes for each set (column) of data from activity sheet 2.

2. Follow the same procedure you used in activity sheet 1 to draw box plots for each of the three sets of data using the number line below.

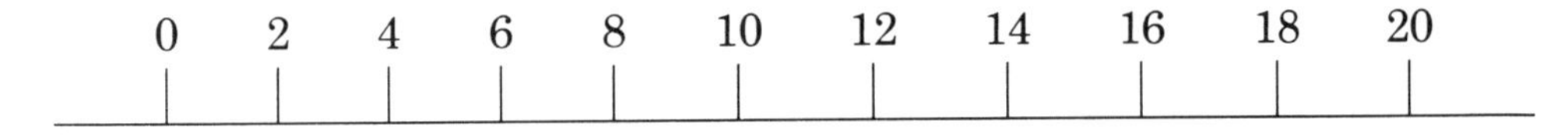

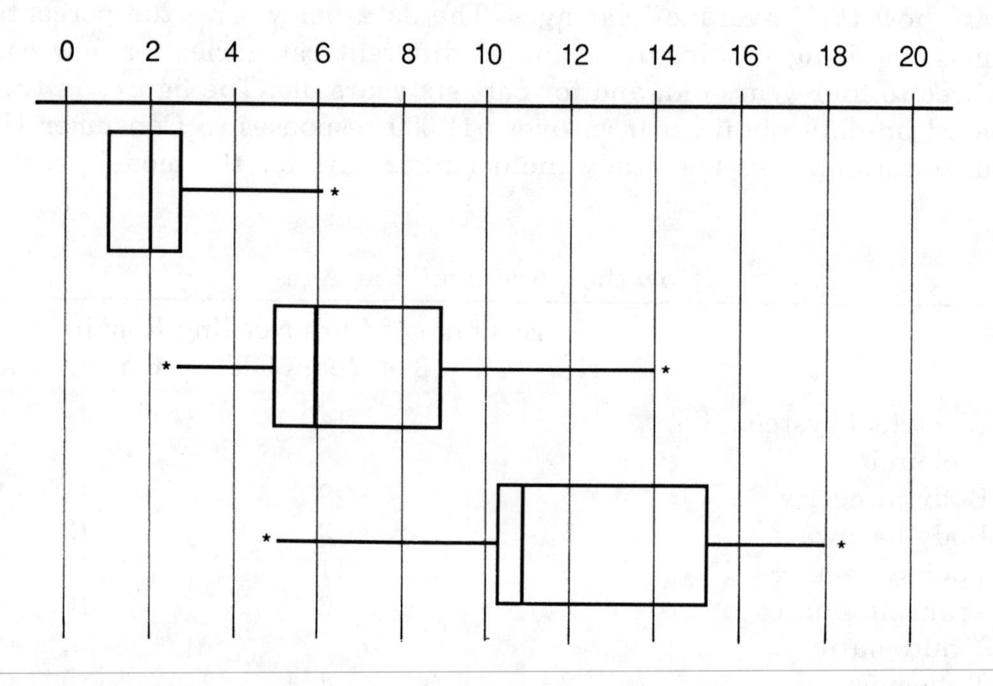

3. How many whiskers does the box plot for the new cars have? What do you think the missing whisker tells us about the new-car data?

4. The length of the rectangle in a box plot is called the *interquartile range* (IQR).

   a) Compute the IQR for each data set.

   b) What does the IQR tell us about the variability of these data sets?

5. If the most reliable age group is the one with the smallest percentage of repairs, which age of car is the most reliable?

6. If the least reliable age group is the one with the largest median, which age of car is the least reliable?

7. Why do box plots furnish a better way to compare the relative reliability ratings of the three age groups of cars than stem-and-leaf plots?

8. If you wanted to compare the reliability of the fuel systems for the three age groups of cars, would the box plot give you the information you needed? What does your answer indicate about what box plots tell us and what they do not tell us?

ACTIVITIES

PLOTTING AND PREDICTING FROM PAIRS

September 1984

By ALBERT P. SHULTE, Oakland Schools, Pontiac, MI 48054, and
JIM SWIFT, Nanaimo Senior Secondary School, Nanaimo, BC V9R 5K3

Teacher's Guide

Introduction: The collection of data, the selection of appropriate displays that illuminate the data, and the interpretation of those displays are all basic skills in today's information-oriented world. Recognizing the need for these skills in the information age, recent curriculum recommendations have called for an increased emphasis on such activities as locating and processing quantitative information; collecting, organizing, presenting, and interpreting data; and drawing inferences and predicting from data. Not only are such processes basic skills needed by all students, but they also provide necessary mathematical methods useful in problem solving. The worksheets included here involve *bivariate* data (associated pairs of numbers) and provide experiences with all the activities listed above.

The ability to work with bivariate data is an important skill, seldom treated in conventional mathematics courses. The examples in the following worksheets center around the use of a *scatter plot* to examine relationships shown by bivariate data. As such, the activities show real-life applications of coordinate graphing and provide a bridge between algebra and statistics.

Grade levels: 7–12

Materials: Graph paper, rulers, and a set of worksheets for each student. Data can be collected from printed sources or by a survey. Reference materials (e.g., the *World Almanac*) should be available.

Objectives: (1) To investigate relationships shown by bivariate data (pairs of numbers representing values of two variables); (2) to graph bivariate data and interpret the graphs; (3) to use graphs of bivariate data to aid in prediction; (4) to collect and analyze bivariate data from various sources

Procedure: The worksheets are designed to be done in sequence over a period of several days and should be distributed one at a time. The students can work independently to complete the worksheets or with the teacher's assistance when requested. Students may wish to work in pairs when gathering data.

Each worksheet should be discussed thoroughly in class before students are assigned the next worksheet. Particular attention should be given to the data collected by students. One way of highlighting the students' efforts is by using their graphs for bulletin-board displays.

Students should also be encouraged to bring in clippings from newspapers and magazines that use graphical displays of bivariate data. Opportunities should be provided for discussion of the relationships depicted.

Answers:

1.

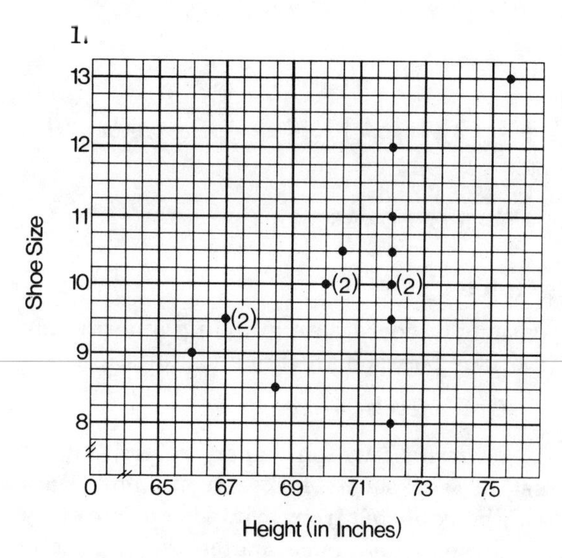

2. The shoe size also tends to increase.

3, 4. Answers will vary. (Possible variables include weight versus height for a class of students or number of points scored versus number of fouls by a school basketball team.)

5.

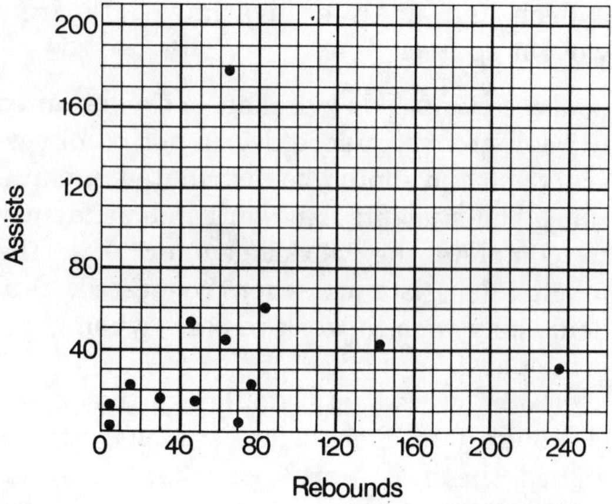

6. Players with more rebounds tend to have fewer assists. (However, if the data for Laimbeer and Thomas are removed, a slightly *positive* relationship, if anything, seems to develop.)

7, 8. Answers will vary. (Possible variables include gas mileage versus weight of an automobile.)

9.

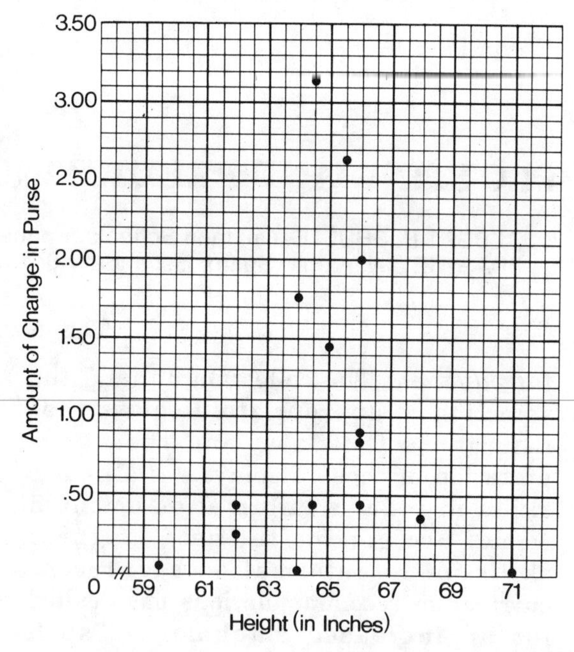

10. No particular relationship is evident. (Some students may say that people in the middle of the height range appear to have more money.)

11, 12. Answers will vary. (Possible variables include a student's height versus month of birth.)

13. 4

14. 1960–1965

15. 1967

16. 5760 ounces

17. Answers will vary. (Possible data might be school enrollments over the last twenty-five years or record-breaking times for running a mile run since 1900 graphed in increments of five years.)

BIBLIOGRAPHY

Haylock, Derek W. "A Simplified Approach to Correlation." *Mathematics Teacher* 76 (May 1983):332–36.

Holmes, Peter, ed. *Statistics in Your World.* Twenty-seven booklets and teacher's guides. Developed by the Schools Council Project on Statistical Education (Ages 11–16). Slough, Berks, England: W. Foulsham & Co., 1980.

Landwehr, James M., and Anne Watkins. "Exploring Data." Prepared for the ASA-NCTM Joint Committee on the Curriculum in Statistics and Probability, 1983.

National Council of Teachers of Mathematics. *An Agenda for Action: Recommendations for School Mathematics of the 1980s.* Reston, Va.: The Council, 1980.

Shulte, Albert P., and James R. Smart, eds. *Teaching Statistics and Probability,* 1981 Yearbook. Reston, Va.: National Council of Teachers of Mathematics, 1981. ■

PLOTTING AND PREDICTING FROM PAIRS SHEET 1

1. Table 1 gives the height and shoe size for fifteen men. Plot the data on graph 1.

TABLE 1

Man	Height (Inches)	Shoe Size
1	70	10
2	75½	13
3	72	11
4	72	12
5	70	10
6	67	9½
7	72	9½
8	72	10
9	70½	10½
10	72	10
11	72	8
12	67	9½
13	66	9
14	68½	8½
15	72	10½

GRAPH 1

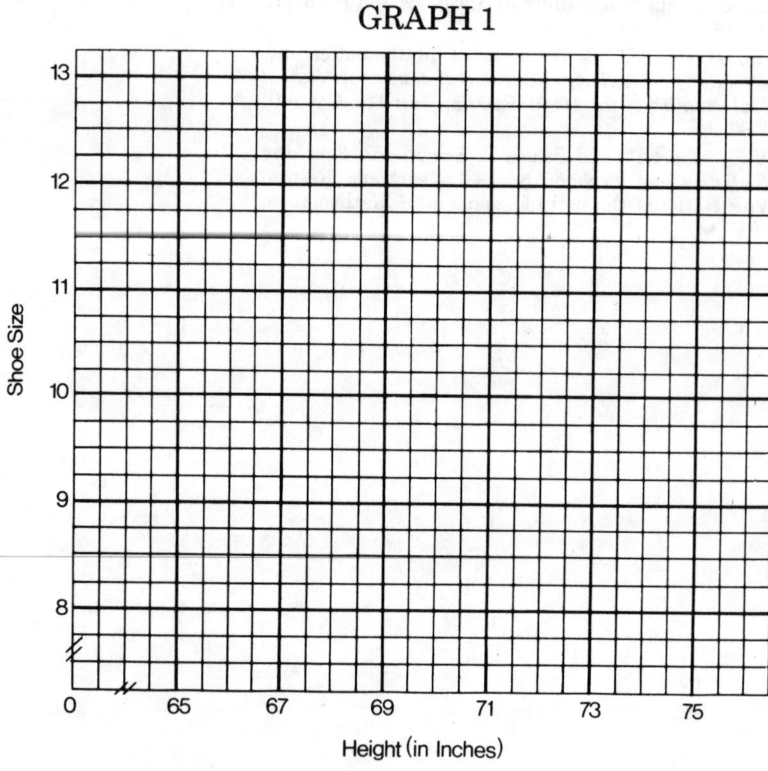

2. As the height increases, what tends to happen to the shoe size?

____

____

3. The two variables, height and shoe size, increase together. They have a *positive relationship*. What are some other pairs of variables that you think have a positive relationship?

____

____

4. Collect data for one set of pairs from question 3. Graph the data to see whether the relationship appears to be positive.

5. Table 2 gives the number of rebounds and assists made by the Detroit Pistons basketball team during their first nineteen games in 1983–84. Plot the data on graph 2.

TABLE 2

Player	Rebounds	Assists
Tripucka	81	60
Thomas	65	179
Laimbeer	237	30
Long	63	47
Tyler	78	21
Levingston	141	42
V. Johnson	45	54
Benson	70	7
Tolbert	49	15
Russell	18	21
Cureton	30	8
Thirdkill	4	11
*Austin	3	1
Totals	884	496

*No longer with team
(Data courtesy of Detroit Pistons)

GRAPH 2

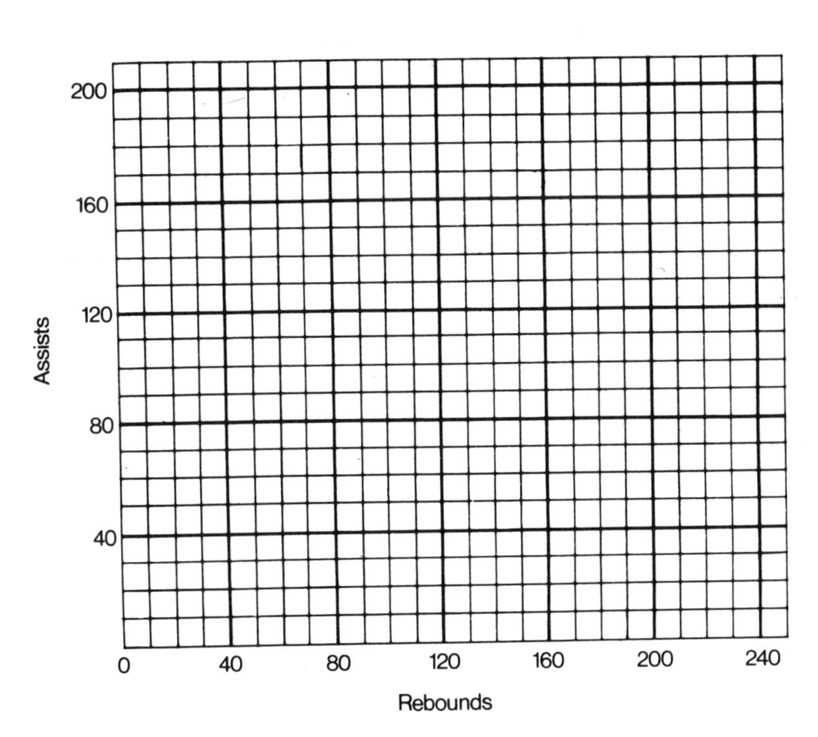

6. Do players with more rebounds tend to have more or fewer assists?

____

When one variable increases and the other decreases, they are said to have a *negative relationship.*

7. Name some pairs of variables that you think have negative relationships.

____

8. Collect data for one set of pairs from question 7. Graph the data. Does the relationship appear to be negative? ________

9. Table 3 gives the height of fifteen women and the amount of change they had in their purse on the morning of 6 December 1983. Plot the data on graph 3.

TABLE 3

Woman	Height (Inches)	Change in Purse
1	64	$0.01
2	66	0.90
3	62	0.27
4	64	1.77
5	65	1.47
6	64½	3.13
7	66	2.01
8	59½	0.05
9	71	0.01
10	68	0.38
11	66	0.46
12	65½	2.62
13	64½	0.47
14	62	0.42
15	66	0.85

GRAPH 3

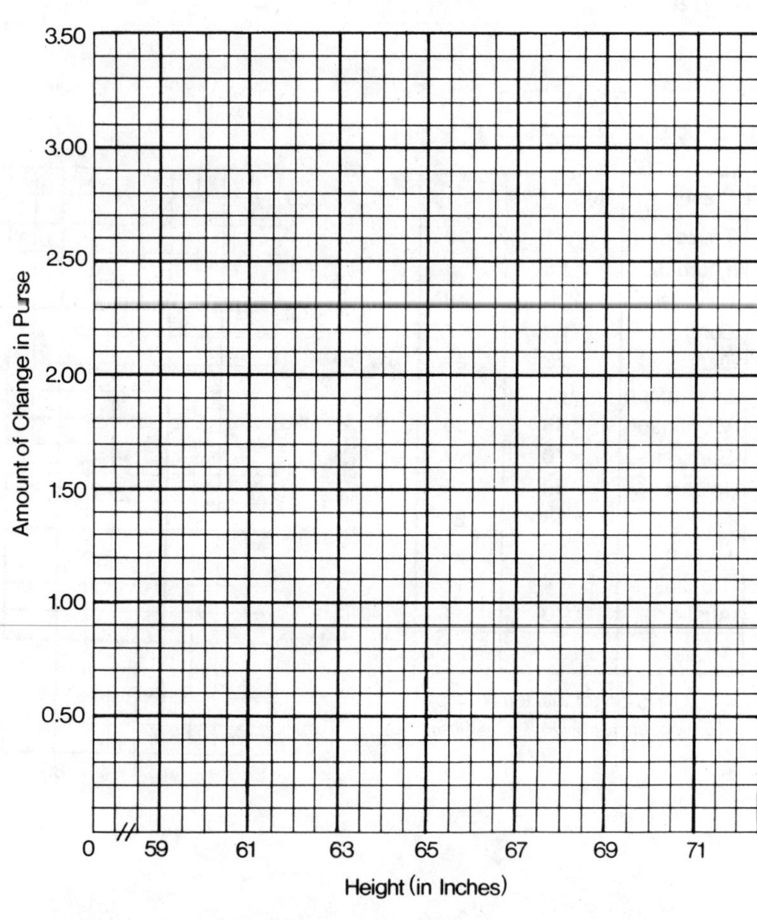

10. How would you describe the relationship between the women's height and the amount of change they had?

____

In this case, the data tend to pile up in the middle. Height and amount of change do not vary together. We can say that these variables seem to have *no relationship* or *are not related.*

11. Name some pairs of variables that you think are not related.

____

12. Collect data for one set of pairs from question 11. Graph the data. Do the pairs appear to have a relationship? ________

Some relationships change over time. Graph 4 shows the number of ounces of soft drinks consumed by the average person in the United States each year from 1945 to 1980.

GRAPH 4

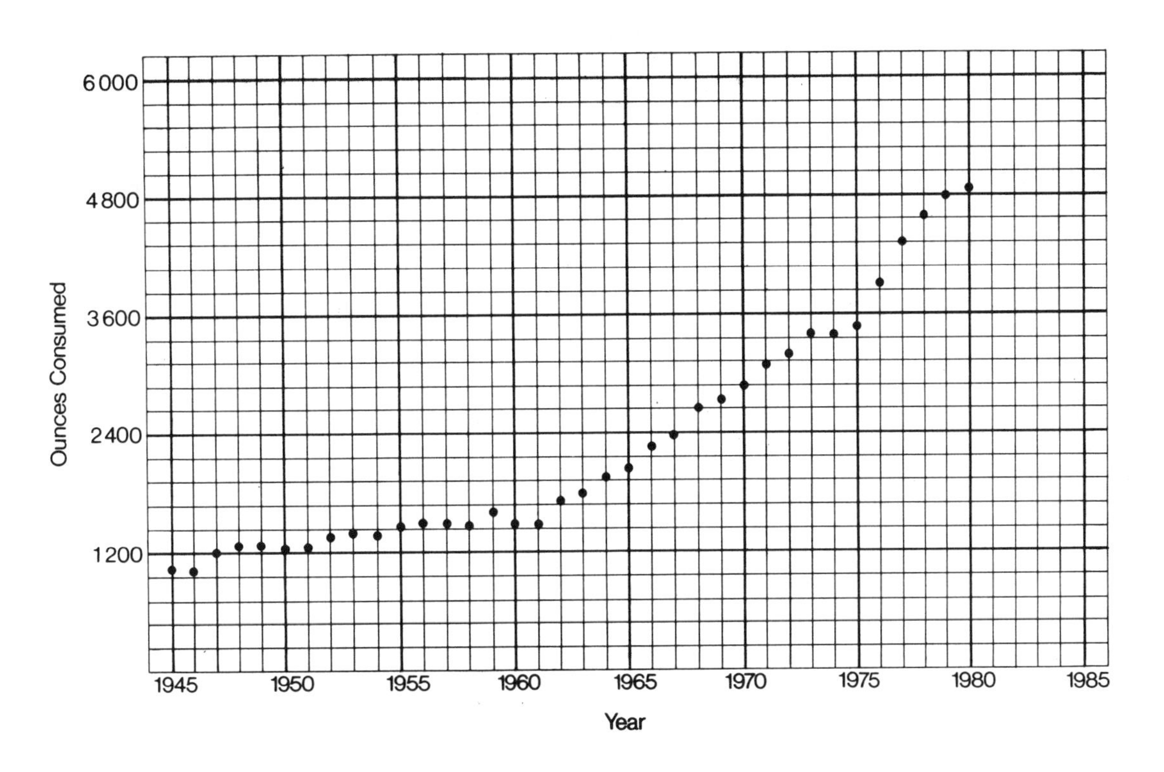

13. In 1980 people drank about ______ times as many soft drinks as they did in 1945.

14. In what five-year period did the amount consumed start to increase sharply? ______

15. In what year did people drink twice as many soft drinks as they did in 1945? ______

16. Connect the points for 1960 and 1980 with a line. Extend the line (and graph) to 1985. About how much should people be drinking then if the trend continues? ______

17. Find some time-related data. Graph the data and draw some conclusions from them. Try to answer such questions as the following:

    a) Is the relationship increasing, decreasing, or just fluctuating up and down?

    b) Can you give a possibe reason for the type of change observed?

    c) Do you expect the trend to continue in the near future?

ACTIVITIES

DATA FITTING WITHOUT FORMULAS

April 1986

By ALBERT P. SHULTE, Oakland Schools, Pontiac, MI 48054
JIM SWIFT, Nanaimo District Secondary School, Nanaimo, BC V9R 5K3

Teacher's Guide

Introduction: Recent curriculum recommendations have stressed the importance of teaching students methods of collecting, organizing, and displaying data. These recommendations have also urged the development of students' facility in interpreting data, drawing inferences from data, and making predictions based on the data. The activity "Plotting and Predicting from Pairs" (Shulte and Swift 1984) provided a student-oriented introduction to scatterplots as a means of organizing, displaying, and interpreting bivariate data. The present activity builds on that one and develops the idea of making predictions on the basis of the information presented in a scatterplot. A simple method of fitting a straight line to a set of ordered pairs—the median-fit technique—is described. A more extensive development of this technique, which was created by James McBride of Princeton University, can be found in Landwehr and Watkins (1984).

Grade levels: 8–12

Materials: Calculators, graph paper, rulers, and a set of worksheets for each student

Objectives: (1) To investigate relationships shown by bivariate data; (2) to graph bivariate data and interpret the graphs; and (3) to fit lines to scatterplots to aid in making predictions from the data

Procedures: Reproduce a set of worksheets for each student. The five worksheets are designed to be completed in sequence, but they can be treated as related subactivities in several instructional periods as follows: worksheets 1 and 2, worksheet 3, worksheets 4 and 5. Students can complete the worksheets independently or with the assistance of the teacher as needed. Each subactivity should be discussed thoroughly in class before pupils are assigned the next worksheet(s).

Worksheets 1 and 2: The data given have been collected over several years by Stuart A. Choate, coordinator of mathematics for the Midland, Michigan, public schools. Students will need calculators to compute the totals and averages for the jar-and-candle experimental data given on worksheet 1. You may need to emphasize that the points to be graphed on worksheet 2 are of the following form: (jar capacity, average time to extinguish the candle). Note that although the individual times vary widely, the averages lie almost on a line. Fitting a line to the data plotted allows students to make predictions. To complete exercise 4, students will need to be familiar with the slope-intercept form, the point-slope form, or the two-point form of an equation for a line. Caution pupils that different scales are used on the two axes.

Worksheet 3: "Fitting" a line to a set of coordinate pairs can be done in many ways. The method introduced here is a simple one that works well if many pairs are involved. Depending on the level and background of your class, you may wish to review the method for finding the median of an ordered set of data. (If an odd number of values occurs, the median is the value appearing in the middle when the numbers are arranged in order. If an even number of values occurs, the median is the average of the two middlemost values when the numbers are arranged in order.) For some classes it may even be helpful to work through an example of finding a *median point* for a set of ordered pairs. To find the median point for the following five points, which are arranged in ascending order of the x-coordinates, (a) determine the median of the x-coordinates (5) and then (b) find the median of the y-coordinates (3):

x	2	3	5	6	8
y	1	3	5	2	6

The point (5, 3) is the *median point* of the set of five given points.

The median-fit technique requires little, if no, arithmetic and can be used to indicate if "fitting" a straight line to a set of data is appropriate. If the three median points do not lie close to a straight line, then it may not be appropriate to describe the relationship by a linear equation.

Most students discover that they can find the median point most easily with a table of coordinate pairs. But it is also easy to find the median point directly from the graph. Cover up all the points except the ones for which the particular median point is being found. Slide a transparent ruler up from the x-axis until the same number of points are above it as below it. Draw a horizontal line across the graph at this point. Repeat the procedure with the ruler parallel to the y-axis. The point where the two lines cross is the median point.

Worksheets 4 and 5: The purpose of these two worksheets is to have students consider the errors created when a scatterplot is used to predict values. After completion of these two sheets, students should be encouraged to include an estimate for the error in their prediction equations. For predicting heights, the equation can take the form $H = 2h \pm 3$. Pupils should be encouraged to compare the physical law derived from these data with the rules of thumb given at the top of worksheets 4 and 5.

BIBLIOGRAPHY

Haylock, Derek W. "A Simplified Approach to Correlation." *Mathematics Teacher* 76 (May 1983):332–36.

Holmes, Peter, ed. Statistics in Your World. Series of 27 booklets and teacher's guides. Developed by the Schools Council Project on Statistical Education (ages 11–16). Slough, Berks, England: W. Foulsham & Co., 1980.

Landwehr, James M., and Ann E. Watkins. *Exploring Data.* Washington, D.C.: American Statistical Association, 1984.

National Council of Teachers of Mathematics. *An Agenda for Action: Recommendations for School Mathematics of the 1980s.* Reston, Va.: The Council, 1980.

Shulte, Albert P., and James R. Smart, eds. *Teaching Statistics and Probability,* 1981 Yearbook of the National Council of Teachers of Mathematics. Reston, Va.: The Council, 1981.

Shulte, Albert P., and Jim Swift. "Activities: Plotting and Predicting from Pairs." *Mathematics Teacher* 77 (September 1984):442–47, 464.

Answers

Worksheet 1: 1. (a) gallon jar, 2574; half-gallon jar, 1924; quart jar, 801; pint jar, 448; half-pint jar, 225. (b) gallon jar, 73.5; half-gallon jar, 40.7; quart jar, 22.9; pint jar, 12.8; half-pint jar, 7.3.

Worksheet 2: 1.

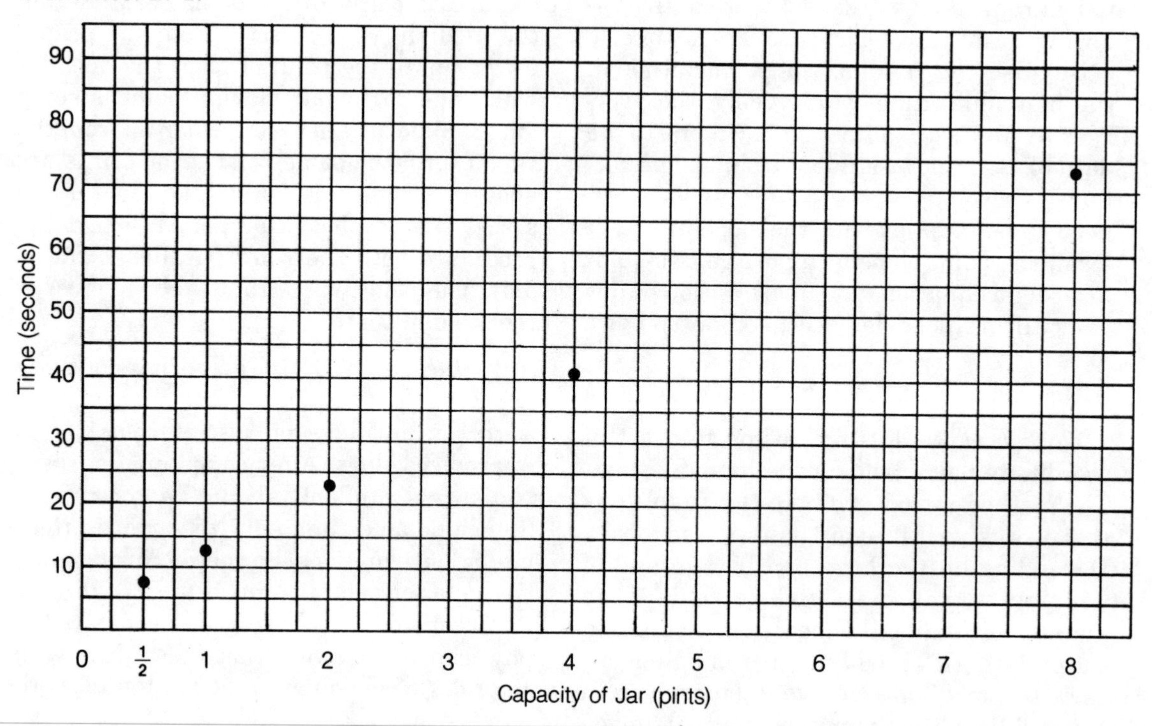

2., 3., 4. Answers will vary.

Worksheet 3: 1. (22.7, 24.2), (24.7, 26.2), (27.4, 28.9)
2., 3.

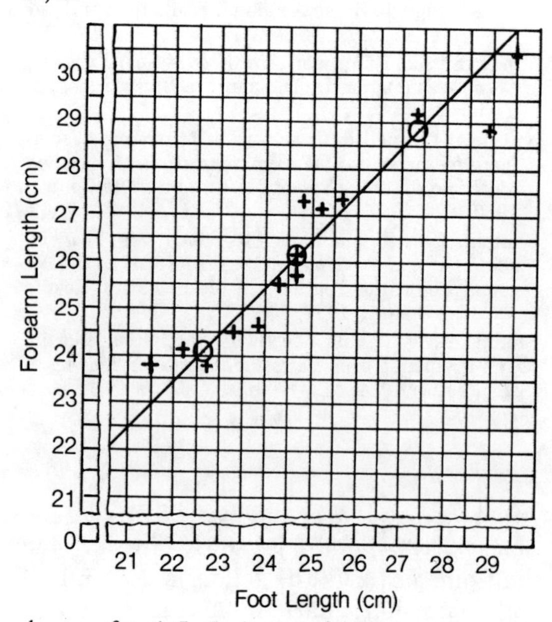

4. $a = f + 1.5$; 5. Answers will vary.

Worksheet 4: 1. (90.3, 179.6), (91.2, 182.0), (94.7, 187.4)
2.

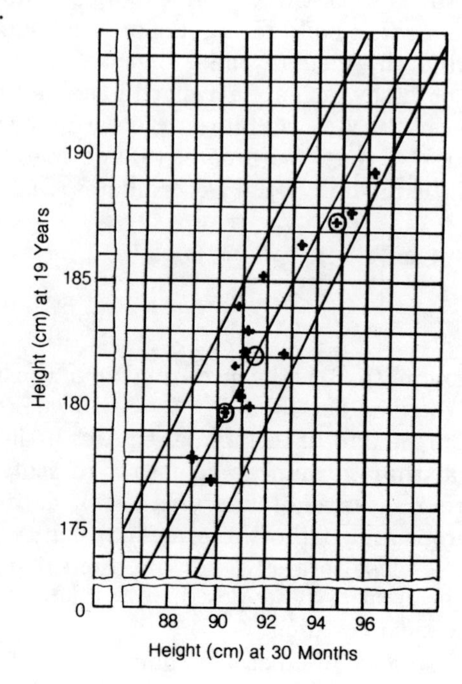

3. $H = 2h - 1$; 4. $H = 2h - 3.8$; 5. $H = 2h + 2.2$; 6. (*a*) 185 cm, (*b*) 182.2 cm, (*c*) 188.2 cm.

Worksheet 5: 1. (80.4, 160.4), (83.2, 166.5), (88.7, 176.9).

2.

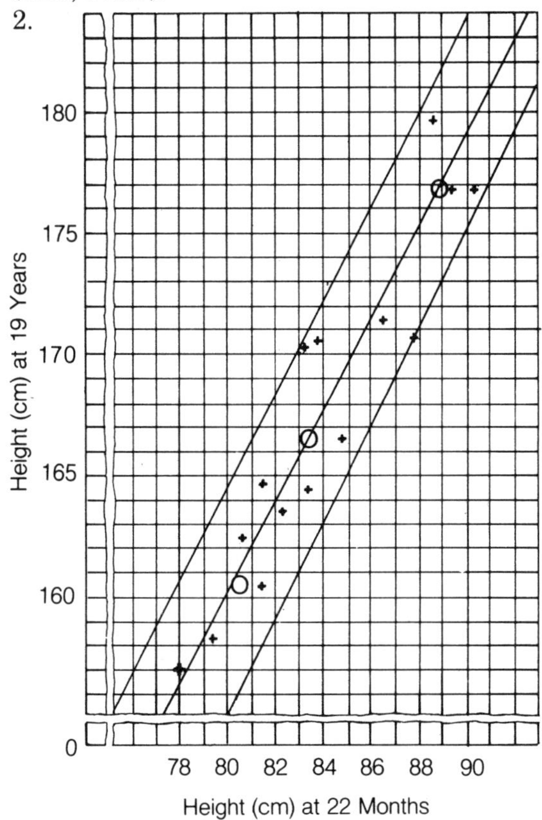

3. $H = 2h - 0.4$; 4. $H = 2h - 5.1$; 5. $H = 2h + 3.8$; 6. (*a*) 169.6 cm, (*b*) 164.9 cm, (*c*) 173.8 cm.

1. This table contains the time in seconds it took for a candle to go out after being covered with a jar of a certain size. The experiment was carried out thirty-five times with each size of jar.
   a) Find the total for each size of jar.
   b) Find the average for each size of jar.

Experiment	Time (Seconds)				
Number	Gallon Jar	½-Gallon Jar	Quart Jar	Pint Jar	½-Pint Jar
1	91	50	24	12	7
2	128	54	23	14	6
3	64	55	28	15	5
4	55	35	20	10	5
5	74	34	24	11	5
6	66	36	26	13	7
7	70	50	18	15	8
8	95	50	29	11	9
9	52	33	16	8	5
10	100	46	28	14	9
11	90	48	25	13	9
12	70	42	21	13	8
13	80	44	24	12	7
14	70	40	25	14	10
15	53	33	20	12	7
16	86	53	28	17	10
17	52	24	15	9	7
18	100	40	20	10	7
19	85	42	27	12	7
20	50	30	22	11	7
21	98	63	23	17	9
22	86	42	23	10	8
23	60	43	24	15	11
24	55	38	12	10	5
25	55	35	20	13	6
26	90	50	27	13	7
27	83	45	27	17	9
28	65	40	21	12	7
29	65	33	22	13	8
30	67	43	28	18	11
31	57	23	23	11	5
32	65	32	25	13	5
33	54	25	15	11	5
34	70	40	25	14	10
35	73	33	23	15	4
Total					
Average					

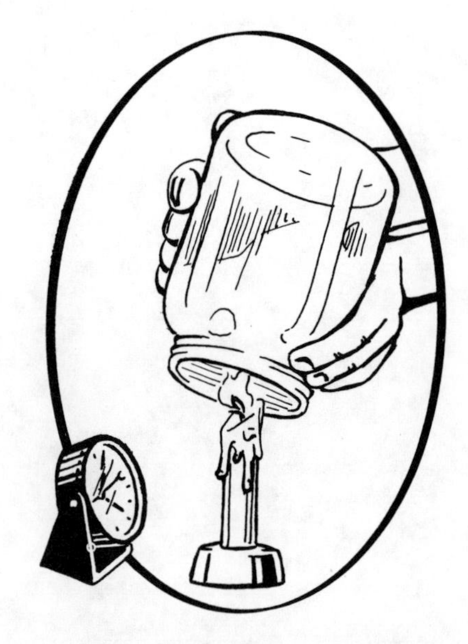

GRAPHING AND FITTING CANDLE DATA SHEET 2

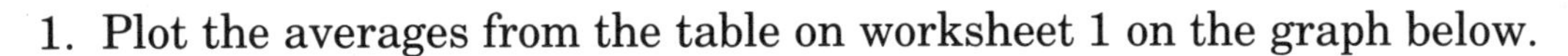

1. Plot the averages from the table on worksheet 1 on the graph below.

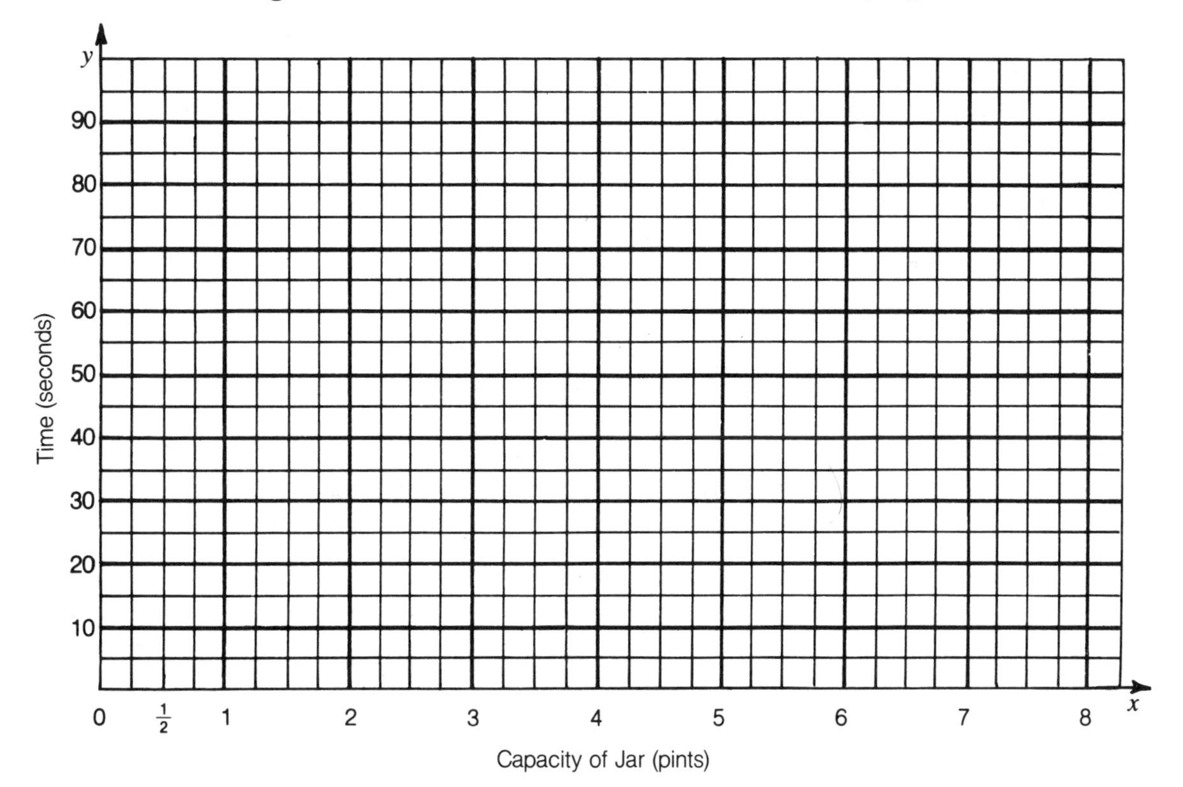

2. The points appear to lie approximately on a line. Use a ruler to draw a line that you think best fits the points.

3. Use the line you drew to predict the time it will take to extinguish a candle when covered by a jar of each of the following sizes:

   a) pint jar ________

   b) quart jar ________

   c) three-pint jar ________

   d) half-gallon jar ________

   e) three-quart jar ________

   f) gallon jar ________

4. Write an equation for the line you drew in exercise 2: ____________________

Studying relationships in this manner is the way scientists discover physical laws. Your line and its equation represent your attempt to express as a physical law the relationship between the jar's size and time necessary to extinguish a candle.

Below are the lengths, in centimeters, of one foot and forearm of each of fifteen students. They have been arranged first in order of ascending foot length and then into three groups of five points.

Lower group:	foot length	21.5	22.3	22.7	23.4	23.9
	arm length	23.8	24.2	23.8	24.5	24.6
Middle group:	foot length	24.3	24.7	24.7	24.8	25.2
	arm length	25.5	25.8	26.2	27.3	27.2
Upper group:	foot length	25.6	26.5	27.4	28.9	29.5
	arm length	27.4	26.8	29.1	28.9	30.5

1. For each of the three groups, find the median of the foot lengths and the median of the arm lengths and enter them as coordinates in the spaces below. These three points are called the *median points* of each group.
   Lower group (____, ____) Middle group (____, ____) Upper group (____, ____)

2. Plot each of the individual points and the median points for each group on the following graph.

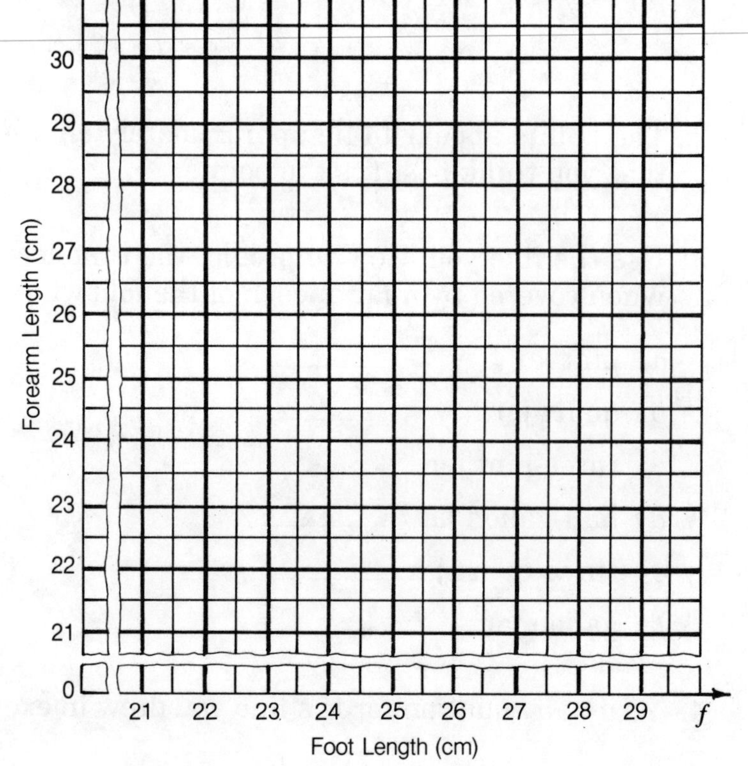

3. Use a ruler to draw a straight line through the three median points.

4. Use your line to find an equation that relates forearm length (a) to foot length (f)

   ____

5. Repeat exercises 1–4 for similar data collected from students in your class. If you cannot divide the class into three groups of equal size, make the upper and lower groups the same size.

The height of a boy at the age of 2 years and 6 months (30 months) is said to be one-half of his mature height. Below are the heights, in centimeters, of 15 boys at 30 months and at 19 years.

1. Find the median points for each of the three groups.

						Median point
Height (h) at 30 months	89.0	89.9	90.3	90.8	90.9	(____, ____)
Height (H) at 19 years	178.0	177.1	179.6	181.8	184.0	
Height (h) at 30 months	91.0	91.1	91.2	91.2	91.9	(____, ____)
Height (H) at 19 years	180.5	182.0	183.1	180.1	185.1	
Height (h) at 30 months	92.9	93.3	94.7	95.4	96.1	(____, ____)
Height (H) at 19 years	182.0	186.3	187.4	187.9	189.4	

2. Plot the points (including the median points) on the grid at the right and then draw a straight line through the median points.

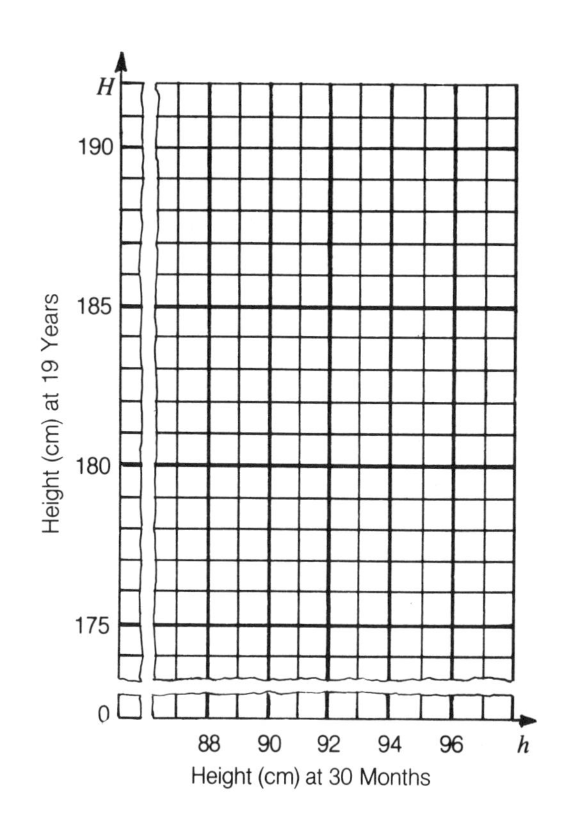

3. Use the equation of the line to express the relationship between h and H.

____

4. Draw a second line, parallel to this median-fit line, so that all the points are either on this line or *above* it. Find the equation of this line.

____

5. Draw a third line, parallel to the median-fit line, so that all the points are either on this line or *below* it. Find the equation of this line.

____

6. A boy was 93 cm tall when he was 30 months old.
   a) Use your equation from exercise 3 to predict the height of this boy when he is 19 years old. __________
   b) Use the equation from exercise 4 to find a *least* estimate of his height at 19 years. __________
   c) Use the equation from exercise 5 to find a *greatest* estimate of his height at 19 years. __________

PREDICTING FEMALE HEIGHTS SHEET 5

The height of a girl at the age of 22 months is said to be one-half of her mature height. Below are the heights, in centimeters, of 15 girls at 22 months and at 19 years.

1. Find the median points for each of the three groups.

						Median point
Height (h) at 22 months	78.0	79.4	80.4	81.3	81.3	(____, ____)
Height (H) at 19 years	157.0	158.4	161.4	164.7	160.4	
Height (h) at 22 months	82.1	83.2	83.2	83.9	84.9	(____, ____)
Height (H) at 19 years	163.7	164.4	170.2	170.5	166.5	
Height (h) at 22 months	86.2	87.9	88.7	89.4	90.1	(____, ____)
Height (H) at 19 years	171.3	170.7	179.7	176.9	176.9	

2. Plot the points (including the median points) on the grid at the right and then draw a straight line through the median points.

3. Use the equation of the line to express the relationship between h and H for this set of data.

____

4. Draw a second line, parallel to this median-fit line, so that all the points are either on this line or *above* it. Find the equation of this line.

____

5. Draw a third line, parallel to the median-fit line, so that all the points are either on this line or *below* it. Find the equation of this line.

____

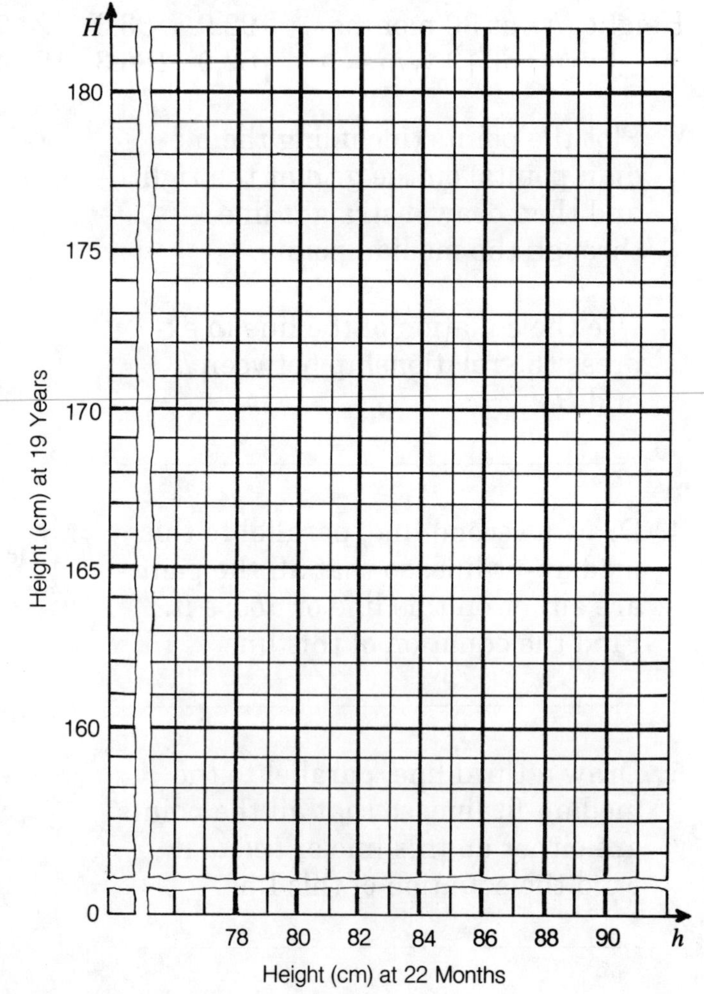

6. A girl was 85 cm tall when she was 22 months old.
   a) Use your equation from exercise 3 to predict the height of this girl when she is 19 years old. __________
   b) Use the equation from exercise 4 to find a *least* estimate of her height at 19 years. __________
   c) Use the equation from exercise 5 to find a *greatest* estimate of her height at 19 years. __________

ACTIVITIES

AREA MODELS FOR PROBABILITY

March 1987

By GLENDA LAPPAN, ELIZABETH PHILLIPS, M. J. WINTER, and WILLIAM M. FITZGERALD, Michigan State University, East Lansing, MI 48824

Teacher's Guide

Introduction: Probability plays an important role in our daily lives. In addition to their use in games of chance, probability concepts are also used as aids to decision making in such diverse fields as scientific research, weather forecasting, military operations, the design and quality control of consumer products, business predictions, insurance calculations, political predictions, and social services.

Teaching probability concepts with games and experiments offers enjoyable activities for all students, even those who have not previously experienced much success in mathematics. The basic ideas are accessible to all, yet the richness of the subject easily lends extra challenges.

The present activity employs an area model to analyze probabilities associated with games of chance. The area model makes these situations accessible to students before a more formal approach is appropriate. The area model also enhances students' understanding of fraction-related concepts and their facility with operations on fractions.

Grade levels: 7–12

Materials: Copies of the activity sheets for each student, a set of transparencies for class discussion, and bobby pins or paper clips to use as spinners

Objective: To develop students' facility in using an area model to analyze one-step and multistep probability situations

Prerequisites: Students should understand the part-to-whole concept of a fraction and be able to find equivalent fractions and sums and products of fractions. They should have some familiarity with the topic of probability, including (1) the idea that the probability of a particular event A occurring is

$$P(A) = \frac{\text{the number of ways } A \text{ can occur}}{\text{the total number of possible events}}$$

and (2) the idea that the notation $P(B)$ means "the probability that event B will occur." Some prior experience with gathering data from simulations would be helpful.

Directions: Distribute copies of the activity sheets one at a time to each student. Depending on the level of the class, you may wish to review the definition of probability. An easy way to review and set the scene for experimentation is to draw a simple spinner on an overhead transparency. Insert a pencil or pen point into the head of a bobby pin at the center of the spinner and then spin it (fig. 1).

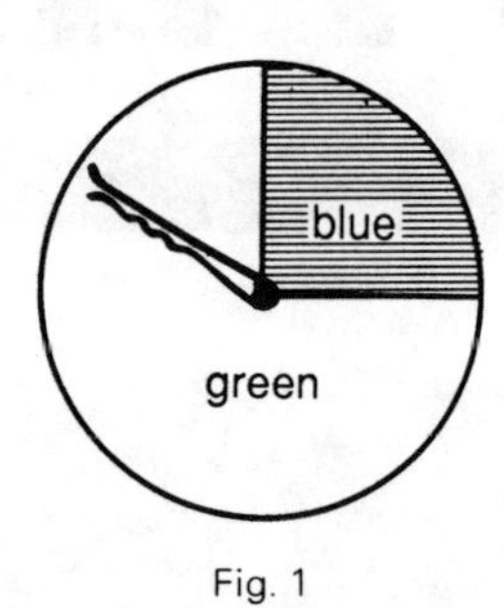

Fig. 1

Ask the class, "What is the probability that the pointer will land in the area marked blue?" Discuss the relationship between the size of the area of the sector of the circular spinner and the probability of landing in that area. Here $P(\text{blue}) \neq \frac{1}{2}$ because the area of this sector is less than that of a semicircle. Stress the need to subdivide the area of the spinner into equal parts so that the part of the whole can be named correctly. By adding the dotted line segments to subdivide the green region (fig. 2), we see that the blue area is one of four equal parts. So, $P(\text{blue}) = \frac{1}{4}$. Alternately, by adding only the horizontal segment, we can reason that the area of the blue region is $\frac{1}{2}$ of $\frac{1}{2}$, or $\frac{1}{4}$, of the total area. Students should now be able to complete sheet 1 with little difficulty; note that problem 4 has a slight twist to it.

On completion of the first sheet, students can be instructed to make a spinner

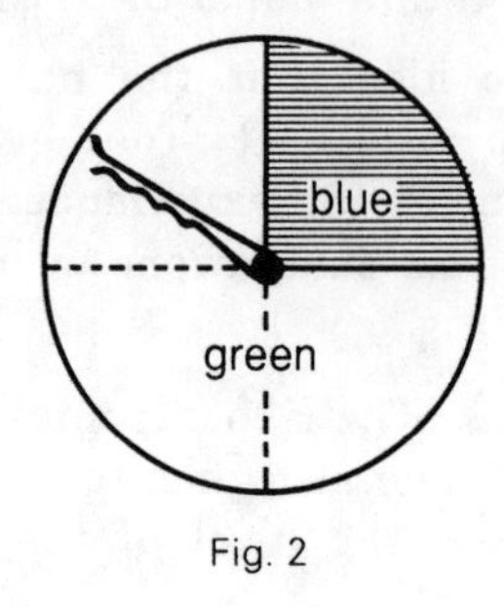

Fig. 2

as depicted in problem 2, spin it 100 times, and then compare the experimental results to the expected values in problem 2*b*. Pupils will find that a chart such as the following will be helpful for recording the results of their trials:

Area	Tally	Experimental Probability
A		$\frac{}{100}$
B		$\frac{}{100}$
C		$\frac{}{100}$
D		$\frac{}{100}$

Next students should compare their experimental probabilities with the corresponding theoretical probabilities. To make these comparisons easier, they should first express their probability ratios as decimals. Note that a (theoretical) probability statement is a prediction about what will happen over *many* trials of an experiment. The more times the spinner is spun, the closer the agreement between the corresponding theoretical and experimental probabilities.

Sheet 2 extends the area model to situations in which area is not an attribute. Depending on the students' ability, more structure can be imposed on the first two problems by having students subdivide the large squares into 36 smaller squares (fig. 3). Each of the smaller squares then represents 1/36th of the whole. In this context, the theoretical probability can be determined simply by counting the small squares marked red or blue.

The area models for problems 1–3 on sheet 3 can be further structured for less able pupils. These students can initially subdivide the first large square into a four-by-four array of smaller squares, the second large square into a six-by-six array of smaller squares, and the third square into a six-by-

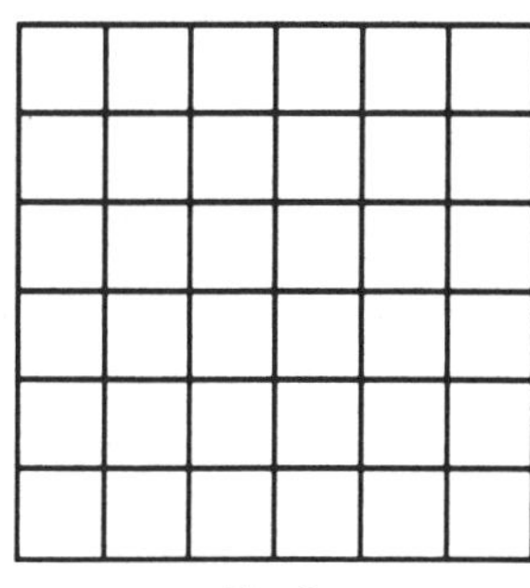

Fig. 3

four array of small *rectangles*.

After they complete sheet 3, encourage students to (*a*) work in pairs to make the spinners, (*b*) simulate Mary's attempt fifty times by spinning first spinner A and then spinner B, (*c*) record the number of times a mix for purple is obtained, (*d*) determine the experimental probability of Mary's getting a mix for purple, and (*e*) compare how well the experimental results fit their theoretical analysis of Mary's chances.

Answers

Sheet 1: 1.*a*. $P(B) = \frac{1}{2}$; 1.*b*. $P(D) = \frac{1}{3}$, $P(E) = \frac{1}{3}$, $P(F) = \frac{1}{3}$; 1.*c*. $P(G) = \frac{1}{4}$, $P(R) = \frac{1}{4}$, $P(Y) = \frac{2}{4}$, or $\frac{1}{2}$. 2.*a*. $P(A) = \frac{1}{6}$, $P(B) = \frac{1}{6}$, $P(C) = \frac{1}{6}$, $P(D) = \frac{3}{6}$ or $\frac{1}{2}$; 2.*b*. approximately $\frac{1}{6} \times 100$, or about 17 times; approximately $\frac{1}{2} \times 100$, or about 50 times

3.*a*.

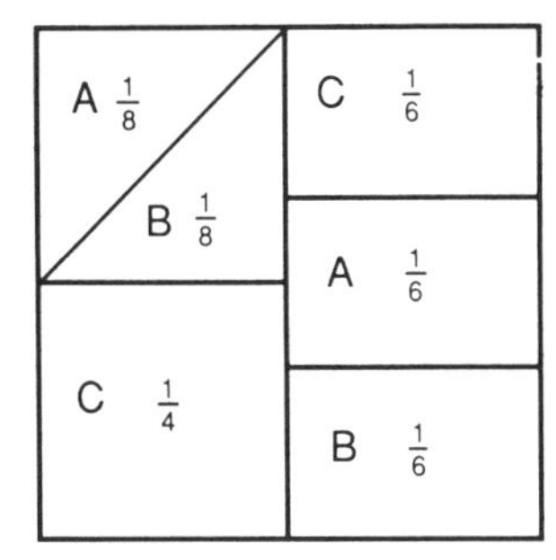

3.*b*. $P(A) = \frac{1}{8} + \frac{1}{6} = \frac{7}{24}$; $P(B) = \frac{1}{8} + \frac{1}{6} = \frac{7}{24}$; $P(C) = \frac{1}{4} + \frac{1}{6} = \frac{10}{24}$, or $\frac{5}{12}$; $P(R) = \frac{1}{18} + \frac{1}{9} + \frac{1}{6} = \frac{6}{18}$, or $\frac{1}{3}$; $P(B) = \frac{1}{18} + \frac{1}{9} = \frac{3}{18}$, or $\frac{1}{6}$; $P(G) = \frac{1}{18} + \frac{1}{6} = \frac{4}{18}$, or $\frac{2}{9}$; $P(Y) = \frac{1}{6} + \frac{1}{9} = \frac{5}{18}$; 3.*c*. 1: The sum of the probabilities is the area of the whole region. 4. Answers will vary. One possible answer is shown: $P(\text{blue}) = \frac{1}{2}$.

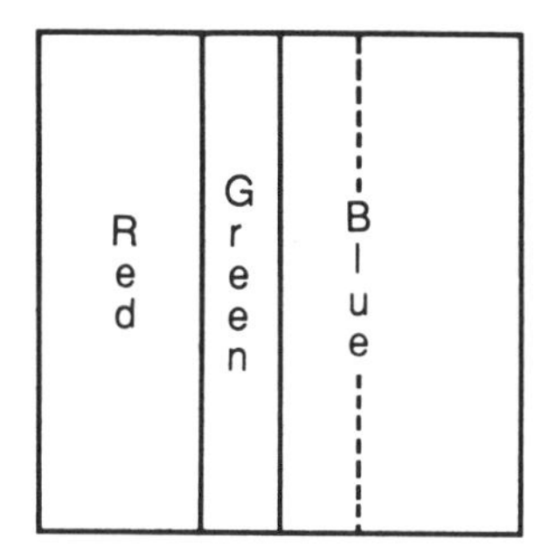

Sheet 2: 1.*d*. Equal probabilities for the two possible paths occur at the first intersection on the right branch;

1.*e*.

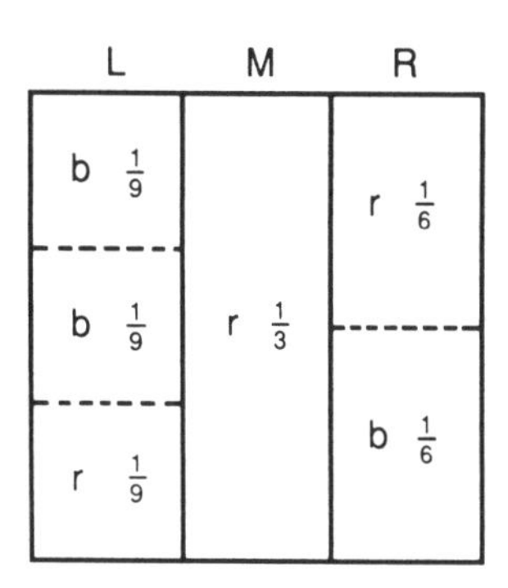

1.*f*. $P(\text{red}) = \frac{1}{9} + \frac{1}{3} + \frac{1}{6} = \frac{11}{18}$. 2.*a*. Answers will vary;

2.*b*.

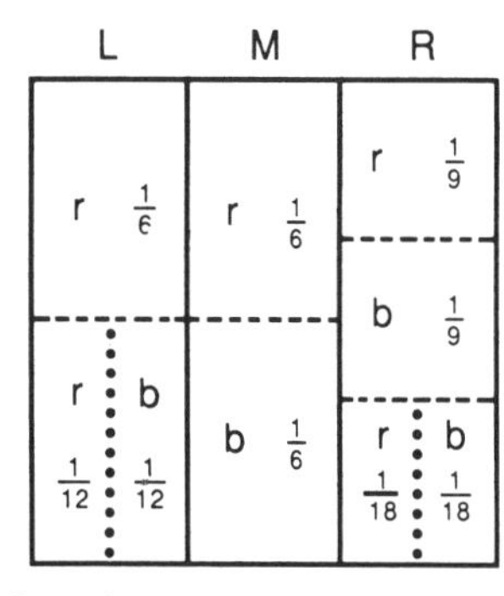

$$P(\text{red}) = \frac{1}{6} + \frac{1}{12} + \frac{1}{6} + \frac{1}{9} + \frac{1}{18} = \frac{21}{36}, \quad \text{or} \quad \frac{7}{12}$$

$$P(\text{blue}) = \frac{1}{12} + \frac{1}{6} + \frac{1}{9} + \frac{1}{18} = \frac{15}{36}, \quad \text{or} \quad \frac{5}{12}$$

3. Answers will vary.

Sheet 3: 1.

$$P(\text{purple}) = \frac{1}{16} + \frac{1}{16} = \frac{2}{16}, \quad \text{or} \quad \frac{1}{8}$$

2.

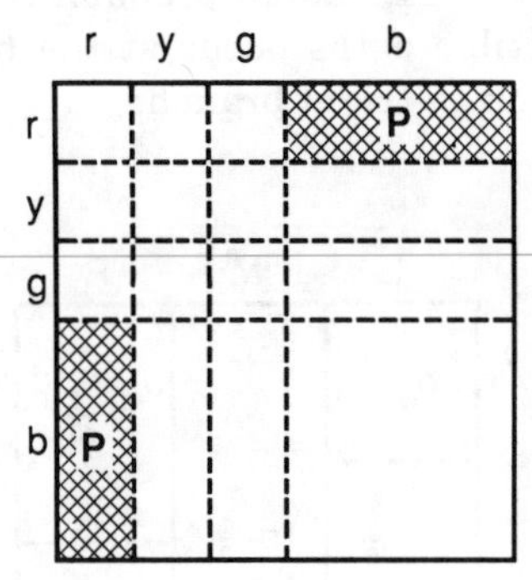

$$P(\text{purple}) = \frac{1}{12} + \frac{1}{12} = \frac{2}{12}, \quad \text{or} \quad \frac{1}{6}$$

3.

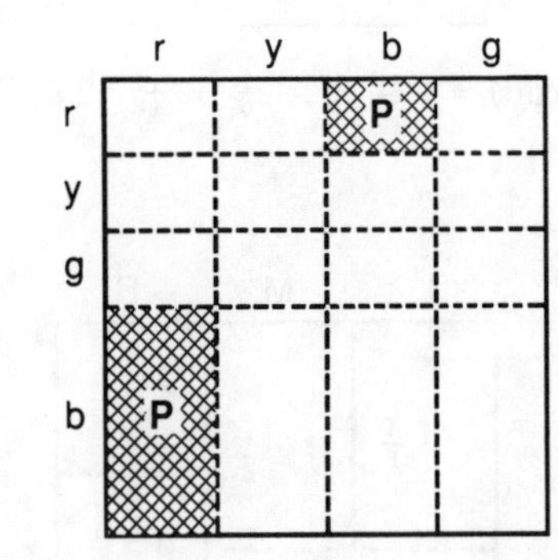

$$P(\text{purple}) = \frac{1}{24} + \frac{1}{8} = \frac{4}{24}, \quad \text{or} \quad \frac{1}{6}$$

4.*a*. John and Mary have the best chance.
4.*b*. Her chances are unchanged. Reversing the order of the spinners merely rotates the square we use for analyzing the probability.

BIBLIOGRAPHY

Armstrong, Richard D. "An Area Model for Solving Probability Problems." In *Teaching Statistics and Probability*, 1981 Yearbook of the National Council of Teachers of Mathematics, edited by Albert P. Schulte, pp. 135–42. Reston, Va.: The Council, 1981.

Dahlke, Richard, and Robert Fakler. "Geometrical Probability." In *Teaching Statistics and Probability*, 1981 Yearbook of the National Council of Teachers of Mathematics, edited by Albert P. Shulte, pp. 143–53. Reston, Va.: The Council, 1981.

Lappan, Glenda, and M. J. Winter. "Probability Simulation in Middle School." *Mathematics Teacher* 73 (September 1980):446–49.

Phillips, Elizabeth, Glenda Lappan, Mary Jean Winter, and William M. Fitzgerald. *Probability, Middle Grades Mathematics Project.* Menlo Park, Calif.: Addison-Wesley Publishing Co., 1986.

Shulte, Albert P., and Stuart A. Choate. *What Are My Chances?* Books A and B. Palo Alto, Calif.: Creative Publications, 1977.

ANALYZING GAMES OF CHANCE SHEET 1

1. Many games of chance use spinners. On a spinner, the probability that the pointer will stop in a particular section depends on the relative area of that section. For the first spinner below, the probability of landing in the area marked A is $\frac{1}{2}$ because this area is one-half that of the whole region. We write $P(\text{A}) = \frac{1}{2}$. Determine the remaining probabilities for the areas of each spinner.

a.

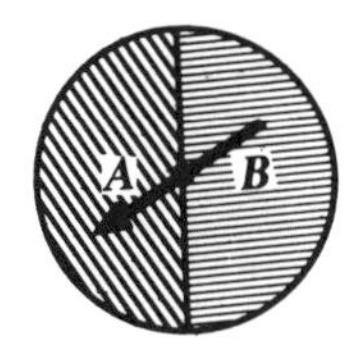

$P(\text{A}) = \frac{1}{2}$

$P(\text{B}) =$ ______

b.

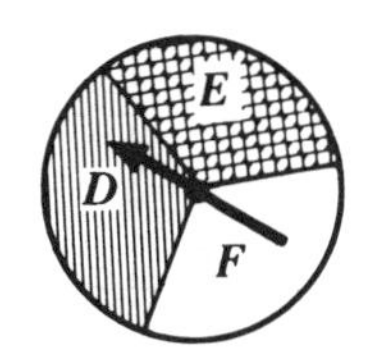

$P(\text{D}) =$ ______

$P(\text{E}) =$ ______

$P(\text{F}) =$ ______

c.

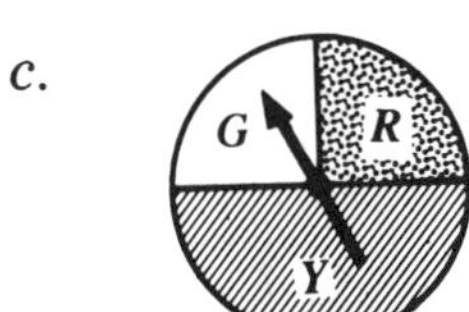

$P(\text{G}) =$ ______

$P(\text{R}) =$ ______

$P(\text{Y}) =$ ______

2. *a.* Analyze the spinner below and determine the probability that the pointer will stop in the indicated section.

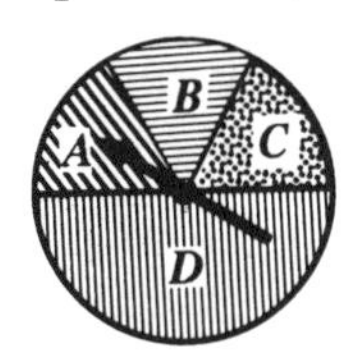

$P(\text{A}) =$ ______ $P(\text{B}) =$ ______

$P(\text{C}) =$ ______ $P(\text{D}) =$ ______

b. If you spin the spinner 100 times, how often would you expect the pointer to land in the area marked B? ______ In the area marked D? ______

3. Shown below are two square dart boards.

a. Label each of the smaller sections with the fraction that represents its part of the whole region.

b. Suppose you threw a dart randomly at each of the boards. Assuming that the dart strikes the board each time, determine the probability of its landing in the area indicated.

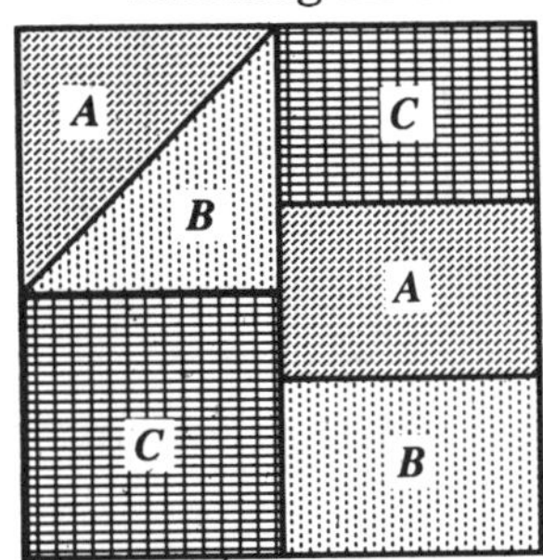

$P(\text{A}) =$ ______

$P(\text{B}) =$ ______

$P(\text{C}) =$ ______

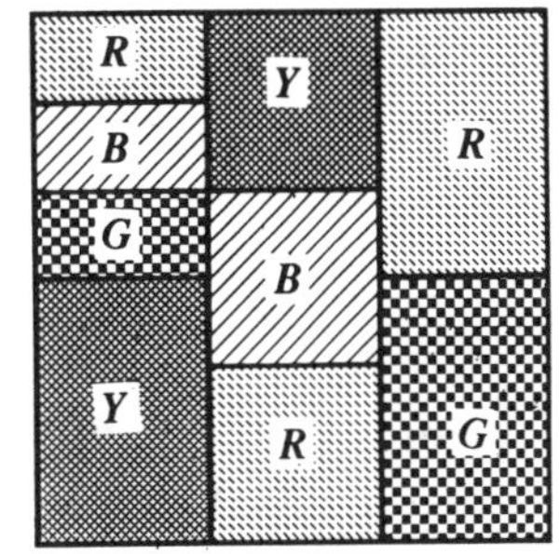

$P(\text{R}) =$ ______

$P(\text{B}) =$ ______

$P(\text{G}) =$ ______

$P(\text{Y}) =$ ______

c. For each board, what should be the sum of the probabilities? ______ Explain your answer.

4. Create a square dart board with areas marked "red," "green," and "blue" so that $P(\text{red}) = \frac{1}{3}$ and $P(\text{green}) = \frac{1}{6}$. Determine $P(\text{blue})$: ______

ANALYZING GAMES OF CHANCE — SHEET 2

1. When a marble is dropped randomly into a track, it rolls down and ends up in either a red (r) slot or a blue (b) slot. At each intersection the marble has an equal chance of rolling down each path. If the marble ends up in a red slot, the player wins a prize.

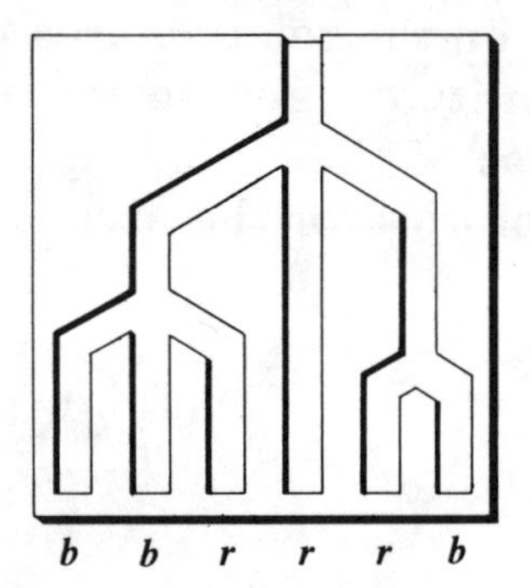

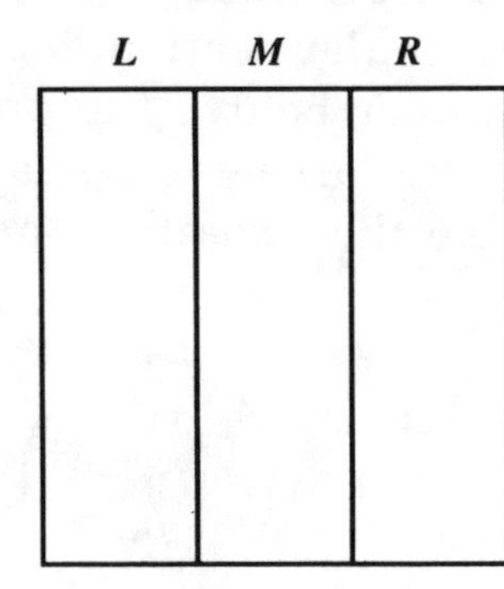

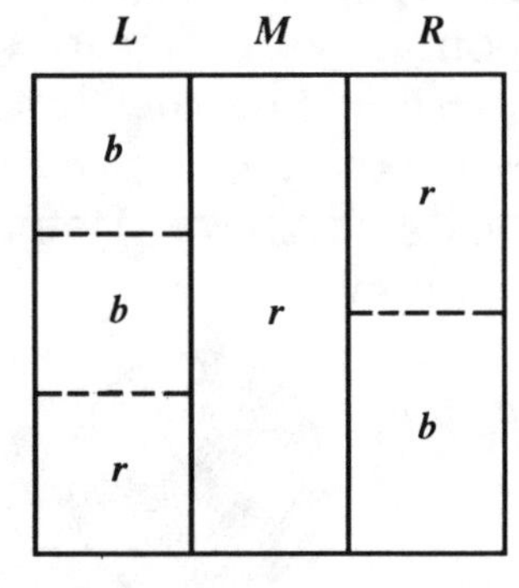

To analyze this game you can use a square whose area represents the total probability, 1, of the marbles ending up in one of the colored slots.

a. First subdivide the area into three equal parts (see the second diagram above) to represent the equal probabilities that the marble will go into each path at the first branch—L (left), M (middle), R (right).

b. Next subdivide the section marked L into three equal parts (third diagram) to correspond with the equal probabilities that the marble will fall into each path at the next branch. Label two regions b and one r.

c. The middle section (M) is not subdivided, since the marble must end up at a red slot if it enters this track.

d. Why is the right section (R) subdivided into two equal parts? ____________

____

e. Label each subregion with the fraction that represents its part of the whole area.

f. Use your completed diagram to help determine the probability that you win a prize, that is, find *P*(red): ______

2. *a.* Do you think the chances of the marble rolling into a red slot are better or worse on the track at the right than on the first track? ______

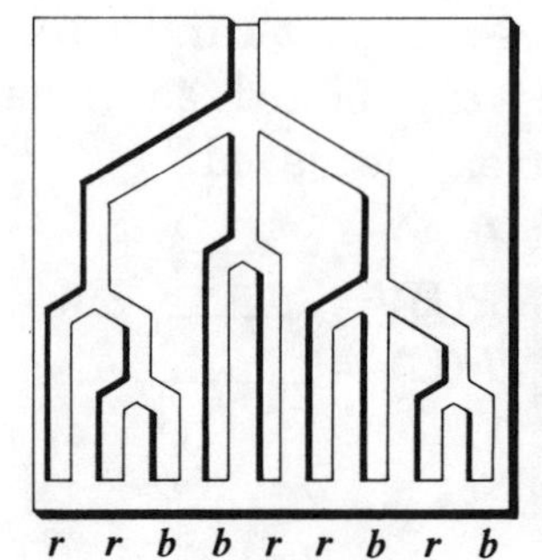

b. Using the given square region and a procedure similar to that outlined above, find these probabilities:

P(red) = ______ *P*(blue) = ______

3. Design a track for a marble-rolling game that has slots of three colors. Analyze your game using an area model. For each color chosen, find the probability that the marble will end up in a slot of that color.

The final game entails two spinners, A and B, for mixing paint colors. To play, you spin twice on any combination of the spinners. If you can make purple (red and blue in either order), you win a prize. Sarah, John, and Mary try the game.

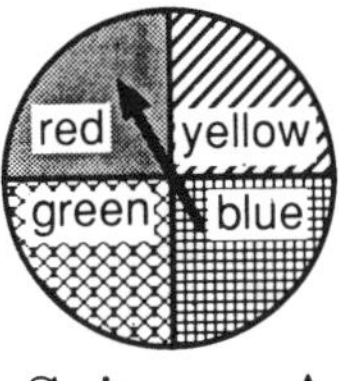

Spinner A

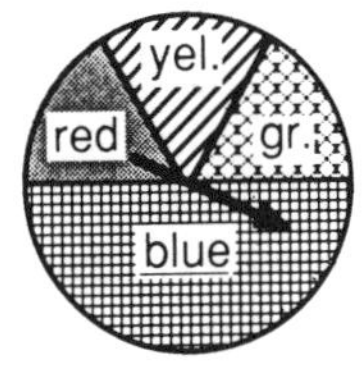

Spinner B

1. Sarah spins spinner A twice. Complete the area model at the right to help you find the probability that she gets the paint she needs to make purple (either red then blue or blue then red).

   After two spins, P(purple) = ______.

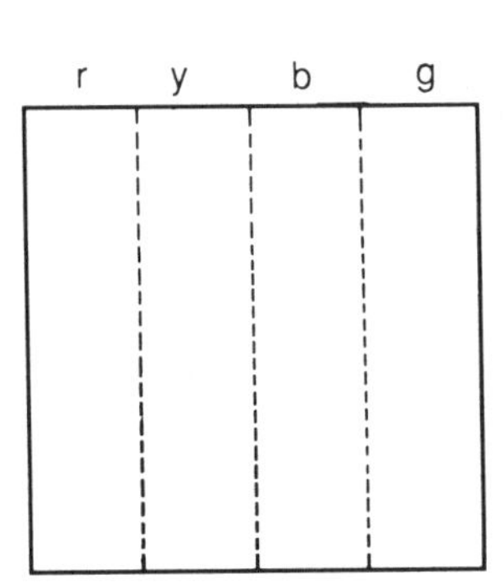

2. John spins spinner B twice. With the given square region, help determine the probability that he gets the paint he needs to make purple.

   After two spins, P(purple) = ______.

3. Mary first spins spinner A and then spinner B. What is the probability that she can make purple after her two spins?

   After two spins, P(purple) = ______.

4. *a.* Who has the best chance of making purple—Sarah, John, or Mary? ______

   b. Suppose Mary spins the spinners in reverse order. First spinner B, then spinner A. Would she have a better chance to get a mix for purple? ______ Explain.

ACTIVITIES

AREA MODELS AND EXPECTED VALUE

November 1987

By GLENDA LAPPAN, ELIZABETH PHILLIPS, WILLIAM M. FITZGERALD, and M. J. WINTER,
Michigan State University, East Lansing, MI 48824

Teacher's Guide

Introduction: In the "Activities" section of the March 1987 *Mathematics Teacher* the authors presented a model based on area for analyzing games of chance that involved dependent events. In this set of activities we again use the area model but extend the questions to include expected value. *Expected value* is the average payoff in points over a very long run of trials. Students have a natural affinity for games that involve chance. These activities build on this interest to reinforce understanding of dependent probabilities and to introduce the concept of expected value in an experimental way. The activities provide the motivation and concrete foundation for a more formal study of these ideas at a later stage.

The first activity introduces the concept of expected value by analyzing the average points per spin on a variety of spinners over a given number of trials. In the last three activities the area model is used to determine the probabilities of dependent events, and then the expected value is calculated.

The activities are designed to be used with minimal instruction. However, a richer experience is obtained if for each activity the teacher launches the challenge, discusses some preliminary guesses about the outcomes, and then allows students to explore the problem in small groups with a final whole-class summary of results and generalizations.

Grade levels: 7–12

Materials: Copies of the activity sheets for each student, a set of transparencies for class discussion, and bobby pins or paper clips to use as spinners

Objective: To practice using area models in analyzing compound situations and calculating the expected value or long-term average

Prerequisites: Students should have some familiarity with the topic of probability, including

1. the idea that the probability of a particular event A occurring is—

$$P(A) = \frac{\text{the number of ways } A \text{ can occur}}{\text{the total number of possible events}},$$

2. the idea that the notation $P(B)$ means "the probability that event B will occur," and

3. some facility with using the area model to determine probabilities of dependent events. (See Lappan et al., 1987.)

Directions: Distribute the activity sheets one at a time. Introduce these activities by posing some simple problems involving probabilities, such as flipping a coin, tossing a die, or drawing colored marbles from a bag to review the definition of probability.

On sheet 1 the probability of spinning and obtaining a particular number is easily determined by subdividing the circular spinner into congruent sectors. If students have never computed experimental probabilities, this is a good time to instruct them to make the spinners, spin each twenty times, record their results, and determine the experimental probabilities.

To calculate the expected value or the long-term average, we employ the definition of probability and equivalent fractions. For example, in problem 1 $P(5) = 1/4$ and $P(-2) = 3/4$. Since $1/4 = 25/100$, we would *expect* the number 5 to occur 25 times in 100 spins. Similarly, since $3/4 = 75/100$, we would *expect* the number -2 to occur 75 times. At the end of 100 spins, the total is

$$25 \times 5 + 75 \times (-2) = -25 \text{ points.}$$

The average points for 100 spins is $-25/100$, or -0.25 points per spin. The average points per spin is called the expected value.

In each problem the number of spins is different. The number of spins is a multiple of the denominators of the probabilities of the events, making it easier to calculate the expected value. The expected value is independent of the number of trials. For example, in problem 1 you should also ask the students to calculate the expected value for 136 spins:

$P(5) = 1/4 = 34/136$, so we expect 5 to occur 34 times for 170 points

$P(-2) = 3/4 = 102/136$, so we expect -2 to occur 102 times for -204 points

Thus, the total number of points is -34 and the expected value is $-34/136$, or -0.25 points per spin.

In general, for a situation with two events A and B, the expected value is

$$\frac{[P(A) \times (\text{total trials} \times \text{pts. for } A) + P(B) \times (\text{total trials} \times \text{pts. for } B)]}{\text{total trials}} = P(A) \times (\text{pts. for } A) + P(B) \times (\text{pts. for } B).$$

More advanced students may discover this formula, but for the majority of students simply using the definition of probability is the best way to determine expected value.

Before doing the problems on sheet 2 you should review the area model for analyzing dependent events. The spinner from problem 1 on sheet 1 can be used. Change the directions to read "add the numbers you obtain by spinning twice." Ask the questions, "What are all the possible sums?" and "What are the probabilities for obtaining these sums?"

Use the area of the grid that has been subdivided into sixteen congruent squares to represent the probability of 1. Separate the area of the grid vertically to represent the probabilities of 1/4 and 3/4 corresponding to obtaining the numbers 5 or -2 on the first spin (fig. 1). Separate the grid horizontally to represent the probabilities of obtaining 5 or -2 on the second spin (fig. 2). Analyze the grid to find the probabilities of obtaining the sums -4, 3, and 10: $P(-4) = 9/16$, $P(3) = (3/16) \times 2 = 6/16$, $P(10) = 1/16$.

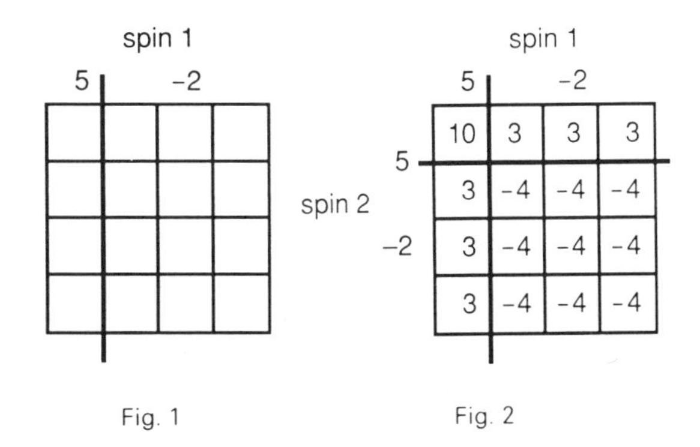

Fig. 1

Fig. 2

The sum of -4 is the most likely to occur.

After introducing the game on sheet 2, have the students predict which outcome they think is the most likely to occur. Most students will guess the score of 1 and are surprised to find that the score of 0 is the most likely and the score of 1 is the least likely. However, the expected value is 0.96. Many students will enjoy knowing that this problem is identical to simulating a one-and-one free throw situation for a player with a 60-percent free throw shooting average. An extension problem would be to analyze the outcome for a two-shot foul for a 60-percent free throw shooter. The analysis would be as follows (fig. 3):

$$P(2 \text{ pts.}) = 0.36$$
$$P(1 \text{ pt.}) = 0.48$$
$$P(0 \text{ pts.}) = 0.16$$

The expected value is $P(2 \text{ pts.}) \times 2 + P(1 \text{ pt.}) \times 1 + P(0 \text{ pts.}) \times 0 = 0.72 + 0.48$, or 1.20 points per turn.

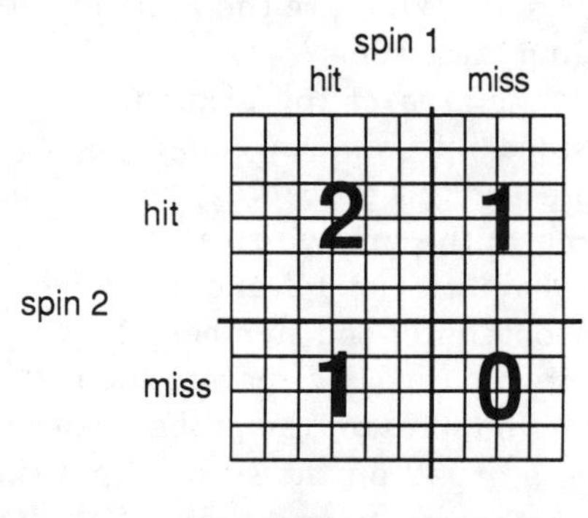

Fig. 3

On sheet 3 the students will need help in determining all the arrangements for placing the coins in the two cans. Each arrangement has a different expected value. This problem is easily accessible using the area grids to analyze the dependent events of selecting a can and then selecting a coin from the can. Some students may need help in subdividing the grids. A 6×6 grid works nicely. The students are to calculate the expected value for 100 trials. This is the first time the number of trials is not a common multiple of all the denominators.

The game on sheet 4 should be played first to get an idea of the outcomes before analyzing the theoretical probabilities. Note that we are not actually computing an expected value for the game but expected total points for each player. However, the game can be analyzed for an expected value if we ask "What is the average number of points scored on a round?"

In the process of doing these activities the students may discover that the probability of event A followed by event B occurring is $P(A) \times P(B)$. If this discovery occurs, use the area model for multiplying fractions to support this conjecture.

Answers

Sheet 1: 1. $P(5) = 1/4$; -2 occurs 75 times for -150 points; the expected total points is -25; the expected value is -0.25 points per spin.

2. $P(5) = 1/3$, $P(10) = 1/3$, $P(-12) = 1/3$; 5 occurs 20 times for 100 points, 10 occurs 20 times for 200 points, -12 occurs 20 times for -240 points; the expected total points is 60; the expected value is 1 point per spin.
3. $P(-1) = 5/12$, $P(2) = 1/6$, $P(3) = 1/4$, $P(-4) = 1/6$; the expected total points in 72 spins is 0 points and the expected value is 0 points per spin.
4. For 240 spins the expected value for each spinner is exactly the same as computed in problems 1, 2, and 3.

Sheet 2: 1 and 2. Answers will vary.

3. See figure 4.

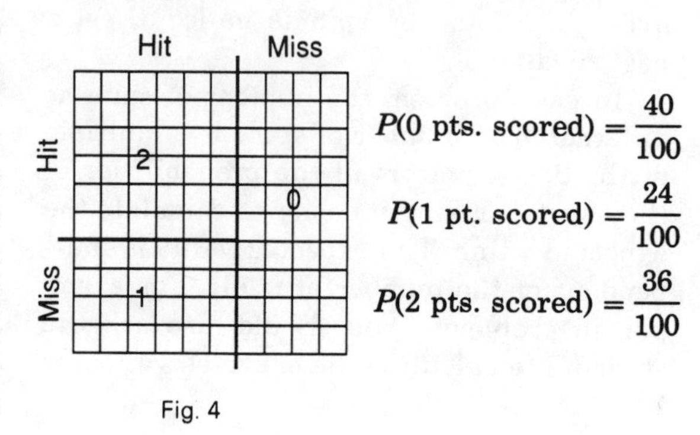

Fig. 4

In 100 turns we expect 0 points to occur 40 times for 0 points, 1 point to occur 24 times for 24 points, and 2 points to occur 36 times for 72 points. The total expected points is 96, and the expected value is 0.96 points per turn.

4. For a 20-percent "hit" area: $P(0 \text{ pts.}) = 80/100$, $P(1 \text{ pt.}) = 16/100$, $P(2 \text{ pts.}) = 4/100$. The expected value is 0.24 points per turn. For a 40-percent "hit" area: $P(0 \text{ pts.}) = 60/100$, $P(1 \text{ pt.}) = 24/100$, $P(2 \text{ pts.}) = 16/100$. The expected value is 0.56 points per turn. For an 80-percent "hit" area: $P(0 \text{ pts.}) = 20/100$, $P(1 \text{ pt.}) = 16/100$, $P(2 \text{ pts.}) = 64/100$. The expected value is 1.44 points per turn.

Sheet 3: 1.

Can 1: *GG*	*GS*	*G*	*S*	
Can 2: *SS*	*GS*	*GSS*	*GGS*	*GGSS*

2. *GG-SS* (fig. 5a)
   $P(G) = 1/2$
   $P(S) = 1/2$
   The expected value = \$3 per turn

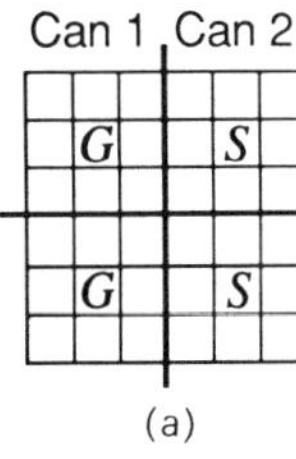

(a)

G-GSS (fig. 5b)
$P(G) = 2/3$
$P(S) = 1/3$
The expected value = \$3.67 per turn

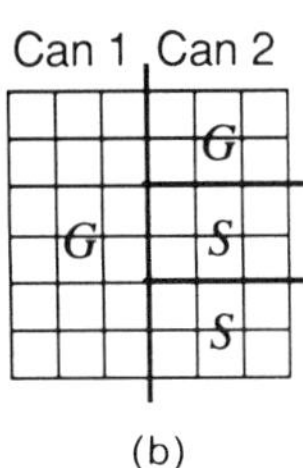

(b)

S-GGS (fig. 5c)
$P(G) = 1/3$
$P(S) = 2/3$
The expected value = \$2.33 per turn

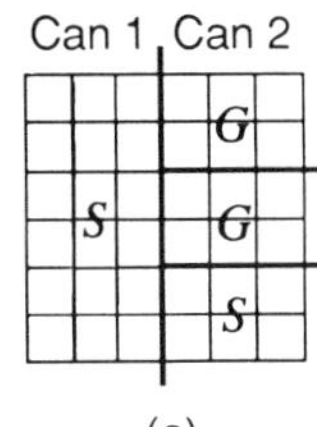

(c)

-GGSS (fig. 5d)
$P(G) = 1/4$
$P(S) = 1/4$
The expected value = \$1.50 per turn

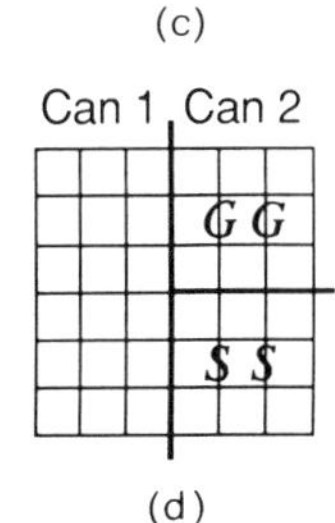

(d)
Fig. 5

The best arrangement is to put a gold coin alone in one of the cans and a gold and two silver coins in the other can.

3. The best arrangement is to put a gold coin alone in two of the cans and a gold and three silver coins in the third can. Then $P(G) = 3/4$, $P(S) = 1/4$. The expected value is \$4 per turn.

Sheet 4: 1 and 2. Answers will vary.

3. $P(A \text{ scores}) = 3/4$, $P(B \text{ scores}) = 1/4$ in figure 6. The expected number of points for each player on 20 rounds: *A* gets 15 points and *B* gets 10 points; for 100 rounds: *A* gets 75 points and *B* gets 50 points.

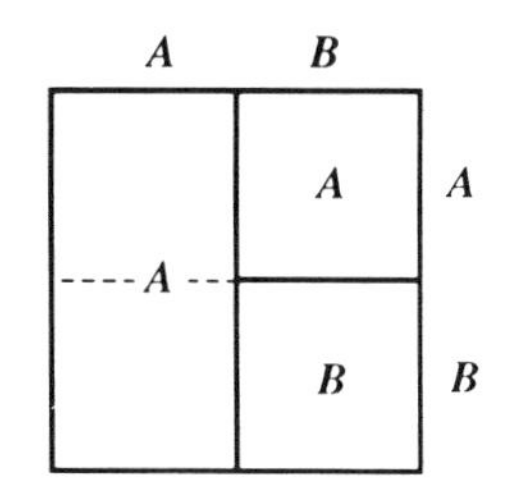

Fig. 6

4. To make the game fair, player *B* should get 3 points for scoring.
5. Answers will vary.
6. $P(\$20) = 1/16$, $P(\$10) = 9/16$, $P(\$15) = 6/16$ in figure 7. The expected value is \$12.50 per week. It should make no difference which method you choose for obtaining your allowance; the amount is the same.

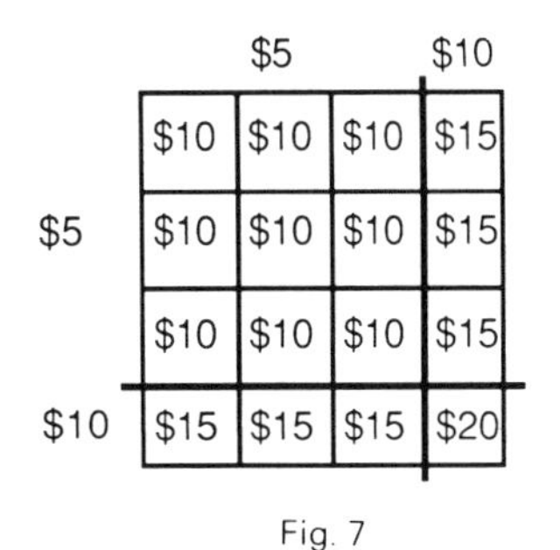

Fig. 7

BIBLIOGRAPHY

Armstrong, Richard D. "An Area Model for Solving Probability Problems." In *Teaching Statistics and Probability*, 1981 Yearbook of the National Council of Teachers of Mathematics, edited by Albert P. Shulte, pp. 135–42. Reston, Va.: The Council, 1981.

Dahlke, Richard, and Robert Fakler. "Geometrical Probability." In *Teaching Statistics and Probability,* 1981 Yearbook of the National Council of Teachers of Mathematics, edited by Albert P. Shulte, pp. 143–53. Reston, Va.: The Council, 1981.

Lappan, Glenda, Elizabeth Phillips, M. J. Winter, and William M. Fitzgerald. "Area Models for Probability." *Mathematics Teacher* 80 (March 1987): 217–23.

Lappan, Glenda, and M. J. Winter. "Probability Simulation in Middle School." *Mathematics Teacher* 73 (September 1980): 446–49.

Phillips, Elizabeth, Glenda Lappan, M. J. Winter, and William M. Fitzgerald. *Probability. Middle Grades Mathematics Project.* Menlo Park, Calif.: Addison-Wesley Publishing Co., 1986.

Shulte, Albert P., and Stuart A. Choate. *What Are My Chances?* Books A and B. Palo Alto, Calif.: Creative Publications, 1977.

In games of chance involving points it is often interesting to know what the *average* points per turn would be over many turns. This long-term average is called the *expected value* for the game. The following games involve spinners. For each turn a player spins and gets the points indicated in the area in which the spinner lands.

1. For this spinner the probability of landing in the area marked -2 is $\frac{3}{4}$. We write $P(-2) = \frac{3}{4}$.

   Find $P(5)$. ________

   In 100 spins the number of times you would expect

   5 to occur is __25__ for a total of __125__ points.

   -2 to occur is ________ for a total of ________ points.

   The expected total points for 100 spins is ________.

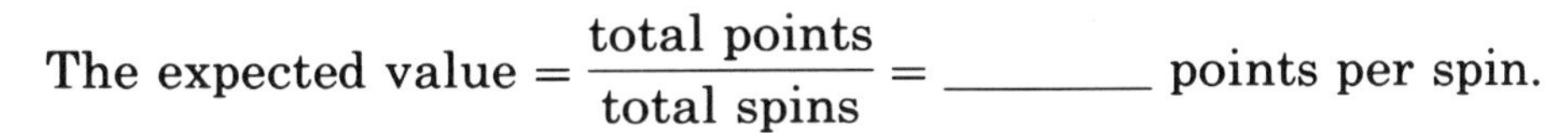

   The expected value $= \dfrac{\text{total points}}{\text{total spins}} =$ ________ points per spin.

2. Determine the probabilities for the spinner below.

   $P(5) =$ ________ $P(10) =$ ________ $P(-12) =$ ________

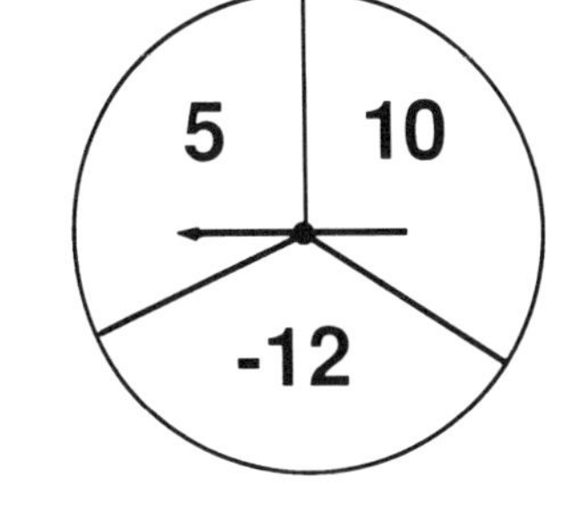

   In 60 spins the number of times you would expect

   5 to occur is ________ for a total of ________ points.

   10 to occur is ________ for a total of ________ points.

   -12 to occur is ________ for a total of ________ points.

   The expected total points for 60 spins is ________.

   The expected value $= \dfrac{\text{total points}}{\text{total turns}} =$ ________ points per spin.

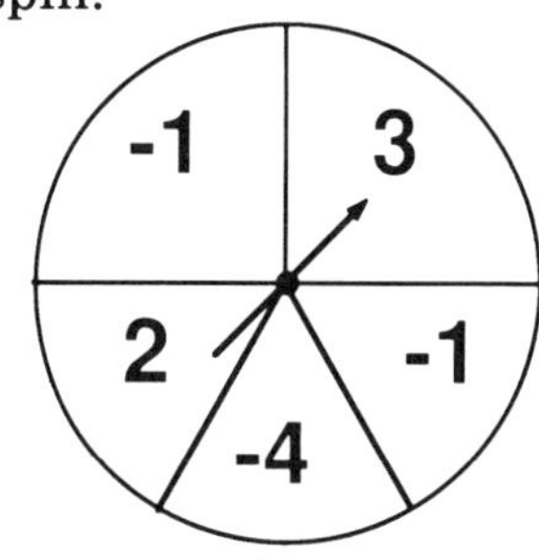

3. Determine the probabilities for this spinner.

   $P(-1) =$ ____ $P(2) =$ ____ $P(3) =$ ____ $P(-4) =$ ____

   For 72 turns, the expected total points is ________.

   The expected value is ________.

4. Calculate the expected value for 240 spins on each of these spinners. Discuss your results.

The spinner for this game is divided into two parts representing 60 percent and 40 percent of the total area. A player has a 60-percent chance of landing in the "hit" area. If a player hits on the first spin, he or she takes a second spin on the spinner. Two points are given if both spins are hits; one point is given if the first spin is a hit and the second spin is a miss; and zero points are given if the first spin is a miss. A player spins either once or twice to complete a turn depending on whether a "miss" or "hit" was obtained on the first spin.

1. Which score do you think is most likely to occur on a turn: zero, one, or two? ________

2. Insert a pencil point into the head of a bobby pin at the center of this spinner. Spin the spinner 50 times. Record your results in the chart below:

Points	Frequency	Total	Experimental Probability
0	________	______	______
1	________	______	______
2	________	______	______
	Total turns	50	

3. Analyze the game using this 100 grid to represent the total probability of one. Determine each of the following probabilities.

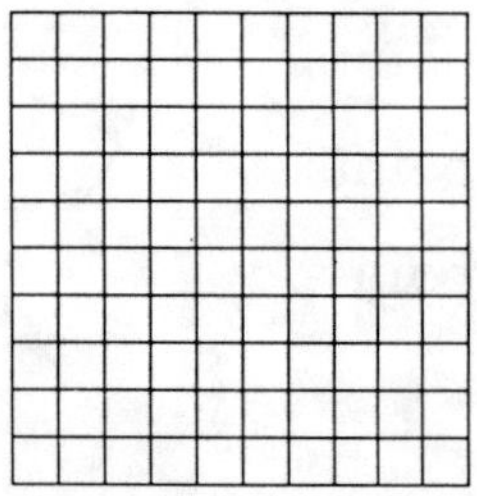

P(0 pts. scored) = ________ .

P(1 pt. scored) = ________ .

P(2 pts. scored) = ________ .

In 100 turns the number of times you would expect

0 points to occur is ________ times for a total of ________ points.

1 point to occur is ________ times for a total of ________ points.

2 points to occur is ________ times for a total of ________ points.

The expected total points for 100 turns is ________ .

The expected value $= \dfrac{\text{total points}}{\text{total turns}} =$ ________ points per turn.

4. Determine the probabilities and expected value for spinner games with 20-percent, 40-percent, and 80-percent probability of landing in the area marked "hit."

A game of chance involves two gold coins each worth \$5 and two silver coins each worth \$1. The coins are identical in size and shape. While your partner is blindfolded you place the coins into two identical cans in any way you want (one can could be empty). Your partner then selects a can at random and reaches in and pulls out a coin. The two of you get to keep the coin.

1. What are all the possible ways that the four coins could be placed into the two cans?

2. Use the area model to help analyze each arrangement. For each arrangement calculate the expected value for 100 trials.

Example: One possible arrangement

Can 1 G, S

Can 2 G, S

can 1	can 2
G	G
S	S

$P(G) = 2/4 = 1/2$

$P(S) = 2/4 = 1/2$

Expected value = \$3 per turn

Can 1 ________

Can 2 ________

$P(G) =$ ________

$P(S) =$ ________

Expected value = ________

Can 1 ________

Can 2 ________

$P(G) =$ ________

$P(S) =$ ________

Expected value = ________

Can 1 ________

Can 2 ________

$P(G) =$ ________

$P(S) =$ ________

Expected value = ________

Can 1 ________

Can 2 ________

$P(G) =$ ________

$P(S) =$ ________

Expected value = ________

What arrangement of coins gives the greatest chance of drawing a gold coin? ________

3. If three cans are used with three gold coins and three silver coins, what arrangement gives the best chance of drawing a gold coin? Calculate the expected value for this arrangement.

A spinner is marked with an *A* and a *B* as shown. Each round will consist of either one or two spins. The player with the highest score wins.

On each round player A spins first:

- If the spinner lands in the area marked *A*, player *A* scores a point (and this turn ends with player *A* spinning to start the second turn).
- If on the first spin the spinner lands in the area marked *B*, then player *B* spins the spinner; player *B* scores 2 points if the spinner lands in *B*, and player *A* scores a point if it lands in *A*.

1. Does this game seem fair? ________
   Explain.

2. With a partner try out the game. Play a total of 20 rounds. Record your results. Calculate the points for each player.
   Who won? ______________
   Now do you think it is fair?
   ____
   Explain.

Player	Points Scored	Total Score
A	________	________
B	________	________

3. Find the probability that player *A* will score and the probability that player *B* will score on a given round. Use the square grid provided to help.

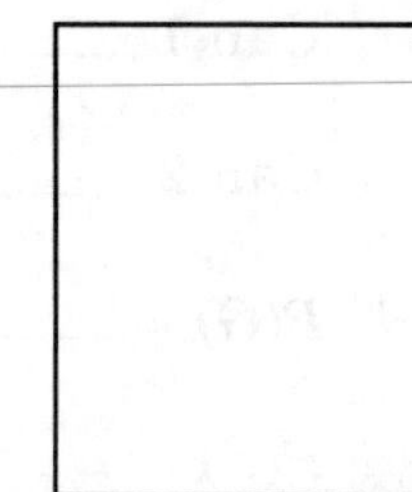

P(*A* will score on this round) = ________ .

P(*B* will score on this round) = ________ .

   Calculate the expected number of points for each player for 20 rounds. For 100 rounds.

4. How could you modify the awarding of points to make this a fair game?

5. Devise another one- or two-person game with points. Analyze the probabilities and the expected points scored for each player.

6. Your father offers you the following choice for obtaining your weekly allowance: you can receive $12.50 per week or you can receive the amount of money you get by taking two spins on the spinner at the right. Which do you choose? ______________ Why? ______________ Analyze how much money you would expect to get per week for thirty-two weeks if you choose the spinner each time. Compare this with the amount of money you would get if you choose the $12.50 per week.

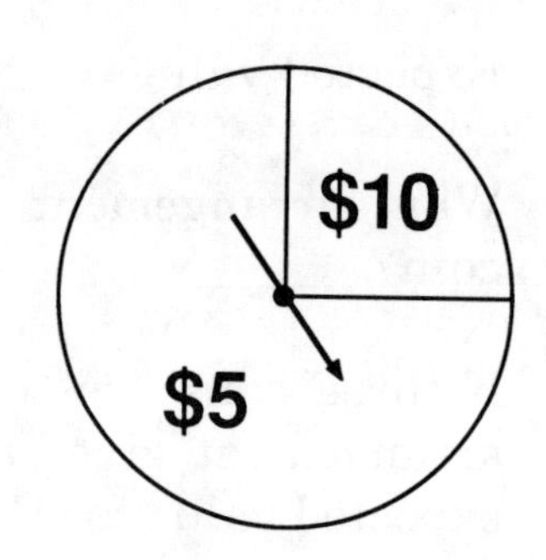

ACTIVITIES

PROBLEM SOLVING WITH SIMULATION

December 1989

By DWAYNE E. CHANNELL, Western Michigan University, Kalamazoo, MI 49008

Teacher's Guide

Introduction: This activity is designed to give students an opportunity to solve an interesting and significant applied problem using a random-number-based simulation and elementary mathematics. The solution of this problem requires students to model the problem mathematically and then to work cooperatively in small groups to collect, organize, and interpret relevant data. In working through the problem presented here, students engage in communication, reasoning, and making connections, in addition to problem solving. Additional experience in the construction and interpretation of stem-and-leaf and box-and-whisker plots is furnished.

Grade levels: 6–10

Materials: One set of activity sheets for each student, a dodecahedral die and calculator for each group of students, and at least one computer and printer for classroom use. Possible alternatives to the dodecahedral die are suggested in the directions to this activity.

Objectives: Students use a simulation technique to solve a real-world problem and gain additional experience in the use of stem-and-leaf and box-and-whisker plots to summarize and interpret data.

Prerequisites: Students should have some experience in the use of stem-and-leaf and box-and-whisker plots. If such experiences are lacking, you may wish to have them complete the activities by Landwehr and Watkins (1985) and Bryan (1988) prior to beginning this activity.

Directions: This activity requires approximately two class periods to complete, depending on the amount of the material assigned as homework.

Sheet 1: Distribute sheet 1 and discuss the problem and its analysis with the students. Allow students to guess at possible solutions. Be certain to discuss that the minimum possible solution is sixty chocolate chips and that no maximum number of chips can *guarantee* at least five chocolate chips in each cookie. Following this discussion, split the class into groups of three or four students, give each group a dodecahedral die, and have them complete problem 1, which involves the simulated placement of *exactly* sixty chocolate chips into a batch of one dozen cookies. If dodecahedral dice are not available, some other random-number-generating technique will need to be used. Some possibilities include rolling a standard six-sided die and flipping a coin, constructing a spinner with twelve equal sectors, or using a computer with a program like the following written in Applesoft BASIC:

```
100  REM DODECAHEDRAL DIE
     SIMULATOR
110  POKE(203), PEEK(78)
120  HOME
130  PRINT INT(12*RND(1)+1)
140  INPUT "ANOTHER TOSS (Y OR
     N)?";A$
150  GET A$
160  IF A$ = "Y" THEN 130
170  END
```

Once all groups have completed the simulated placement of the sixty chocolate chips, you may wish to have each group share their results and discuss the variability across groups.

Sheets 2 and 3: Distribute sheet 2. Have students complete the table in problem 2 using the data collected in problem 1. Have small groups continue the simulated placement of chocolate chips until they have satisfied the five-chip-per-cookie minimum. Work should continue in the small groups until each group has answered the questions in problem 3. As each group completes sheet 2, give each group member a copy of sheet 3 and ask them to perform a second simulation as suggested in problems 4 and 5. Once the second simulation has been completed, have the groups share their results with each other. Use the table in problem 6 to record these results. If time permits, extend this table so that the results from all student groups can be recorded and displayed to the class on an overhead transparency. Once the average (mean) number of chips in all cookie batches has been computed, have students discuss their interpretations of the solution. Although some may accept this average as the "best" solution, others may argue that the maximum chip count obtained during the simulation represents the "best" solution. All interpretations should take into account the seller's desire to keep production costs to a minimum.

Sheet 4: The purpose of sheet 4 is to present students with a large computer-generated data set relevant to this problem and to have them summarize and interpret these data. An Applesoft BASIC program that simulates the placement of chocolate chips into cookies is listed in program 1. Depending on the students' backgrounds, you may wish to discuss this program and the purpose of various statements with your class. If sufficient computer facilities are available, have each group of students run this program and carry out the instructions in problem 8. If only one computer is available, the program could be run once and the output copied and distributed for use in problems 9, 10, and 11.

Extensions: The program simulating this problem has been written to allow the user to chose the number of cookies for each batch, the number of batches to make, and the minimum number of chips desired for each cookie. The problem presented in the activity assumed twelve cookies in a batch with a minimum of five chips in each cookie. You may wish to have your students use the program to investigate these questions:

Question 1: If the baker doubles (triples, etc.) the number of cookies in a batch, does the number of chocolate chips needed for each batch double (triple, etc.)? (In general, the answer to this question is yes.)

Question 2: If the baker doubles (triples, etc.) the number of chocolate chips desired for each cookie, does the number of chocolate chips needed for each batch double (triple, etc.)? (In general, the answer to this question is no.)

For each question it is important to perform a large enough number of simulations to

PROGRAM 1

```
100  REM  CHIPPIES COOKIE PROBLEM
110  HOME : PRINT "CHOCOLATE CHIP
       COOKIE MIXING SIMULATION" : PRINT
120  INPUT "NUMBER OF COOKIES IN A
       BATCH? ";NCOOKIES
130  INPUT "NUMBER OF BATCHES TO
       MAKE? ";NBATCH
140  INPUT "MINIMUM NUMBER OF CHIPS
       PER COOKIE? ";MCHIPS
150  POKE (203), PEEK (78) : REM  SET SEED
       FOR RANDOM NUMBER GENERATION
160  DIM CHIPS(NCOOKIES)
170  HOME : PRINT "NUMBER OF CHIPS
       USED IN EACH BATCH:" : PRINT
180  FOR BATCH = 1 TO NBATCH: REM  FOR
       EACH BATCH
190  FOR COOKIE = 1 TO NCOOKIE:
       REM  SET CHIP COUNTS IN COOKIES
       TO ZERO
200  LET CHIPS(COOKIE) = 0
210  NEXT COOKIE
220  FOR CHIP = 1 TO NCOOKIES * MCHIPS
       - 1: REM  DISTRIBUTE MINIMUM # OF
       CHIPS
230  LET COOKIE = INT (NCOOKIES * RND (1)
       + 1): REM  SELECT A COOKIE
240  LET CHIPS(COOKIE) = CHIPS(COOKIE)
       + 1: REM  ADD A CHIP TO THE
       COOKIE
250  NEXT CHIP
260  LET CUSED = NCOOKIES * MCHIPS - 1:
       REM  CHIPS USED SO FAR IN BATCH
270  LET COOKIE =  INT (NCOOKIES * RND
       (1) + 1): REM  SELECT A COOKIE
280  LET CHIPS(COOKIE) = CHIPS(COOKIE)
       + 1: REM  ADD A CHIP TO THE
       COOKIE
290  LET CUSED = CUSED + 1:
       REM  COUNT CHIP USED IN BATCH
300  FOR COOKIE = 1 TO NCOOKIE:
       REM  FOR EACH COOKIE
310  IF CHIPS(COOKIE) < MCHIPS THEN 270:
       REM  CHECK CHIP COUNT IN COOKIE
320  NEXT COOKIE
330  PRINT CUSED,
340  LET TCHIPS = TCHIPS + CUSED:
       REM  KEEP TRACK OF ALL CHIPS
       USED
350  NEXT BATCH
360  PRINT : PRINT
370  PRINT "AVERAGE NUMBER OF CHIPS
       USED ACROSS"
380  PRINT NBATCH;" BATCHES TO OBTAIN
       AT LEAST ";MCHIPS
390  PRINT "CHIPS PER COOKIE WAS
       ";TCHIPS / NBATCH;" ."
400  END
```

obtain reliable results. Although it takes several minutes to run 100 simulations using the BASIC program, it is desirable to invest the time so as to obtain useful output.

Other problems that can be solved using simulation techniques can be found in Channell (1984), Channell and Hirsch (1984), and Gnanadesikan, Scheaffer, and Swift (1987).

Answers:

Sheets 1, 2, and 3: Since the solution process used in this activity is based on random procedures, the answers to the problems will vary.

Sheet 4: Although answers will vary each time the computer program is run, a sample run of the program is furnished in table 1,

TABLE 1

Program Execution

```
]RUN
CHOCOLATE CHIP COOKIE MIXING
  SIMULATION
NUMBER OF COOKIES IN A BATCH? 12
NUMBER OF BATCHES TO MAKE? 100
MINIMUM NUMBER OF CHIPS PER COOKIE? 5
NUMBER OF CHIPS USED IN EACH BATCH:

 95     113     102      94     125
100     122      96     104      97
 83      83     146     111     121
111      77     123      96     104
110     123      99     115      88
162      99     119     124      97
 99     110      88     133     127
107      89     106     105      96
159     119      96     139     112
 97      97      80     121     124
105      92     118     136     101
110     129     101     108      99
119     123      98     115     106
107     102      97      98     104
119      98     114     147     123
 85     145      87      92     150
103      85      96      94      95
 88     109     112     112      90
134     101     106     103     121
119     132     136     133     111

AVERAGE NUMBER OF CHIPS USED ACROSS
100 BATCHES TO OBTAIN AT LEAST 5
CHIPS PER COOKIE WAS 109.51.
```

and answers to problems 9, 10, and 11 are given for these sample data.

9.

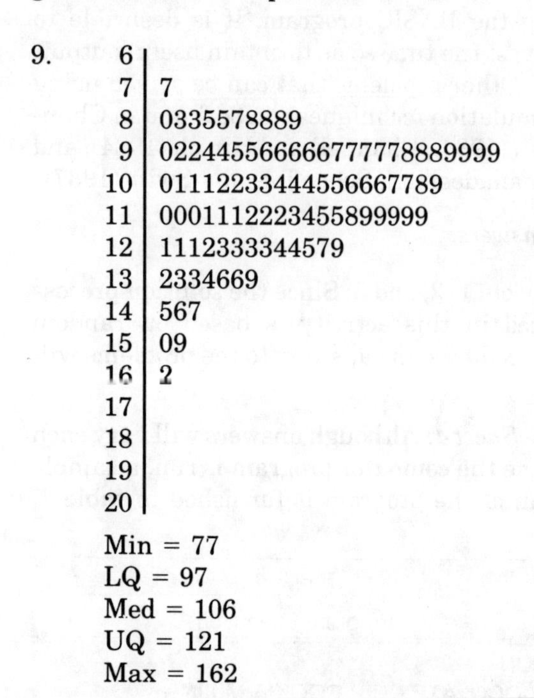

Stem	Leaves
6	
7	7
8	0335578889
9	0224455666667777778889999
10	01112233444556667789
11	0001112223455899999
12	1112333344579
13	2334669
14	567
15	09
16	2
17	
18	
19	
20	

Min = 77
LQ = 97
Med = 106
UQ = 121
Max = 162

10.

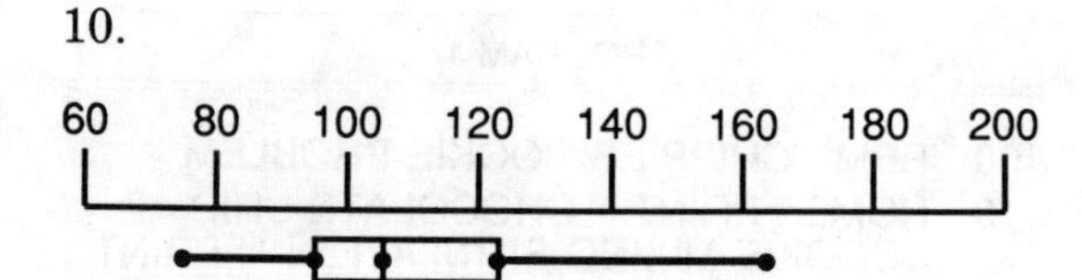

REFERENCES

Bryan, Elizabeth H. "Exploring Data with Box Plots." *Mathematics Teacher* 81(November 1988):658–63.

Channell, Dwayne E. "Problem Solving with Computers." *Mathematics Teacher* 77(October 1984):534–41.

Channell, Dwayne E., and Christian R. Hirsch. "Computer Methods for Problem Solving in Secondary School Mathematics." In *Computers in Mathematics Education,* 1984 Yearbook of the National Council of Teachers of Mathematics, edited by Viggo P. Hansen, 171–83. Reston, Va.: The Council, 1984.

Gnanadesikan, Mrudulla, Richard L. Scheaffer, and Jim Swift. *The Art and Techniques of Simulation.* Palo Alto, Calif.: Dale Seymour Publications, 1987.

Landwehr, James M., and Ann E. Watkins. "Stem-and-Leaf Plots." *Mathematics Teacher* 78 (October 1985):528–38.

Problem: The Cookie Bin is a small shop that bakes and sells cookies at an indoor shopping mall. The owner plans to introduce a new bite-sized chocolate-chip cookie, to be called "Chippies," to her line of products. She would like to have each cookie contain at least five chocolate chips. To minimize the cost of ingredients, she wants to determine the least number of chocolate chips that need to be mixed into a batch of dough for a dozen Chippies so as to meet the five-chip-per-cookie minimum. How many chocolate chips should she use for each batch?

Analysis: Assume that each chip placed into the dough for one dozen cookies has an equally likely chance of ending up in any one of the twelve cookies made from that batch. One approach to solving this problem is to simulate the random placement of chocolate chips into the individual cookies in the batch. Since each cookie should contain at least five chips, at least sixty chocolate chips must be mixed into the cookie dough. A dodecahedral (12-sided) die can be rolled to determine which of the cookies is to get a given chocolate chip. By rolling the die at least sixty times and recording each result, we can determine the number of cookies containing five or more chips.

1. A picture of twelve Chippies, without chocolate chips, is shown. Roll a dodecahedral die exactly sixty times and record each result by drawing a dot (chocolate chip) on the cookie corresponding to that roll. After the sixty chocolate chips have been randomly placed, determine the number of cookies that contain five or more chips.

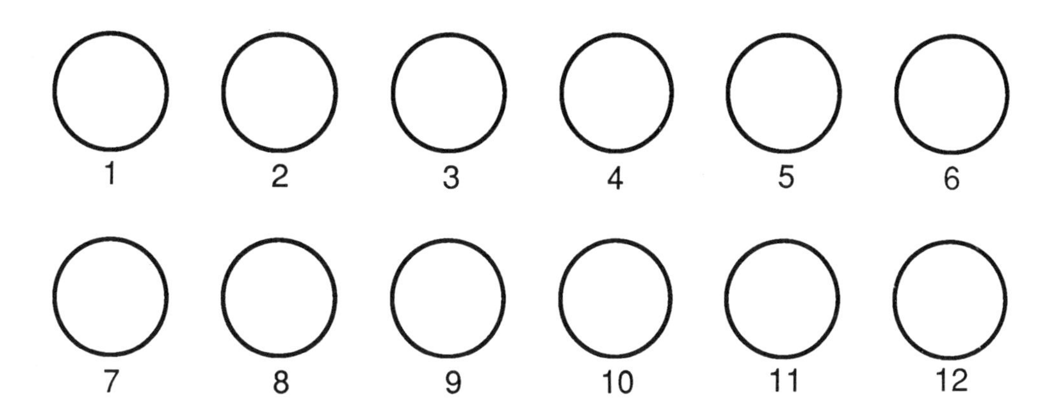

After the random placement of sixty chocolate chips, how many of the twelve cookies have five or more chips? _________

2. After placing sixty chocolate chips into a batch of twelve cookies, you have found that many cookies do not have five or more chips. Obviously, to meet the Cookie Bin owner's requirements, more chips must be added to the cookies. Record your results from problem 1 by filling in the appropriate number of squares in the grid for each of the sixty chocolate chips already placed.

   Continue rolling the dodecahedral die and recording results in the graph until you have used the minimum number of chips to fill each cookie with five or more chips.

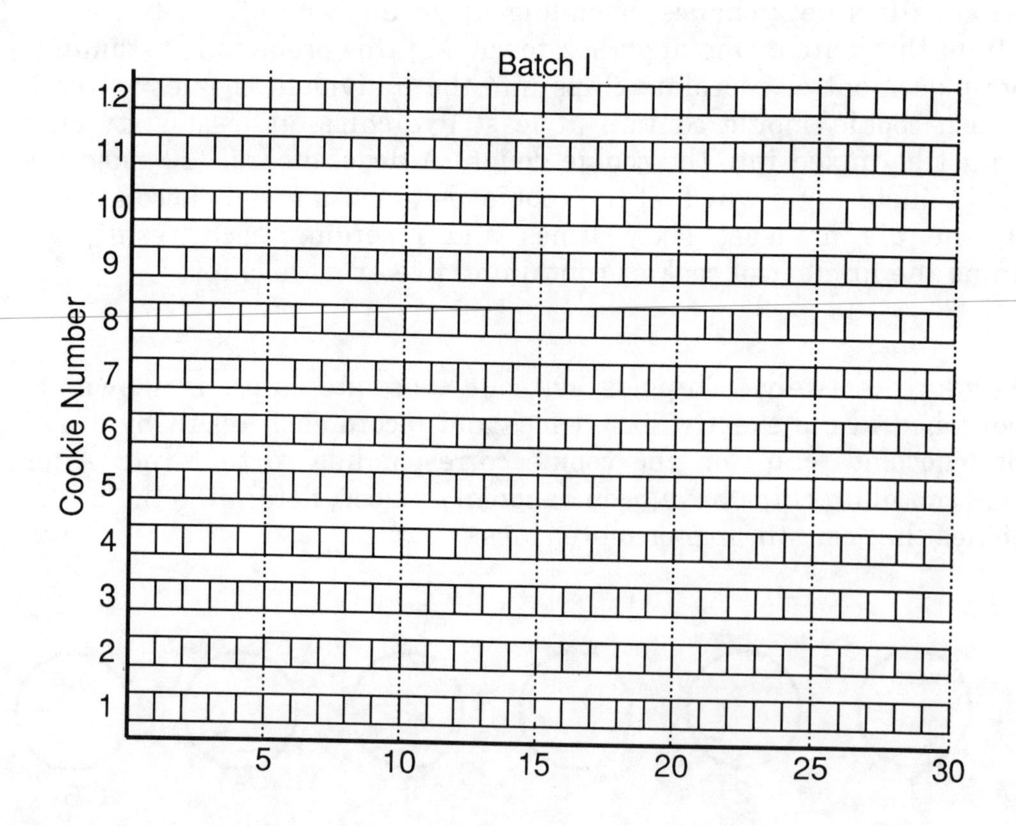

3. Answer the following questions using the information you have recorded in the graph.

   a. Based on your experiment, how many chocolate chips are needed to meet the five-chip-per-cookie minimum? __________

   b. What is the mean number of chocolate chips used per cookie?

   c. What is the maximum number of chocolate chips used in any one cookie? __________

4. The results of a single simulation of this problem do not give a reliable estimate of a solution. Repeat the experiment a second time by "making" a second batch of Chippies. Record your results in the graph and compare this result with those from the first batch.

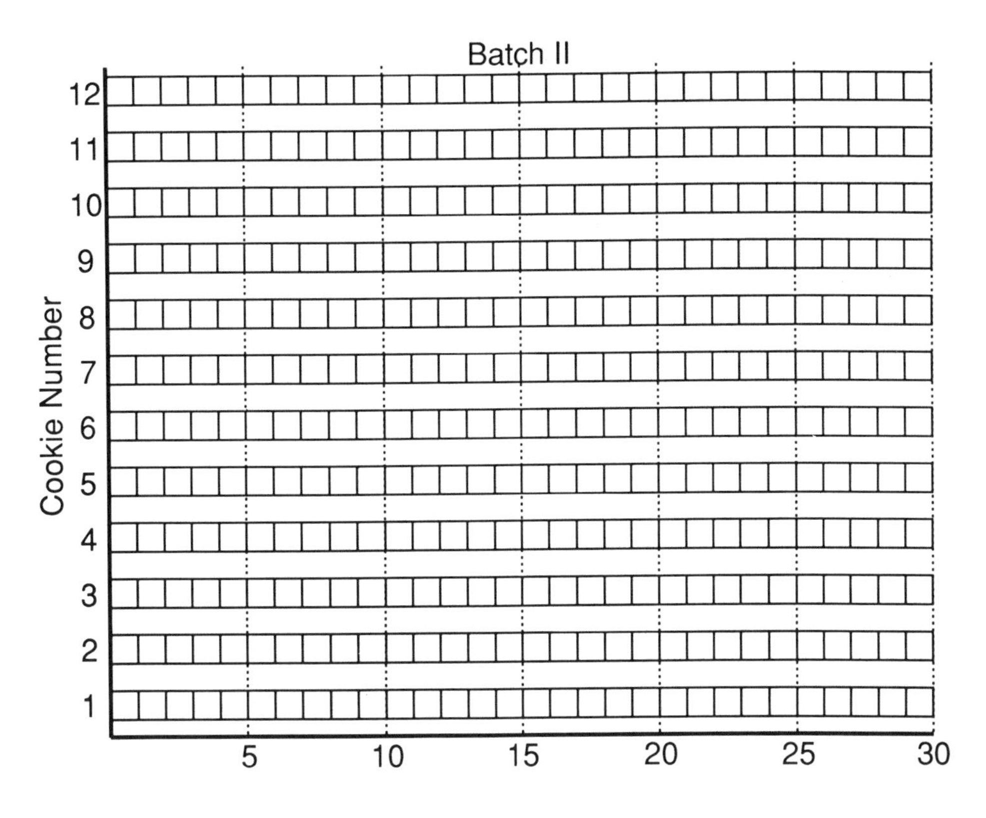

5. How many chocolate chips were needed in the second batch to fill each cookie with at least five chocolate chips? ________

6. Enter the number of chips used in your group's two batches of cookies in the table. Enter similar data from four other groups of students in your class. Compute an overall mean for all ten batches.

Your Group		Second Group		Third Group		Fourth Group		Fifth Group	
1	2	1	2	1	2	1	2	1	2

What is the mean of all ten batches? ________

7. How many chocolate chips do you think the Cookie Bin owner should put into the dough for each batch of one dozen Chippies? ________

8. The mean of the results from the ten batches of Chippies is a more reliable approximation of the solution than the results from a single batch. An even more reliable answer can be obtained by simulating the making of even more batches of cookies. This task can be handled most efficiently by using a computer program to perform the same simulation that you have been doing with the dodecahedral die.

   Obtain from your teacher the computer program that simulates this process. Run the program using input of twelve cookies per batch, 100 batches, and a minimum of five chips per cookie. The program should compute 100 chocolate-chip counts corresponding to the number of chips needed in each of the batches of twelve cookies to obtain a minimum of five chips per cookie.

9. Construct a stem-and-leaf plot of the 100 data values. Order the data from low to high in your plot.

6 |
7 |
8 |
9 |
10 |
11 |
12 |
13 |
14 |
15 |
16 |
17 |
18 |
19 |
20 |

10. Construct a box-and-whisker plot from the data as summarized in the stem-and-leaf plot.

60 80 100 120 140 160 180 200

11. Base your responses to these questions on the computer-generated data:

    a. What number of chips guarantees that every cookie will have five or more chips? __________

    b. What number of chips will result in each cookie's having five or more chips 75 percent of the time? __________

    c. What number of chips will result in each cookie's having five or more chips 90 percent of the time? __________